INSECT PESTS AND FRESH HORTICULTURAL PRODUCTS

Treatments and Responses

Edited by

ROBERT E. PAULL

Department of Plant Molecular Physiology
College of Tropical Agriculture and Human Resources
University of Hawaii

and

JOHN W. ARMSTRONG

United States Department of Agriculture
Agricultural Research Service
Tropical Fruit and Vegetable Research Laboratory
Hawaii

CAB INTERNATIONAL

CABI Publishing is a division of CAB International

CABI Publishing
CAB International
Wallingford
Oxon OX10 8DE
UK

Tel: +44 (0)1491 832111
Fax: +44 (0)1491 833508
Email: cabi@cabi.org
Web site: www.cabi-publishing.org

CABI Publishing
875 Massachusetts Avenue
7th Floor
Cambridge, MA 02139
USA

Tel: +1 617 395 4056
Fax: +1 617 354 6875
Email: cabi-nao@cabi.org

A catalogue record for this book is available from the British Library, London, UK

ISBN 0 85198 872 5

First published 1994
Transferred to print on demand 2004

Printed and bound in the UK by Antony Rowe Limited, Eastbourne.

Contents

Contributors

ARMSTRONG, J.W. *United States Department of Agriculture, Agricultural Research Service, Tropical Fruit and Vegetable Research Laboratory, PO Box 4459, Hilo, HI 96720, USA.*

BRECHT, J.K. *Horticultural Sciences Department, University of Florida, Gainesville, FL 32611-0690, USA.*

BURDITT, A.K. JR *9068-C SW 82 Terrace, Ocala, FL 34481-9438, USA.*

FORNEY, C.F. *Agriculture and Agri-food Canada, Research Station, Kentville, Nova Scotia, Canada, B4N 1J5.*

FRAMPTON, E.R. *Ministry of Agriculture and Fisheries, Lincoln, New Zealand.*

GAUNCE, A.P. *Agriculture Canada, Research Station, Summerland, BC, Canada, V0H 1Z0.*

GREANY, P.D. *United States Department of Agriculture, Agricultural Research Services, 1700 SW 23rd Drive, Gainesville, FL 32604, USA.*

HALLMAN, G.J. *United States Department of Agriculture, Agricultural Research Service, Subtropical Horticulture Research Station, 13601 Old Cutler Road, Miami, FL 33158, USA.*

HARA, A.H. *Department of Entomology, University of Hawaii, Beaumont Agricultural Research Center, 461 W Lanikaula Street, Hilo, HI 96720, USA.*

HAYES, C.F. *Department of Physics and Astronomy, University of Hawaii, Honolulu, HI 96822, USA.*

HOUCK, L.G. *United States Department of Agriculture, Agricultural Research Service, Horticultural Crops Research Laboratory, 2021 S Peach Avenue, Fresno, CA 93727, USA.*

JESSUP, A.J. *NSW Agriculture, Horticultural Research and Advisory Station, Gosford, NSW 2250, Australia.*

KADER, A.A. *Department of Pomology, University of California, Davis, CA 95616, USA.*

KE, DY *Department of Pomology, University of California, Davis, CA 95616, USA.*

MCDONALD, R.E. *United States Department of Agriculture, Agricultural Research Service, Horticultural Research Laboratory, 2120 Camden Road, Orlando, FL 32803, USA.*

MEHERIUK, M. *Agriculture Canada, Research Station, Summerland, BC, Canada V0H 1Z0.*

MORRIS, S.C. *CSIRO, Division of Horticulture, PO Box 52, North Ryde, NSW 2113, Australia.*

NATION, J.L. *Department of Entomology, University of Florida, Gainesville, FL 32611, USA.*

PAULL, R.E. *Department of Plant Molecular Physiology, College of Tropical Agriculture and Human Resources, University of Hawaii, 3190 Maile Way, Honolulu, HI 96822, USA.*

PREISLER, H.K. *United States Department of Agriculture, Forest Service, PO Box 245, Berkeley, CA 94701, USA.*

ROBERTSON, J.L. *United States Department of Agriculture, Forest Service, PO Box 245, Berkeley, CA 94701, USA.*

STARK, J.D. *Washington State University, Puyallup Research and Extension Center, 7612 Pioneer Way East, Puyallup, WA 98371-4998, USA.*

WILLIAMSON, M. *Department of Agricultural Engineering, University of Hawaii, 3050 Maile Way, Honolulu, HI 96822, USA.*

WINKELMAN, P. *Department of Agricultural Engineering, University of Hawaii, 3050 Maile Way, Honolulu, HI 96822, USA.*

Foreword

Considerable effort has been expended on the postharvest physiology, storage and technology of horticultural crops to improve the ability to deliver high quality product to the consumer. However, until the loss of ethylene dibromide and concerns with potential loss of methyl bromide, little attention was focused on alternative methods to control the insect pests without these fumigants which have been used as the primary control treatment for protection of crops moved in the multibillion dollar export trade. Many of these pests, especially the tephritid fruit flies, can seriously disrupt produce trade among countries and even among states within the US. Thus, effective insect control treatments which are not harmful to the commodity, workers, or the consumer are essential for allowing unrestricted movement of fresh horticultural crops in domestic or international commerce.

Making a connection between a commodity treatment which is effective in eliminating some pest, and yet preserving the quality of fresh horticultural products is often a difficult and complicated problem. On 1 October 1982, in recognition of the need to accelerate research and development of effective and acceptable quarantine treatments as alternatives to fumigation, the Western Directors approved a regional research project, W-164 'Postharvest Technology and Quarantine Treatments for Insect Control in Horticultural Crops'. The purpose of this regional effort was to bring together state Agricultural Experiment Station and USDA scientists with expertise in postharvest horticulture and those in entomology to establish a coordinated research agenda with the following specific objectives:

1. To develop and evaluate effective insect control procedures for use in postharvest storage and in quarantine treatments of produce for domestic and/or international markets.
2. To determine the effects of insect control procedures on market quality and consumer acceptance of fresh horticultural commodities.

Participants in this effort included representation from the state Agricultural Experiment Stations in California, Hawaii, Oregon, Washington and Florida, with USDA scientists from USDA-ARS Fresno, California; Hilo, Hawaii; and Yakima, Washington. In addition scientists from Agriculture Canada Research Stations in Summerland, British Columbia and Kentville, Nova Scotia; DSIR, Auckland, New Zealand; Queensland Department of Primary Industries, Brisbane, and CSIRO, North Ryde, Australia were involved. This collaborative regional research project provided an effective and efficient use of resources and a set of standardized treatment conditions for a given quarantine treatment were developed which could then be assessed for both efficacy in achieving the required insect control and the ability of the commodity to withstand such a treatment and yet maintain quality in the market system. Through this process a 'catalogue' of possible postharvest insect control treatments was generated which included cold/heat treatments, modified atmospheres, radiation and fumigants, a catalogue which systematically described the effectiveness of each of these treatments.

W-164 reached its tenth anniversary on 30 September, 1992, and the chapters presented here summarize the accomplishments realized from this regional and international effort, and bring together discussions on various aspects of pest infestation of horticultural products. We hope this book will prove useful, not only for workers in the field and those in the regulatory arena responsible for establishing quarantine protocols, but also for students who will be entering this complex area.

J.M. Lyons
Co-Administrative Advisor, W-164
University of California, Davis, CA
October 1982–September 1992

Introduction 1

ROBERT E. PAULL[1] and JOHN W. ARMSTRONG[2]
[1]*Department of Plant Molecular Physiology, College of Tropical Agriculture and Human Resources, University of Hawaii, 3190 Maile Way, Honolulu, HI 96822, USA;*
[2]*United States Department of Agriculture, Agricultural Research Service, Tropical Fruit and Vegetable Research Laboratory, PO Box 4459, Hilo, HI 96720, USA.*

Introduction

Postharvest treatments are required to disinfest fresh fruits, vegetables, nuts, dried fruits and vegetables, flowers, and ornamentals of economically important quarantined insects. These insect pests are of quarantine importance because they are absent from the importing country (e.g. various tephritid fruit flies, codling moth, and mango seed weevil), or the importing country has a 'zero tolerance' for all live insects whether or not they are economically important (Table 1.1). While there are many insects and arthropods of quarantine importance, fruit flies of the family Tephritidae represent a major group of destructive pests that attack a wide range of fruit (Table 1.1).

While quarantine treatments are used to kill insects in fruits, insects on flowers and plant foliage generally are killed by preharvest insecticide use and/or removal by hand. In either case, treatment monitoring and/or commodity inspection by regulatory personnel are required to ensure quarantine treatment procedures and regulations, including handling requirements to preclude reinfestation after treatment, are followed. Failure to control the spread of insect pests can result in expensive quarantine and eradication procedures (Carey, 1991), product losses due to infestation, and costly new quarantine treatment requirements (Dowell, 1983).

Quarantine problems are compounded by the rapid expansion of world trade in fresh horticultural products and international passenger traffic, along with shorter transit times. Greatest risk may be associated with contraband fruit smuggled by airline passengers rather than by

From: Paull, R.E. and Armstrong, J.W. (eds) (1994) *Insect Pests and Fresh Horticultural Products: Treatments and Responses.* CAB INTERNATIONAL, Wallingford, UK.

Table 1.1. Examples of insects and mites that can be carried by fruit and vegetables, regarded as quarantine pests by various countries.

Common name	Insect	Common hosts	Approximate distribution
Codling moth	*Cydia pomonella* (L.)	Apple, pear, peach, quince, *Prunus* spp., walnut	Worldwide, except Japan
Sweet potato weevil	*Cylas formicarius* (Fab.)	Sweet potato	Africa, Asia, Pacific Islands, North and South America
Mango seed weevil	*Sternochaetus mangiferae* (Fab.)	Mango	Africa, Asia, Australia, Hawaii
Red-legged earth mite	*Halotydeus destructor* (Tucker)	Leafy vegetables	Australia, New Zealand, Africa
Fruit flies			
Mexican fruit fly	*Anastrepha ludens* (Lw.)	Citrus, other tropical and subtropical fruits	Central America, Mexico
Caribbean fruit fly	*A. suspensa* (Loew)	Tropical and subtropical fruits	Caribbean, Southern Florida
Mediterranean fruit fly	*Ceratitis capitata* (Wied.)	Deciduous, subtropical and tropical fruits	Southern Europe, Africa, Central America, South America, Western Australia, Hawaii
Melon fly	*Bactrocera cucurbitae* Coq.	Cucurbits, tomato, many other fleshy fruits	Asia, Hawaii, Papua New Guinea, Africa
Oriental fruit fly	*B. dorsalis* (Hendel)	Most fleshy fruits or vegetables	Asia, Hawaii
Queensland fruit fly	*B. tryoni* (Froggatt)	Deciduous, subtropical and tropical fruits	Australia, Pacific Islands
European Cherry fruit fly	*Rhagoletis cerasi* (L.)	Cherry, *Lonicera* spp.	Europe
Cherry fruit fly	*R. cingulata* (Lw.)	Wild and cultivated cherry, *Prunus* spp.	North America
Apple maggot fly	*R. pomonella* (Walsh)	Apple, blueberry	USA, Canada
Mealybugs			
Pineapple mealy bug	*Dysmicoccus brevipes* (Ckll.)	Pineapple	Africa, Asia, Australia, Pacific Islands, South America

commercial shipments. For example, a special inspection effort by regulatory personnel of all passenger baggage coming through Los Angeles International Airport from selected Central and South American countries during one week in 1990 resulted in 677 interceptions of contraband fruits with a total weight of almost 1 ton. The intercepted fruit contained 61 live fruit fly larvae (Anonymous, 1990). These special inspection results contrast with an average interception rate of 434 contraband fruits for one week of random baggage inspections for all international arrivals. Increasing travel and trade means that quarantine barriers are becoming increasingly more difficult to maintain and new fruit fly introductions occur at an alarming rate (Carey and Dowell, 1989).

Quarantined insect pests, especially tephritid fruit flies, have seriously disrupted fresh fruit marketing between countries, and even between geographical areas within countries (e.g. Florida to California, Hawaii to the Mainland US, Queensland to Victoria in Australia, Okinawa to Japan) unless accepted quarantine treatments are available. Therefore, effective postharvest quarantine treatments (Table 1.2) that are not harmful to either the fruit or people coming in contact with or consuming the fruit are essential to the unrestricted trade of fresh horticultural commodities through domestic and international marketing channels.

Disinfestation and Quarantine

Many disinfestation procedures using chemical or physical treatments, or combinations thereof, are accepted by the regulatory agencies of most importing countries. However, prohibitions on the use of ethylene dibromide (Anonymous, 1984; Ruckelshaus, 1984) and potential restrictions on the use of methyl bromide as postharvest disinfestation fumigants in some countries necessitate alternative treatments to toxic fumigants (Armstrong, 1992). Current research efforts are directed to the use of physical disinfestation treatments, such as heat or cold, that may be more costly and difficult to apply than chemical treatments, such as fumigants. Furthermore, many commercially important commodities attacked by quarantined pests have no approved disinfestation treatments.

Some basic concepts of quarantine need to be defined:

1. *Quarantine host* is any horticultural product that, at one or more of its growth stages, can be naturally infested in the field, and support development.
2. *Quarantine treatment* is any individual action or treatment that, by itself, can be used to disinfest host horticultural products.

Table 1.2. Currently approved and possible postharvest insect control treatments for fruits, vegetables, and flowers. This lists the usefulness of various potential quarantine procedures for horticultural commodities. It does not indicate their effectiveness against specific insect pests.

Commodity	Fumigation: Methyl bromide	Fumigation: Phosphine	Fumigation: Other fumigants	Cold treatments	Heat treatments	Modified atmospheres	Radiation	Other procedures, treatments, or combinations of treatments	
I. ***Fresh fruits***									
Pome fruit									
Apple	T	O		T		X	X	X	Certification (infestation level and handling system)
Pear	T	O		T		X	X		
Stone fruit									
Apricot	T	O		T			X	T	Cold and MB
Cherry	T	O		T			X	T	Cold and MB
								X	Certification (infestation level and handling system)
Nectarine	T	O		T			X	T	Cold and MB
Peach	T	O		T			X	T	Cold and MB
Plum	T	O		T			X	T	Cold and MB
Soft fruit									
Strawberry	T	O			O		X		
Bush berries							X		
Grapes	T			T				T	Cold and MB
Kiwifruit				T					
Persimmon				T					
Pomegranate				T					
Fig, fresh							X		
Date	T	T		T		T	X		
Avocado	O				X		O	T	Cold and MB
Citrus fruits			T(HCN)		X				
Grapefruit	X	X		T			X	X	Conditioning before cold treatment

Lemon/Lime	T			X/O					
Mandarin/Tangelo	T						X		
Orange	T			T			X		
Banana									
Cavendish	O	O		O	X			X	Certification
Brazilian									
Litchi	O			X	X				
Papaya	X	O		X	T		X	T	Picking maturity + hot water + cooling
Mango	X/O	O			X		X		
Cherimoya				O					
Carambola				T	O				
II. *Fresh vegetables*									
Tomato/Pepper/Eggplant	T/O	O			T(VH)			T/X	Dimethoate dip
Melons/Cucumber	T/O				T(VH)				
III. *Dried fruit and vegetables*									
Raisins, prunes, etc.	T	T		X	X	X	X		
IV. *Tree nuts*									
Almond	T	T		X	X	T	X		
Filbert	X	X							
Macadamia	X	X							
Pecan	X	X							
Pistachio	X	X			X				
Walnut	T	T		X	X	T	X		
V. *Flowers*	T		X (HCN)	X	X		O		

Source: This table was an outcome of US Dept of Agriculture Research Service and Western Land Grant College regional project and was assembled by Dr G. Mitchell, University of California, Davis (used with permission).

T = treatment registered for use against certain pests; X = possible treatment; O = host intolerance expected based on past tests or observations; blanks indicate no information available; HCN = hydrogen cyanide; VH = vapour heat.

3. *Quarantine system* is a systems approach, utilizing various individual actions or treatments sequentially so that the combined effects of all the actions or treatments are required to disinfest the product.
4. *Quarantine security* is that degree of statistical probability needed by quarantine treatments or systems to disinfest host products so that, upon transport or shipment of the treated products, the targeted pests cannot become established in any area where they do not already exist.

Insects – the Problem

The reproduction rate, short life cycle, and adaptability of pest insects, as well as their destructive potential, underscore the need for quarantine treatments and barriers. The reproduction rates for many insects of quarantine importance are rapid, especially under optimum conditions, such as warm average temperatures and the absence of low winter temperatures (Fletcher, 1987). The time required for life cycle completion differs with environmental factors and species, e.g. oriental fruit flies produce 3 to 18 generations per year (Saeki *et al.*, 1980; Vargas *et al.*, 1984) depending upon temperature, and the adaptability of insects allows them to move into new ecological niches, adapt to new conditions, and even expand their host range. The damage caused by insects includes small surface blemishes, destruction of the edible flesh, fruit drop, seed destruction, and spoilage from decay organisms entering the fruit aided by insect activity (Hagen *et al.*, 1981).

It is more difficult to develop quarantine treatments or systems for pest organisms when more than one life stage is present on or in the host commodity because the stage that is most difficult to kill must be determined. For example, the egg and pupal stages may be more tolerant to fumigants than larval or adult stages because of lower metabolic and respiration rates. A diapausing stage, when part of the target pest's life cycle, may exhibit even greater tolerance because of lower metabolic and respiration rates.

Pest organisms' life stages can be on the surface or inside the host commodity, and the location of the pest organism also affects the type and severity of quarantine treatment or system required to disinfest the host commodity. Pests with life stage(s) only on the surface of the host commodity are generally more susceptible to treatments because they are more exposed. Insects with life stage(s) inside the host commodity require more severe treatments because penetration by the treatment into the commodity is needed to cause insect mortality.

Quarantine Security

One of the difficult problems in developing quarantine treatments and systems is appropriate statistical criteria that will assure quarantine security. In the case of fruit fly disinfestation treatments, the United States generally uses the probit 9 concept (Baker, 1939). Probit 9 corresponds to a mortality rate of 99.9968%, which infers that no more than three individuals from a population of 100 000 will survive a quarantine treatment. Using 95% confidence limits, probit 9 could have 29 to 136 survivors per million with an average of 32 (Couey and Chew, 1986). Japan often uses a variation of the probit 9 concept that requires no survivors from a treated population of 30 000 target pests. The probit 9 concept has come under increased scrutiny (Landolt *et al.*, 1984) because it does not consider several factors that directly impact quarantine situations, such as the actual infestation per volume of treated fruit, the potential for culling infested fruits during processing, the infestation variability between good hosts and poor hosts, or the probability of a potential normal mating pair occurring at the same time at the market destination.

The assumption is that probit 9 ensures insect control. The alternative of zero tolerance is meaningless because insects are not detected and you are attempting to prove a negative. Additionally, probit 9 cannot be used to predict treatment time due to the large confidence levels required and, in many cases, mortality is curvilinear, not logarithmic. Risks are clearly different for different crops: apples generally have only one codling moth larva per fruit, whereas there may be 20 or more fruit fly eggs and larvae in a papaya, providing different degrees of risk. Tolerance or susceptability to a treatment often changes with the different insect development stages (Jang, 1986, 1991). A caveat to the probit 9 'no live insect' idea occurs with irradiation that is designed to sterilize rather than kill the target insects.

Alternative quarantine statistics that challenge the probit 9 concept have appeared in the literature (Landolt *et al.*, 1984; Couey and Chew, 1986), but few changes in quarantine statistics have occurred. An exception is New Zealand, which developed a maximum pest limit concept for imported produce (Baker *et al.*, 1990; Harte *et al.*, 1992). Maximum pest limit statistics include a sampling model for the accurate assessment of infestation levels in the host commodity, and a limit on the number of immatures that may be present in consignments (e.g. three live fruit fly larvae per day) imported during a specified time to a specified location (Baker *et al.*, 1990). This assumes that the emerging adults will not be able to find a mate because the commercial distribution would scatter the arriving fruit over a large area. A higher limit may be appropriate for Lepidoptera due to low numbers in each fruit (Baker *et al.*, 1990).

The difficulty with this system is the lack of good biological data on infestation indices available, and the limitation of sampling methods for accurate assessment. Application of a treatment of known efficiency ensures that the limit is not exceeded if the infestation, determined by the sampling model, is below a predetermined value (Baker *et al.*, 1990).

Biotic Potential and Life Cycles

Quarantine pests of fresh horticultural products can vary greatly in destructiveness and biotic potential. The biotic potential of any pest is directly related to its life cycle, which is affected by the availability of hosts (both non-economic and economic) and environmental factors, especially temperature and humidity. Biotic potential varies widely between various pests. The following examples taken from the insect orders, Diptera, Lepidoptera, and Coleoptera, illustrate the variation in biotic potential.

Tephritid fruit flies

Development times are somewhat similar for many fruit fly species. Required times for the different developmental stages at 25°C, for example, are about 1–3 days for egg hatch to occur, 7–9 days for larval development and pupation, and 10–11 days for adult emergence (Fletcher, 1987). The time required for completion of life cycles in the field is determined primarily by temperature (O'Loughlin *et al.*, 1984), but is also affected by fruit ripeness, moisture content, and degree of larval crowding (Ibrahim and Rahman, 1982; Vargas *et al.*, 1983a; Couey *et al.*, 1985). Adult fruit flies require about 6–10 days to become sexually mature, and fecundity is usually high (Vargas *et al.*, 1984).

The abundance of larval hosts is one of the major factors regulating fruit fly populations (Tan and Lee, 1982; Vargas *et al.*, 1983b). Major environmental mortality factors include high and low temperatures, wet or arid conditions, parasitism, and frugivorous birds (Fletcher, 1987). Fruit fly egg development has an upper temperature limit of 36°C and a lower limit of 12°C (Messenger and Flitters, 1958). These life cycle characteristics are suited to exploiting resources, especially if they occur intermittently. This ability combined with good adult mobility, relatively long life spans (50–93 days), high fecundity (800–1600 eggs per female), and multivoltine reproduction (10–12 generations per year) (Vargas *et al.*, 1984), explain the threat posed by these economic pest insects (Carey and Dowell, 1989).

Tortricid fruit moths

Two fruit moths of major quarantine importance include the codling moth (*Cydia pomonella* (L.)) and the oriental fruit moth (*Grapholita molesta* (Busck)) that attack mainly pome fruits and stone fruits. These moths have a lower fecundity than fruit flies, producing from 50–150 eggs per female (Metcalf *et al.*, 1962), and fewer generations (1–7) per year (Metcalf *et al.*, 1962). Unlike fruit flies that cannot survive the low winter temperate climates, codling moth and oriental fruit moth can survive temperatures below 0°C by over-wintering as diapausing larvae in cacoons under tree bark or in rubbish, weeds, and mummied fruit (Metcalf *et al.*, 1962). In the spring, the larvae break diapause, pupate, and emerge as adults after 2–4 weeks (Metcalf *et al.*, 1962). The eggs require 3–20 days to hatch and development from first to fifth larval instar requires about 3–5 weeks, depending on temperature (Metcalf *et al.*, 1962; Yokoyama and Miller, 1987). Major environmental mortality factors include extreme cold and wet conditions, parasitism, disease, and insectivorous birds (Metcalf *et al.*, 1962).

Mango weevils

The mango weevil, *Cryptorhynchus mangiferae* (F.), is an important pest of mangoes, the only known host of this pest. Except for North and South America, this cucurlionid pest is found throughout the mango growing areas of the world (Balock and Kozuma, 1954; Shukla and Tandon, 1985). Eggs are laid on the outside of developing fruit and the enclosing larvae burrow through the flesh to penetrate the young seed, where the weevil completes its development (Rutherford, 1914; Jarvis, 1946). Mango weevil oviposition may occur in less than 8 days (Shukla and Tandon, 1985) or up to 90 days (Balock and Kozuma, 1954), and usually occurs when the fruits are marble size (Shukla and Tandon, 1985). As the fruits mature, the seed covering becomes hard and the first instars cannot penetrate the endocarp. The duration of the larval period may be influenced by climate, location, host cultivar, and non-biotic site characteristics, such as soil chemistry or humidity (Hansen *et al.*, 1989). Larval development requires from 3–10 weeks (Balock and Kozuma, 1954; Hansen *et al.*, 1989; Shukla and Tandon, 1985) with evidence of much individual variation (Hansen *et al.*, 1989). The number of larval instars (5–7) is disputed and actually may not be constant (Hansen *et al.*, 1989). The pupal stage requires about a week (Subramanyam, 1925; Balock and Kozuma, 1954; Shukla and Tandon, 1985). The fruits, after falling to the ground, are highly susceptible to rodent and insect attack and the seeds may become so damaged that weevils move out of the seeds and seek hiding places where they over-season (Hansen *et al.*, 1989). Although an occasional weevil may be found in the fruit flesh, the weevil is primarily a pest of the seed with one seed supporting up to five larvae (Balock and Kozuma, 1954; Hansen *et al.*, 1989). Although the mango weevil is rather innocuous when compared to the tephritid fruit flies and tortricid fruit moths, it remains a quarantine pest because of its absence from North and South America and the Caribbean.

Plant Resistance to Insects

Both fruit, cultivar, and stage of ripeness can significantly influence host status (Greany, 1989). 'Starking Delicious' is more resistant to apple maggot than 'Jonathan' or 'Golden Delicious', with some variation in apple selections to codling moth (Goonewardene *et al.*, 1979). Green fruit are not frequently hosts under normal conditions. Mature green 'Brazilian', 'Valery', and 'William's' bananas are not infested or are resistant to fruit flies (Armstrong, 1983; Armstrong *et al.*, 1983), although these fruits are fruit fly hosts when ripe; 'Eureka' and 'Lisbon' lemons are not hosts for the Mediterranean fruit fly at commercial maturity (Spitler *et al.*, 1984); and papaya with less than 25% yellow are not preferred fruit fly hosts (Seo *et al.*, 1982; Couey and Hayes, 1986). Reasons for these temporal non-host status have not been determined.

Pineapples with 50% or more 'Smooth Cayenne' parentage are not fruit fly hosts (Seo *et al.*, 1970a, 1973a; Armstrong *et al.*, 1979; Armstrong and Vargas, 1982). However, each new pineapple cultivar has to be tested in the field to determine its fruit fly host status. The reason for the non-host status of Smooth Cayenne is unclear (Macion *et al.*, 1968) although oviposition and egg hatch occur, fruit water content appears to be a major biophysical factor in larval mortality. In a number of cases, the fruit skin intactness can alter host status such as for 'Sharwil' avocado (Armstrong *et al.*, 1983).

There are occassional disagreements between regulatory agencies of different countries over the acceptance of non-host status for certain fruits. To date, the non-host status of mature-green 'Brazilian', 'Valery', and 'William's' bananas from Hawaii has not been accepted by the US, while Japan imports mature-green bananas from Central American countries where the Mediterranean fruit fly and several species of *Anastrepha* occur. Conversely, the US determined that strawberries were not a host for Mediterranean fruit fly, based on more than 30 years of inspections of imported strawberries (Armstrong *et al.*, 1984), whereas Japan regulates strawberries as a Mediterranean fruit fly host based on the works of Thiem (1937), Bohm (1958), and Baas (1959).

Certified Pest-free Zones or Growing Periods

Preharvest quarantine systems, such as certified pest-free areas or growing periods, are accepted by various quarantine enforcement agencies throughout the world. The use of preharvest quarantine systems generally precludes the need for, or modifies the severity of, postharvest quarantine treatments or systems.

The US accepts certain fruits without quarantine treatment from Chile provided that they are from certified Mediterranean fruit fly-free zones (Animal and Plant Health Inspection Service, 1982). These fly-free zones were established by large-scale insecticide control programmes and they are maintained by approved Mediterranean fruit fly trapping and detection protocols. Any area losing its fly-free zone certification due to the trapping of Mediterranean fruit fly must undergo an eradication programme and recertification.

In Australia, fly-free growing periods are established for the Queensland fruit fly, *Dacus tryoni* (Froggatt), and for the Queensland fruit fly complex found in Queensland and the Northern Territory. During the cold winter months of May through August, all fruits grown in these two areas can be shipped to the state of Victoria without treatment for fruit flies (Heather *et al.*, 1987).

Chemical Control

Fumigation

Fumigation was the primary method for disinfesting fruit flies from fresh fruits since the development of methyl bromide (MB) fumigation in the early 1940s (USDA, 1940) and ethylene dibromide (EDB) fumigation in the early 1950s (USDA, 1951). Fumigants used or tested include methyl bromide (Seo *et al.*, 1971; Moran *et al.*, 1974, 1975; Anthon *et al.*, 1975, 1977; Benschoter, 1979, 1980; Gaunce *et al.*, 1980; Hatton and Cubbedge, 1980; Yokoyama *et al.*, 1987), ethylene dibromide (Havens *et al.*, 1979; Seo *et al.*, 1972, 1979b; Wade and Rigney, 1979), phosphine (Spalding *et al.*, 1977; Seo *et al.*, 1979a,c), ethyl formate (Vincent and Lindgren, 1972), hydrogen cyanide (Houck *et al.*, 1989; Hansen *et al.*, 1991a,b), acetaldehyde (Aharoni *et al.*, 1979), and others (Burditt *et al.*, 1963). Pressure and temperature directly affect the fumigant concentration and fumigation time required for effective results. Vacuum or negative-pressure atmosphere fumigation improves the penetration permitting the use of lower fumigant concentrations and treatment times. A major problem with many of these fumigants is phytotoxicity, which varies with commodity, variety, seasonal conditions, temperature and humidity, as well as fumigant dose and time.

Chemical fumigants provide flexibility, as the gaseous condition provides the property of diffusion that allows penetration of the fruit mass (Seo *et al.*, 1979b). All insecticidal fumigants are toxic to humans, acting rapidly (hydrogen cyanide) or slowly (methyl bromide). Additionally, some of these fumigants are highly flammable (hydrogen cyanide).

Insecticides

Insecticides incorporated into sprays, dips, and fruit waxes have potential for disinfesting fruit flies from host fruits, vegetables and, especially, flowers and foliage ornamentals. The insect growth regulator methoprene, incorporated in fruit waxes and applied to infested papayas and peaches, was shown to cause high levels of mortality to Mediterranean fruit fly (Saul *et al.*, 1985, 1987), oriental and melon fruit flies (Saul and Seifert, 1990). Postharvest dipping in water containing either of the deep-penetrating insecticides dimethoate or fenthion adequately disinfested Queensland fruit fly or banana fruit fly, *D. musae* Tryon, from tomatoes or bananas, respectively (Saunders and Elder, 1966; Smith, 1977; Swaine *et al.*, 1984a; Heather *et al.*, 1987). However, dimethoate dips to disinfest Queensland fruit fly from the 'Kensington' mango was not efficacious (Swaine *et al.*, 1984b). Insecticides for postharvest use are of concern as to residues and consumers.

Controlled atmospheres

Controlled atmospheres (CA) (reduced O_2, elevated CO_2, and/or added CO) have been evaluated as possible insect control treatments in a number of studies (Klaustermeyer *et al.*, 1973; Moran and Gaunce, 1975; Davis and Jay, 1977; Soderstrom, 1977; Aharoni *et al.*, 1981; Lidster *et al.*, 1981; Gaunce *et al.*, 1982; Benschoter, 1987; Delate *et al.*, 1990; Soderstrom and Brandl, 1990). However, most fresh horticultural crops do not tolerate those CA treatments found effective for insect control in dry cereals and legumes. There is also limited information on insect responses to CA.

Physical Control

Temperature

Heat treatments offer a viable alternative to other quarantine treatment technologies (Couey, 1989). Hot-water immersion, vapour heat and forced hot-air treatments have been the heating technologies of choice against insects (Couey, 1989). Heat treatments have the merit of effective fungicidal and insecticidal action, ease of application, and leave no chemical residues. The disadvantages are the potential for fruit damage, that must be researched on a commodity-by-commodity basis (Armstrong and Couey, 1989; Paull, 1990b), and a relatively higher cost of application compared to fumigation. Insect responses to heat treatment and thermal death relationships have been developed for some insects of quarantine importance (Jang, 1986; 1991; Yokoyama *et al.*, 1987; 1991; Moss and Jang, 1991). Mortality due to heat is dependent upon oxygen availability, pH, previous temperatures, general energy status of the insects and insect age and stage (Moss and Jang, 1991).

Hot-water immersion treatments were developed to disinfest Mediterranean fruit fly from bananas (Armstrong, 1982), papayas (Hayes *et al.*, 1984; Couey and Hayes, 1986), guavas (Gould and Sharp, 1990), and mangoes (Sharp and Spalding, 1984; Sharp, 1986, 1989; Sharp *et al.*, 1988, 1989; Sharp and Picho-Martinez, 1990; Animal and Plant Health Inspection Service, 1992a). Vapour heat treatments have been developed for bell pepper, Chinese peas, cucumber, eggplant, green beans, lima beans, litchis, mango, papaya, tomato, grapefruit, and yellow wax beans (Jones, 1940; Balock and Starr, 1946; Balock and Kozuma, 1954; Seo *et al.*, 1974a; Sugimoto *et al.*, 1983; Furasawa *et al.*, 1984; Hallman *et al.*, 1990). A high-temperature (or hot) forced-air quarantine treatment to disinfest Mediterranean fruit fly, melon fly and oriental fruit fly from papaya was recently developed (Armstrong *et al.*, 1989; Armstrong, 1990; Hansen *et al.*, 1990; Animal and Plant Health Inspection Service, 1992b). This treatment is essentially a vapour heat treatment without the vapour, and may be effective in disinfesting a number of fruit fly hosts including atemoya, beans, some citrus, eggplant, mangoes, peppers, and squashes. Heat treatments are being considered for codling moth (Yokoyama *et al.*, 1991), oriental fruit moths (Yokoyama and Miller, 1987), Caribbean fruit fly in stone fruit (Sharp, 1990) and carambola (Hallman and Sharp, 1990).

Cold treatments, like heat treatments, are relatively easy to apply and leave no chemical residues. Cold treatments have been recommended for disinfestation of fruits for many years following the observations that refrigeration was an effective quarantine treatment for Mediterranean fruit fly (Hooper, 1907; Lounsbury, 1907; Back and Pemberton, 1916). Today, cold or refrigeration quarantine treatments are used to disinfest a wide range of fruits (Rippon and Smith, 1979; Benschoter, 1984, 1988; Hill *et al.*, 1988; Animal and Plant Health Inspection Service, 1992c). The eggs and larval stages of fruit flies are killed by exposure to temperatures below 10°C. However, only temperatures below 3°C are practical because of the long exposure times involved (10 days and longer at 0°C) (Burditt and Balock, 1985; Armstong and Couey, 1989).

Microwaves and ultrasound

Microwaves, or dielectric heating, is limited by non-uniform heating and heating from the centre of the fruit outward. Microwave treatment to disinfest Mediterranean fruit fly, melon fly and oriental fruit fly from mangoes and papayas damaged the fruits before any effect on the fruit flies was observed (Armstrong, unpublished data). Interrupted or pulsated microwave treatments may be effective against mango seed weevil with minimal fruit injury (Seo *et al.*, 1970b).

The use of ultrasound for disinfestation of papaya fruits may merit further investigations (Horner, 1979). Mortality to oriental fruit fly eggs and larvae was shown to be dependent on treatment intensity, temperature, and duration, and no injury was observed on papayas treated with 20 kHz (at 85 W l^{-1}). Hayes (1982) and Hayes *et al.* (1983) showed that ultrasound just below the threshold for papaya fruit damage killed fruit flies no deeper than 1 mm. Since fruit fly eggs in papaya are in the first 4 mm of the fruit, ultrasound alone does not appear to be a viable disinfestation treatment for papayas.

Irradiation

The use of up to 1.0 kilogray (kGy) for insect disinfestation in fresh horticultural commodities was approved by the US Food and Drug Administration (FDA, 1987). However, there are four major problems facing the application of irradiation technology. The first relates to efficacy. Dosages that will kill insects may also damage fruit. Therefore, lower dosages that sterilize insects, rather than kill them, are recommended. Unfortunately, a dose leaving live larvae may cause regulatory problems such as determining whether the commodity has been treated correctly. The second problem deals with overall response of the commodity to these doses in a commercial situation. Treatment studies under actual or simulated harvesting, packing, treating, storing, and shipping conditions must be completed to ensure that the treatment is not in any way irreversibly detrimental to the fruits (Maxie *et al.*, 1971). The third and fourth concerns are the economic feasibility (Maxie *et al.*, 1971) and consumer acceptance (Schulz *et al.*, 1989). If these criteria are met, then irradiation can be a viable quarantine treatment against many quarantine insects.

Irradiation has potential as a quarantine treatment against fruit flies infesting avocado, mango, citrus, tomato, litchis, banana, papaya, pome fruits, stone fruits, and strawberry (Burditt *et al.*, 1971; Burditt and Seo, 1971; Heather, 1986). Doses of 1.8 to 2.4 kGy will kill target insects or prevent the normal completion of the next successive life stage (Anon., 1971; Burditt and Seo, 1971; Moy *et al.*, 1971; Dollar and Hanaoka, 1972; Moy, 1985). Lower doses of 0.15 to 1 kGy of gamma radiation would disinfest fruit flies from papayas, tomatoes, cucumbers, bananas, litchis, bell peppers, eggplant, and mangoes by allowing the emergence of only sterile adults (Balock *et al.*, 1966; Burditt *et al.*, 1971; Burditt and Seo, 1971; Seo *et al.*, 1973b). Fuller rose beetle on citrus and San Jose scale and codling moths on apples are potential targets (Burditt and Hungate, 1989; Angerilli and Fitzgibbon, 1990; Johnson *et al.*, 1990). Furthermore, irradiation was shown to be effective against mango weevil infesting the seeds of commercial mangoes (Burditt and Seo, 1971; Seo *et al.*, 1974b). Cut flowers and other ornamentals with a diverse population of insects are another group of commodities where irradiation may be successful (Wit and van de Vrie, 1985a).

Washing

The simple technique of washing may be used as a quarantine treatment for pest organisms found on the surface of host commodities. The addition of a detergent (insecticidal soaps) increases the effectiveness of washing by dissolving the waxy epicuticle of insect pests and causes rapid drowning (Waller, 1990; Hata *et al.*, 1992). Additionally, the use of hot-water washing may kill surface pest organisms if the thermal death-point of the target organism is reached.

High and low pressures

High pressure and vacuum treatments can be used in conjunction with fumigants and controlled or modified atmospheres. This treatment increases penetration of the host commodity and enhances the effect on the target pest organisms (Aharoni *et al.*, 1980; Steward *et al.*, 1980). Reduced pressure has been suggested to control green peach aphid, *Myzus persicae* (Aharoni *et al.*, 1986).

Postharvest Quarantine Systems

A number of quarantine systems exist for disinfesting tephritid fruit flies from tropical fruits. Quarantine systems are combinations of two or more separate quarantine treatments, such as those previously described. Some examples of quarantine systems include:

1. *Fumigation combined with heat or refrigeration.* Higher than ambient (>21°C) fumigation temperatures generally allow for decreased EDB or MB concentrations and/or decreased fumigation time to achieve efficacy because of increased insecticidal activity and penetration into the commodity (Armstrong and Couey, 1984; Couey *et al.*, 1985; Williamson

et al., 1985; Armstrong *et al.*, 1988). Higher fumigation temperatures increase the respiration rate of the target pest organism which increases the rate of fumigant uptake, and results in less time required to kill the insect (Sun, 1946).

Fumigation with MB followed by a cold treatment is a quarantine system used to disinfest Mediterranean fruit fly, melon fly, or oriental fruit fly, from avocado, apple, apricot, cherry, grape, pear, plum, peach, and nectarine (Seo *et al.*, 1971; Animal and Plant Health Inspection Service, 1992d). Temperate fruits are better candidates than tropical fruits because they are more cold tolerant.

2. *Heat combined with cold.* A combination heat and cold treatment was developed to disinfest Mediterranean fruit fly, melon fly, and oriental fruit fly from papayas. The treatment consists of fruit selection for papayas less than one-quarter ripe for hot-water immersion at 49°C for 20 min followed by refrigeration at 8–9°C for 10 days during surface transit (Couey *et al.*, 1984).

3. *Fruit selection combined with heat.* Fruit selection of less than one-quarter-ripe papayas followed by a two-stage hot-water immersion to disinfest Mediterranean fruit fly, melon fly, and oriental fruit fly eggs from papayas was used in Hawaii (Couey and Hayes, 1986). This treatment has been withdrawn due to treatment failures associated with a genetic blossom-end defect (Zee *et al.*, 1989).

Horticultural Effects of Control Measures

Chilling

Temperature limits plant growth and development (Couey, 1982). The growth and development of plants that evolved in the tropics and to a lesser extent subtropics are severely limited by temperatures between 0°C and up to 10–12°C (Paull, 1990a). Chilling injury limits the postharvest life of these commodities. Low temperatures, <10–12°C for tropical and <7°C for subtropical crops, cannot be used for insect disinfestation without caution (Paull, 1990a). The extent of postharvest loss due to chilling injury is difficult to determine because chilling injury predisposes all commodities to pathogenic attack, especially saprophytes, as well as disrupting ripening of climacteric fruits. Symptoms of chilling injury generally develop after removal from refrigeration to ambient temperatures. The symptoms most commonly seen are surface lesions and failure to ripen (Morris, 1982).

The stage of fruit ripeness and age of leaf tissue at the time of low temperature storage significantly influences the fruit sensitivity to visible chilling injury symptom development (Paull, 1990a). The preclimacteric stage is generally more sensitive to cold than the postclimacteric stage for avocados, papayas, honey dew melons, tomatoes, and mangoes. The extreme sensitivity of bananas to chilling does not allow any difference to be determined. However, banana fruit in the less mature apical hands are less susceptible to chilling injury than the more mature distally borne fruit (Marriott, 1980). Avocados increase in chilling injury sensitivity up to the climacteric, then sensitivity declines (Zauberman *et al.*, 1973).

Heat

Heat treatments are now being widely used though there is a potential for injury to the commodity (Paull, 1990b). The lethal temperature has been reported to be 45°C for tomatoes, 63°C for grapes, and 49 to 52°C for apples (Huber, 1935). Even below these temperatures, growth and metabolism can be adversely affected by heat. Couey (1989) grouped fruits by heat tolerance and heat sensitivety to hot-water immersion treatment. The heat-tolerant fruits were bananas, papayas, mangoes, pears, and litchis, while the heat-sensitive fruits were peaches, raspberries, bell peppers, and cantaloupes. The difficulty with such grouping is that thermotolerance may differ among cultivars, fruit sizes, and maturity and postharvest handling (Jones, 1942; Claypool and Vines, 1956; Sharp *et al.*, 1989).

Heat injury is characterized by a failure to develop normal pigmentation (Goodwin and Jamikorn, 1952), abnormal softening (Paull and Chen, 1990) and by a marked decline in ethylene production (Field, 1984). Respiration rate and ethylene synthesis are affected by exposure to high temperatures (Jones, 1939; Akamine, 1966; Chan, 1986a,b, Paull and Chen, 1990). Colour changes associated with ripening in tomato are inhibited with storage at temperatures above 30°C (Goodwin and Jamikorn, 1952). Above 30°C, bananas (Burg, 1962) and papaya (An and Paull, 1990) fail to ripen normally, with the pulp becoming soft and watery.

Bacteria, plants, and animals exposed to higher temperatures for a brief period acquire transient thermotolerance (Lindquist and Craig, 1988). As this thermotolerance response develops, a unique group of proteins, heat shock proteins (HSP), are synthesized. Field induction of HSP has been shown to occur in cotton (Burke *et al.*, 1985), soybeans (Kimpel and Key, 1985), sorghum (Ougham and Stoddart, 1986), and papayas (Paull and Chen, 1990). The extent of protection against softening disruption is dependent upon season, stage of fruit ripeness at exposure, and pretreatment time and temperature (Paull, 1990b).

Many procedures have been empirically developed to reduce the injury caused by heat treatment. Exposure of ripening papaya fruit either to 42°C for 4 h or to 38–42°C for 1 h followed by 3 h at 22°C results in the development of thermotolerance and the production of heat shock proteins (HSPs) (Paull and Chen, 1990). The 8-h approach time to 44°C used for a papaya vapour heat treatment (Seo *et al.*, 1974a) is designed to reduce subsequent injury when the fruit are exposed to 44°C for 8.75 h. This approach time would be expected also to provide the conditions necessary to develop HSPs and associated thermotolerance. The same would be likely to occur in deciduous fruits (Claypool and Vines, 1956). The quick rise of temperature to 49°C using vapour heat always damaged Valencia oranges and grapefruit (Sinclair and Lindgren, 1955), but they were tolerant when heated for 8 h at 43°C before the 49°C treatment. Cucumber can be preconditioned by a 24-h treatment at 32.5°C to tolerate subsequent exposure for 50 min to hot water at 46°C (Chan and Linse, 1989). Though 32.5°C is lower than the temperature normally used to induce HSPs, the 24-h duration would probably induce HSPs.

Controlled atmospheres

Controlled atmospheres (CAs) with O_2 levels below 1% and/or CO_2 levels above 50% for short durations have been tested for possible use in postharvest insect control. Ke and Kader (1989) reported that keeping fruits in 0.25% or 0.92% O_2 – or in 20% to 99% CO_2 – atmospheres for short durations at 0–10°C had beneficial effects that included reduction of respiration and ethylene production rates, retardation of loss of green colour and softening, and inhibition of decay. However, longer exposures were detrimental to flavour and induced physiological disorders (symptoms included external and internal browning and surface pitting), which limited the tolerance of fresh fruits to the stress CA conditions.

Tolerance of fruits to low O_2 levels is much greater than their tolerance to CO_2-enriched atmospheres. Nectarines, peaches, plums, pears, oranges, and strawberries developed injury after exposure to 20–99% CO_2 for 3–10 days at 0, 5, or 10°C. As expected, the higher the temperature and CO_2 concentration, the shorter the time for injury to occur (Ke and Kader, 1989, 1990; Smilanick and Fouse, 1989; Ke *et al.*, 1990). Cured sweet potato roots can withstand 1% oxygen and 40% CO_2 at 25°C for up to 3 weeks (Delate *et al.*, 1990).

Comparison of fruit tolerance data and the limited data available on the efficacy of low-O_2 atmospheres in insect control, indicates that insecticidal O_2 concentration/time/temperature combinations may have a potential use as quarantine treatments for some insects and host commodities.

Gamma radiation

Gamma irradiation dosages below 1 kGy have no detrimental effects on many commodities, but they may be harmful to others such as avocados, lemons, limes, grapefruit, oranges, and grapes (Kader, 1986; Sommer and Mitchell, 1986). There is a lack of information because very few horticultural storage or response studies have been done at the entomological doses proposed for insect sterilization (100 to 250 Gy).

The low doses (250 Gy) have not been found to alter the dry matter, total and reduced ascorbic acid, carotenoids, pH, titratable acidity, total soluble solids, total reducing sugar of papayas (Dollar *et al.*, 1971), and the fruits in these tests were reported to be firmer. However, more recent studies (R.E. Paull, unpublished) have shown that fruit firmness varies with season and stage of fruit ripening when irradiated. Skin scald can also be a problem on papaya that are irradiated, then immediately placed into storage at 12.5°C. Irradiation of banana at 250 or 500 Gy delays ripening about 2 days at normal ripening temperatures (Ferguson *et al.*, 1966). This was confirmed by Maxie *et al.* (1968) using 350 Gy. Irradiation at these dose levels does not have fungistatic effects (Hunter *et al.*, 1969). Mango palatability was only preserved if the fruit received a dose of less than 100 Gy (Hatton *et al.*, 1961), although different mango varieties can withstand higher doses without affecting quality (Akamine and Goo, 1979).

Chemicals

Slight differences in susceptibility to methyl bromide and ethylene dibromide injury exist among different varieties of the same commodity (Claypool and Vines, 1956), it is therefore not possible to generalize on product compatibility with the fumigation treatment (Mitchell, 1981). For example, the threshold for phytotoxic injury in nectarines varied from 48 to 96 g methyl bromide m^{-3}, and all fumigated fruits were significantly firmer (Harvey *et al.*, 1989). Flowers vary widely in their response to methyl bromide (Wit and van de Vrie, 1985b) and hydrogen cyanide fumigation (Hansen *et al.* 1991a). Increasing hydrogen cyanide dosage increased phytotoxicity, with sunlight causing an increase in the injury in ginger flowers immediately after fumigation. Pincushion protea and, to a lesser extent, ginger flowers are more sensitive to hydrogen cyanide than heliconias or anthuriums. Carnations can also be readily damaged by hydrogen cyanide (Wit and van de Vrie, 1985b).

Conclusions

No single quarantine treatment or system can be expected to work equally against all insects or for all host fruits, and the response to any treatment can vary greatly. Entomologists and postharvest physiologists need to develop insect dosage–mortality information following a standardized protocol to determine the treatment levels required to ensure that horticultural commodities can be shipped through domestic and international marketing channels without the risk of transporting unwanted insect pests or loss of the commodity quality attributes of appearance, texture, flavour, shelf life, and nutritive value. The dosage–mortality information consists of testing all life stages of each target insect pest in sufficient numbers to provide statistical reliability based on the relative susceptibilities of the different life stages to each treatment, and the treatment severity required to achieve the desired security level. The dosage–mortality information must be developed both on the naked insects to determine relative susceptibilities and tolerances, and on infested commodities to determine efficacy and phytotoxic effects.

References

Aharoni, Y., Stewart, J.K., Hartsell, P.L. and Young, D.K. (1979) Acetaldehyde – a potential fumigant for control of the green peach aphid on harvested head lettuce. *Journal of Economic Entomology* 72, 493–495.

Aharoni, Y., Stewart, J.K., Guadagni, D.G. and Mon, T.R. (1980) Thrips mortality and strawberry quality after vacuum fumigation with acetaldehyde and ethyl formate. *Journal of the American Society for Horticultural Science* 105, 926–929.

Aharoni, Y., Stewart, J.K. and Guadagni, D.G. (1981) Modified atmospheres to control western flower thrips on harvested strawberries. *Journal of Economic Entomology* 74, 338–340.

Aharoni, Y., Apelbaum, A. and Copel, A. (1986) Use of reduced atmospheric pressure for control of the green peach aphid on harvested head lettuce. *HortScience* 21, 469–470.

Akamine, E.K. (1966) Respiration of fruits of papaya (*Carica papaya* L., var. Solo) with reference to the effect of quarantine disinfestation treatments. *Proceedings of the American Society for Horticultural Science* 89, 231–236.

Akamine, E.K. and Goo, T. (1979) Effects of ionizing irradiation on 'Haden' mangoes. Research Report. Hawaii Agricultural Experimental Station University of Hawaii No. 205, 11pp.

An, R.F. and Paull, R.E. (1990) Storage temperature and ethylene on the ripening of papaya fruit. *Journal of the American Society for Horticultural Science* 115, 949–953.

Angerilli, N.P.D. and Fitzgibbon, F. (1990) Effects of cobalt gamma radiation on San Jose scale (Homoptera: Diaspididae) survival on apples in cold and controlled-atmosphere storage. *Journal of Economic Entomology* 83, 892–895.

Animal and Plant Health Inspection Service (1982) Fruits and vegetables approved for entry, M319.56-2-A, Section II, Chile. Plant Protection and Quarantine Programs Port of Entry Manual. US Deptartment of Agriculture, Hyattsville, Maryland, USA.

Animal and Plant Health Inspection Service (1992a) T102(a) Hot water treatment for mangoes. In: *Plant Protection and Quarantine Treatment Manual*. US Department of Agriculture, Hyattsville, Maryland, USA, pp. 5.42–5.43.

Animal and Plant Health Inspection Service (1992b) T103(b). High-temperature forced-air treatment of papaya for *Ceratitis capitata, Bactrocera dorsalis,* and *B. cucurbitae*. In: *Plant Protection and Quarantine Treatment Manual*. US Department of Agriculture, Hyattsville, Maryland, USA, p. 5.47.

Animal and Plant Health Inspection Service (1992c) T107(a), (b), (c), and (d). Cold treatment for *Ceratitis capitata* (Mediterranean fruit fly), *Anastrepha ludens* (Mexican fruit fly), species of *Anastrepha* other than *A. ludens*, and *Bactrocera tryoni* (Queensland fruit fly). In: *Plant Protection and Quarantine Treatment Manual*. US Department of Agriculture, Hyattsville, Maryland, USA, pp. 5.55–5.59.

Animal and Plant Health Inspection Service (1992d) T108(a) (1) and (2): Fumigation plus refrigeration. In: *Plant Protection and Quarantine Treatment Manual*. US Department of Agriculture, Hyattsville, Maryland, USA, pp. 5.63–5.64.

Anonymous (1971) Disinfestation of fruit by irradiation. Proc. Panel Joint FAO/IAEA Meeting, Honolulu, Hawaii, USA, 7–11 December 1970, IAEA, Vienna, Austria, 177pp.

Anonymous (1984) Ethylene dibromide: Amendment of notice to cancel registration of pesticide products containing ethylene dibromide. Congressional Federal Register. 49:14182-14185. US Govt. Printing Office, Washington, DC, USA.

Anonymous (1990) Fruit flies found in luggage in inspection of LA airport. *The Packer* 9 June, 17A.

Anthon, E.W., Moffitt, H.R., Couey, H.M. and Smith, L.O. (1975) Control of codling moth in harvested sweet cherries with methyl bromide and effects upon quality and taste of treated fruit. *Journal of Economic Entomology* 68, 524–526.

Anthon, E.W., Moffitt, H.R. and Smith, L.O. (1977) Codling moth: Dosage response of larvae in cherries to methyl bromide fumigation. *Journal of Economic Entomology* 70, 381–382.

Armstrong, J.W. (1982) Development of a hot-water immersion quarantine treatment for Hawaiian-grown 'Brazilian' bananas. *Journal of Economic Entomology* 75, 787–790.

Armstrong, J.W. (1983) Infestation biology of three fruit fly species on 'Brazilian,' 'Valery', and 'William's' cultivars of banana in Hawaii. *Journal of Economic Entomology* 76, 539–543.

Armstrong, J.W. (1990) High temperature forced-air quarantine treatments for fresh fruits infested by tephritid fruit flies. *Acta Horticulturae* 269, 449–451.

Armstrong, J.W. (1991) 'Sharwil' avocados: Quarantine security against fruit fly infestation in Hawaii. *Journal of Economic Entomology* 84, 1308–1315.

Armstrong, J.W. (1992) Fruit fly disinfestation strategies beyond methyl bromide. *New Zealand Journal of Crop and Horticultural Science* 20, 181–193.

Armstrong, J.W. and Couey, H.M. (1984) Methyl bromide fumigation treatments at 30°C for California stonefruits infested with Mediterranean fruit fly, *Ceratitis capitata* (Wiedemann). *Journal of Economic Entomology* 77, 1229–1232.

Armstrong, J.W. and Couey, H.M. (1989) Fumigation, heat, and cold. In: Robinson, A.S. and Hooper, G. (eds), *World Crop Pests, Vol. 3A, Fruit Flies, Their Biology, Natural Enemies, and Control*. Elsevier, Rotterdam, pp. 411–424.

Armstrong, J.W. and Vargas, R.I. (1982) Resistance of pineapple variety '59-656' to field populations of oriental fruit flies and melon flies. *Journal of Economic Entomology* 75, 781–782.

Armstrong, J.W., Vriesenga, J.D. and Lee, C.Y.L. (1979) Resistance of pineapple varieties 'D-10' and 'D-20' to field populations of oriental fruit flies and melon flies. *Journal of Economic Entomology* 72, 6–7.

Armstrong, J.W., Mitchell, W.C. and Farias, G.J. (1983) Resistance of 'Sharwil' avocados at harvest maturity to infestation by three fruit fly species in Hawaii. *Journal of Economic Entomology* 76, 119–121.

Armstrong, J.W., Schneider, E.L., Garcia, D.L. and Couey, H.M. (1984) A methyl bromide quarantine fumigation for strawberries infested with Mediterranean fruit fly. *Journal of Economic Entomology* 77, 680–682.

Armstrong, J.W., Harvey, J.M., Garcia, D.L., Menezes, T.D. and Brown, S.A. (1988) Methyl bromide fumigation for control of oriental fruit fly (Diptera: Tephritidae) in California stone fruits. *Journal of Economic Entomology* 81, 963–965.

Armstrong, J.W., Hansen, J.D., Hu, B.K.S. and Brown, S.A. (1989) High-temperature, forced air quarantine treatment for papaya infested with tephritid fruit flies (Diptera: Tephritidae). *Journal of Economic Entomology* 82, 1667–1674.

Baas, J. (1959) The Mediterranean fruit fly in Central Europe. Part 2. *Hoefchen-Briefe Bayer Pflanzenschutz-Nachrichten* 3, 113–140.

Back, E.A. and Pemberton, C.E. (1916) Effect of cold-storage temperatures upon the Mediterranean fruit fly. *Journal of Agricultural Research* 5, 657–666.

Baker, A.C. (1939) The basis for treatment of products where fruit flies are involved as a condition for entry into the United States. US Department of Agriculture Circular No. 551. 8 pp.

Baker, R.T., Cowley, J.M., Harte, D.S. and Frampton, E.R. (1990) Development of a maximum pest limit for fruit flies (Dipteria: Tephritidae) in produce imported into New Zealand. *Journal of Economic Entomology* 83, 13–17.

Balock, J.W. and Kozuma, T. (1954) Sterilization of papaya by means of vapor-heat quick run-up. Special Report No. 7, Fruit Fly Investigations in Hawaii. US Department of Agriculture Entomology Research Branch, Honolulu, Hawaii, USA.

Balock, J.W. and Starr, D.F. (1946) Mortality of the Mexican fruit fly in mangoes treated by the vapor-heat process. *Journal of Economic Entomology* 38, 646–651.

Balock, J.W., Burditt, A.K., Seo, S.T. and Akamine, E.K. (1966) Gamma radiation as a quarantine treatment for Hawaiian fruit flies. *Journal of Economic Entomology* 59, 202–204.

Benschoter, C.A. (1979) Fumigation of grapefruit with methyl bromide for control of *Anastrepha suspensa*. *Journal of Economic Entomology* 72, 401–402.

Benschoter, C.A. (1980) Seasonal variation in tolerance of Florida 'Marsh' grapefruit to a combination of methyl bromide fumigation and cold storage. *Proceedings of the Florida State Horticultural Society* 92, 166–167.

Benschoter, C.A. (1984) Low-temperature storage as a quarantine treatment for the Caribbean fruit fly (Diptera: Tephritidae) in Florida citrus. *Journal of Economic Entomology* 77, 1233–1235.

Benschoter, C.A. (1987) Effects of modified atmospheres and refrigeration temperatures on survival of eggs and larvae of the Caribbean fruit fly (Diptera: Tephritidae) in laboratory diet. *Journal of Economic Entomology* 80, 1223–1225.

Benschoter, C.A. (1988) Methyl bromide fumigation and cold storage as treatments for California stone fruits and pears infested with the Caribbean fruit fly (Diptera: Tephritidae). *Journal of Economic Entomology* 81, 1665–1667.

Bohm, H. (1958) Zum Vorkommen der Mittelmeerfruchtfliege im Weiner Obstbaugebiet. Pflanzenschutzberichte 21, 129–158.

Burditt, A.K., Jr (1982) Food irradiation as a quarantine treatment of fruits. *Food Technology* 36, 51–62.

Burditt, A.K. and Balock, J.W. (1985) Refrigeration as a quarantine treatment for fruits and vegetables infested with eggs and larvae of *Dacus dorsalis* and *Dacus cucurbitae* (Diptera: Tephritidae). *Journal of Economic Entomology* 78, 885–887.

Burditt, A.K. and Hungate, F.P. (1989) Gamma irradiation as a quarantine treatment for apples infested by codling moth (Lepidoptera: Tortricidae). *Journal of Economic Entomology* 82, 1386–1390.

Burditt, A.K., Jr and Seo, S.T. (1971) Dose requirements for quarantine treatment of fruit flies with gamma irradiation. In: *Proceedings of the Panel of Disinfestation of Fruit by Irradiation*. IAEA, Vienna, Austria, pp. 33–41.

Burditt, A.K., Hinman, F.G. and Balock, J.W. (1963) Screening of fumigants for toxicity to eggs and larvae of the oriental fruit fly and Mediterranean fruit fly. *Journal of Economic Entomology* 56, 261–265.

Burditt, A.K., Seo, S.T. and Balock, J.W. (1971) Basis for developing quarantine treatments for fruit flies. In: *Disinfestation of Fruit by Irradiation*. IAEA, Vienna, Austria. 177pp.

Burg, S.P. (1962) The physiology of ethylene formation. *Annual Review of Plant Physiology* 13, 265–302.

Burke, J.J., Hatfield, J.L., Klein, R.R. and Mullet, J.E. (1985) Accumulation of heat shock proteins in field grown cotton. *Plant Physiology* 78, 394–398.

Carey, J.R. (1991) Establishment of the Mediterranean fruit fly in California. *Science* 253, 1369–1373.

Carey, J.R. and Dowell, R.V. (1989) Exotic fruit fly pests and California agriculture. *Californian Agriculture* 43, 38–40.

Chan, H.T. (1986a) Effects of heat treatments on the ethylene-forming enzyme system in papaya. *Journal of Food Science* 51, 581–583.

Chan, H.T. (1986b) Heat inactivation of the ethylene forming enzyme system in cucumber. *Journal of Food Science* 51, 1491–1493.

Chan, H.T. and Linse, E. (1989) Conditioning cucumbers for quarantine heat treatments. *HortScience* 24, 985–989.

Claypool, L.L. and Vines, H.M. (1956) Commodity tolerance studies of deciduous fruits to moist heat and fumigants. *Hilgardia* 24, 297–355.

Couey, H.M. (1982) Chilling injury of crops of tropical and subtropical origin. *HortScience* 17, 162–165.

Couey, H.M. (1989) Heat treatment for control of postharvest disease and insect pests of fruit. *HortScience* 24,198–202.

Couey, H.M. and Chew, V. (1986) Confidence limits and sample size in quarantine research. *Journal of Economic Entomology* 79, 887–890.
Couey, H.M. and Hayes, C.F. (1986) Quarantine procedure for Hawaiian papaya using fruit selection and a two-stage hot-water immersion. *Journal of Economic Entomology* 79, 1307–1314.
Couey, H.M., Linse, E.S. and Nakamura, A.N. (1984) Quarantine procedure for Hawaiian papayas using heat and cold treatments. *Journal of Economic Entomology* 77, 984–988.
Couey, H.M., Armstrong, J.W., Hylin, J.W., Thornburg, W., Nakamura, A.N., Linse, E.S., Ogata, J. and Vetro, R. (1985) Quarantine procedure for Hawaii papaya using a hot-water treatment and high-temperature, low-dose ethylene dibromide fumigation. *Journal of Economic Entomology* 78, 879–884.
Davis, R. and Jay, E.G. (1977) The current status of controlled atmospheres as a method of insect control. In: Dewey, D.H. (ed.), Horticulture Report 28, Michigan State University, East Lansing, Michigan, pp. 207–211.
Delate, K.M., Brecht, J.K. and Coffelt, J.A. (1990) Controlled atmosphere treatments for control of sweet potato weevil (Coleoptera: Airculionidae) in stored tropical sweet potatoes. *Journal of Economic Entomology* 83, 461–465.
Dollar, A.M. and Hanaoka, M. (1972) Commercial disinfestation of fruit by ionizing radiation. *Hawaii Agricultural Experiment Station Research Bulletin* No. 165.
Dollar, A.M., Hanaoka, M., McClish, G.A. and Moy, J.H. (1971) Semi commercial scale studies on irradiated papaya. In: *Disinfestation of Fruit by Irradiation. Proceedings of a Panel on the Use of Irradiation to Solve Quarantine Problems in the International Fruit Trade.* IAEA PL422/14, pp.137–156.
Dowell, R.V. (1983) The Medfly in California: the threat. *HortScience* 18, 40.
Ferguson, W.E., Yates, A.R., MacQueen, K.F. and Robb, J.A. (1966) The effects of gamma irradiation on bananas. *Food Technology* 20, 105–107.
Field, R.J. (1984) The effect of temperature on ethylene production by plant tissue. In: Roberts, J.A. and Tucker, G.A. (eds), *Ethylene and Plant Development*. Butterworths, London, pp. 47–69.
Fletcher, B.S. (1987) The biology of dacine fruit flies. *Annual Review of Entomology* 32, 115–144.
FDA–US Food and Drug Administration (1987) Irradiation in the production, processing and handling of food; Final rules. *Federal Register* 51, 13375–13399.
Furasawa, K., Sugimoto, T. and Gaja, T. (1984) The effectiveness of vapor heat treatment against the melon fly, *Dacus cucurbitae* Coquillett, in eggplant and fruit tolerance to the treatment. *Research Bulletin of the Plant Protection Service Japan* 20, 17–24.
Gaunce, A.P., Madsen, H.F., McMullen, R.D. and Hall, J.W. (1980) Dosage response of the stages of codling moth, *Laspeyresia pomonella* (Lepidoptera: Olethreutidae), to fumigation with methyl bromide. *Canadian Entomologist* 112, 1033–1038.

Gaunce, A., Morgan, C.V.G. and Meheriuk, M. (1982) Control of tree fruit insects with modified atmospheres. *Proceedings of 3rd Controlled Atmospheres Research Conference*, Oregon State University, Corvallis, Oregon, pp. 383–390.

Goodwin, T.W. and Jamikorn, M. (1952) Biosynthesis of carotenes in ripening tomatoes. *Nature* 170, 104–105.

Goonewardene, H.F., Kwolek, W.F., Mouzin, T.E. and Williams, E.B. (1979) A 'No choice' study for resistance of apple fruits to four insect pests. *HortScience* 14, 165–166.

Gould, W.P. and Sharp, J.L. (1990) Hot-water immersion treatment for guava infested with Caribbean fruit fly (Diptera: Tephritidae). *Journal of Economic Entomology* 85, 1235–1239.

Greany, P.D. (1989) Host plant resistance to tephritids: an underexploited control strategy. In: Robinson, A.S. and Hooper, G. (eds), *Fruit Flies – Their Biology, Natural Enemies and Control*. Elsevier, Amsterdam, pp. 353–362.

Hagen, K.S., Allen, W.A. and Tassan, R.L. (1981) Mediterranean fruit fly: the worst may be yet to come. *Californian Agriculture* 35, 5–7.

Hallman, G.J. and Sharp, J.L. (1990) Hot-water immersion quarantine treatment for carambolas infested with Caribbean fruit fly (Diptera: Tephritidae). *Journal of Economic Entomology* 83, 1471–1474.

Hallman, G.J., Gaffney, J. and Sharp, J.L. (1990) Vapor heat treatment for grapefruit infested with Caribbean fruit fly (Diptera: Tephritidae). *Journal of Economic Entomology* 83, 1475–1478.

Hansen, J.D., Armstrong, J.W. and Brown, S.A. (1989) The distribution and biological observations of the mango weevil, *Cryptorhynohus mangiferae* (Coleoptera: Curculionidae), in Hawaii. *Proceedings of the Hawaiian Entomological Society* 29, 31.

Hansen, J.D., Armstrong, J.W., Hu, B.K.S. and Brown, S.A. (1990) Thermal death of third instars of oriental fruit flies (Diptera: Tephritidae) in developing quarantine treatments for papayas. *Journal of Economic Entomology* 83, 160–167.

Hansen, J.D., Chan, H.T., Hara, A.H. and Tenbrink, V.L. (1991a) Phytotoxic reaction of Hawaiian cut flowers and foliage to hydrogen cyanide fumigation. *HortScience* 26, 53–56.

Hansen, J.D., Hara, A.H., Chan, H.T. and Tenbrink, V.L. (1991b) Efficacy of hydrogen cyanide fumigation as a treatment for pests of Hawaiian cut flowers and foliage after harvest. *Journal of Economic Entomology* 84, 532–536.

Harte, D.S., Baker, R.T. and Cowley, J.M. (1992) Relationship between pre-entry sample size for quarantine security and variability of estimates of fruit fly (Diptera: Tephritidae) disinfestation treatment efficacy. *Journal of Economic Entomology* 85, 1560–1565.

Harvey, J.M., Harris, C.M. and Hartsell, P.L. (1989) Tolerance of California nectarine cultivars to methyl bromide quarantine treatments. *Journal of the American Society for Horticultural Science* 114, 626–629.

Hata, T.Y., Hara, A.H., Jang, E.B., Imaino, L.S., Hu, B.K.S. and Tenbrink, V.L. (1992) Pest management before harvest and insecticidal dip after harvest as a systems approach to quarantine security for red ginger. *Journal of Economic Entomology* 85, 2310–2316.

Hatton, T.T. and Cubbedge, R.H. (1980) Phytotoxicity of methyl bromide as a fumigant for Florida citrus fruit. *Proceedings of the Florida State Horticultural Society* 92, 167–169.

Hatton, T.T., Beraha, L. and Wright, W.R. (1961) Preliminary trials of gamma radiation on mature green Irwin and Sensation mangoes. *Proceedings of the Florida Mango Forum* (June 1961), 15–17.

Havens, C., Hu, B.K.S., Farias, G.J., Seo, S.T. and Harris, E.J. (1979) *Dacus dorsalis*: Fumigation of papayas with ethylene dibromide. *Journal of Economic Entomology* 72, 4–5.

Hayes, C.F. (1982) A study of untrasound and microwaves for the control of fruit flies in papaya postharvest. Hawaii Institute of Tropical Agriculture and Human Resources, Honolulu, Res. Ext. No. 20, p.51.

Hayes, C.F., Chingon, H.T.G., McMurdo, M.B., Ikeda, M.B., Sanderson, S.L. and Deaver, J. (1983) Ultrasonic effects on *Dacus dorsalis*. *Ultrasound in Medicine and Biology* 9, 185–189.

Hayes, C.F., Chingon, H.T.G., Nitta, F.A. and Wang, W.J. (1984) Temperature control as an alternative to ethylene dibromide fumigation for the control of fruit flies (Diptera: Tephritidae) in papaya. *Journal of Economic Entomology* 77, 683–686.

Heather, N.W. (1986) Irradiation of fruit and vegetables. *Queensland Agricultural Journal* (March–April), 85–87.

Heather, N.W., Hargreaves, P.A., Corcoran, R.J. and Melksham, K.J. (1987) Dimethoate and fenthion as packing line treatments for tomatoes against *Dacus tryoni* (Froggatt). *Australian Journal of Experimental Agriculture* 27, 465–469.

Hill, A.R., Rigney, C.J. and Sproul, A.N. (1988) Cold storage of oranges as a disinfestation treatment against the fruit flies *Dacus tryoni* (Froggatt) and *Ceratitis capitata* (Wiedemann) (Diptera: Tephritidae). *Journal of Economic Entomology* 81, 257–260.

Hooper, T. (1907) Cold storage and fruit fly. *Journal of the Department of Agriculture, Western Australia* 15, 252–253.

Horner, M.E. (1979) An investigation into the potential use of ultrasonic irradiation as a means of disinfestation of Hawaiian *Carica papaya* L., M.S. Thesis, Department of Entomology, University of Hawaii, Honolulu, Hawaii, 62pp.

Houck, L.G., Jenner, J.F., Moreno, D.S. and Mackey, B.E. (1989) Permeability of polymer film wraps for citrus fruits fumigated with hydrogen cyanide to control California red scale. *Journal of the American Society for Horticultural Science* 114, 287–292.

Huber, H. (1935) Der Warmehaushalt der Pflanzen. *Naturwiss. Landwirtsch.* 17, 1–148.

Hunter, J.E., Buddenhagen, I.W. and Kojima, E.S. (1969) Efficacy of fungicides, hot water and gamma-irradiation for control of postharvest fruit rots of papaya. *Plant Diseases Report* 53, 279–284.

Ibrahim, A.G. and Rahman, M.D.A. (1982) Laboratory studies on the effects of selected tropical fruit on the larvae of *Dacus dorsalis* Hendel. *Peranika* 5, 90–94.

Jang, E.B. (1986) Kinetics of thermal death in eggs and first instars of three species of fruit flies (Diptera: Tephritidae). *Journal of Economic Entomology* 79, 700–705.

Jang, E.B. (1991) Thermal death kinetics and heat tolerance in early and late third instars of the oriental fruit fly (Diptera: Tephritidae). *Journal of Economic Entomology* 84, 1298–1303.

Jarvis, H. (1946) Pests of mango. *Queensland Agricultural Journal* 62, 10–14.

Johnson, J.A., Soderstom, E.L., Brandl, D.G., Houck, L.G. and Wofford, P.L. (1990) Gamma radiation as a quarantine treatment for Fuller rose beetle eggs (Coleoptera: Curculionidae) on citrus fruit. *Journal of Economic Entomology* 83, 905–909.

Jones, W.W. (1939) The influence of relative humidity on the respiration of papaya at high temperature. *Proceedings of the American Society for Horticultural Science* 37, 199–124.

Jones, W.W. (1940) Vapor heat treatment for fruits and vegetables grown in Hawaii. *Hawaii Agricultural Experimental Station Circular* No. 16. 8 pp.

Jones, W.W. (1942) Respiration and chemical changes of the papaya fruit in relation to temperature. *Plant Physiology* 17, 481–486.

Kader, A.A. (1986) Potential applications of ionizing radiation in postharvest handling of fresh fruits and vegetables. *Food Technology* 40, 117–121.

Ke, D. and Kader, A.A. (1989) Tolerance and responses of fresh fruits to oxygen levels at or below 1%. In: *Proceedings of 5th International CA Research Conference*, Wenatchee, WA, 14–16 June, 1989, vol.2, pp. 209–216.

Ke, D. and Kader, A.A. (1990) Tolerance of 'Valencia' oranges to controlled atmospheres as determined by physiological responses and quality attributes. *Journal of the American Society for Horticultural Science* 115, 779–783.

Ke, D., van Gorsel, H. and Kader, A.A. (1990) Physiological and quality responses of 'Bartlett' pears to reduced-O_2 and enhanced-CO_2 atmospheres and storage temperature. *Journal of the American Society for Horticultural Science* 115, 435–439.

Kimpel, J.A. and Key, J.L. (1985) Presence of heat shock mRNAs in field grown soybeans. *Plant Physiology* 79, 672–678.

Klaustermeyer, J.A., Kader, A.A., Morris, L.L. and France, R. (1973) Preliminary studies related to insect control in harvested lettuce. University of California, Davis, Department of Vegetable Crops Report, 18 pp.

Landolt, P.J., Chambers, D.L. and Chew, V. (1984) Alternative to the use of Probit 9 mortality as a criteria for quarantine treatments of fruit fly (Diptera: Tephritidae) infested fruit. *Journal of Economic Entomology* 77, 285–287.

Lidster, P.D., Sanford, K.H. and McRae, K.B. (1981) Effects of modified atmosphere storage on overwintering populations of the apple rust mite and European red mite eggs. *HortScience* 16, 328–329.

Lindquist, S. and Craig, E.A. (1988) The heat shock proteins. *Annual Review of Genetics* 22, 631–677.

Lounsbury, C.P. (1907) The fruit fly (*Ceratitis capitata*). *Agricultural Journal of the Cape of Good Hope* 31, 186–187.

Macion, E.A., Mitchell, W.C. and Smith, J.B. (1968) Biophysical and biochemical studies on the nature of resistance of pineapples to the oriental fruit fly. *Journal of Economic Entomology* 61, 910–911.

Marriott, J. (1980) Banana physiology and biochemistry of storage and ripening for optimum quality. *CRC Critical Reviews of Food Science and Nutrition* 13, 41–88.

Maxie, E.C., Amezquita, R., Hassan, B.M. and Johnson, C.F. (1968) Effect of gamma irradiation on the ripening of banana fruit. *Proceedings of the American Society for Horticultural Science* 92, 235–254.

Maxie, E.C., Sommer, N.F. and Mitchell, F.G. (1971) Infeasibility of irradiating fresh fruits and vegetables. *HortScience* 6, 202–204.

Messenger, P.S. and Flitters, N.E. (1958) Effect of constant temperature environments on the egg stage of three species of Hawaiian fruit flies. *Annual Entomological Society of Amererica* 51, 109–119.

Metcalf, C.L., Flint, W.P. and Metcalf, R.L. (1962) *Destructive and Useful Insects, Their Habits and Control*, 4th edn. McGraw-Hill, New York, 1087 pp.

Mitchell, F.G. (1981) The Mediterranean fruit fly in California—a status report of quarantine treatments and their possible impact on postharvest handling procedures. *Perishables Handling* 47, 2–7.

Moran, C.V.G. and Gaunce, A.P. (1975) Carbon dioxide as a fumigant against the San Jose scale (Homoptera: Diaspididae) on harvested apples. *Canadian Entomologist* 107, 935–936.

Moran, C.V.G., Gaunce, A.P. and Jong, C. (1974) Control of codling moth larvae in harvested apples by methyl bromide fumigation and cold storage. *Canadian Entomologist* 106, 917–920.

Moran, C.V.G., Gaunce, A.P. and Madsen, B.J. (1975) Control of the San Jose scale on harvested apples by methyl bromide fumigation and cold storage. *Canadian Entomologist* 107, 367–368.

Morris, L.L. (1982) Chilling injury of horticultural crops: an overview. *HortScience* 17, 162–163.

Moss, J.I. and Jang, E.B. (1991) Effects of age and metabolic stress on heat tolerance of Mediterranean fruit fly (Diptera: Tephritidae) eggs. *Journal of Economic Entomology* 84, 537–541.

Moy, J.H. (ed.) (1985) *Radiation Disinfestation of Food and Agricultural Products*. University of Hawaii, Honolulu, HI, 424 pp.

Moy, J.H., Akamine, E.K., Brewbaker, J.L., Buddenhagen, I.W., Ross, E., Spielman, H., Upadhya, M.D., Wenkam, N., Heller, D., Dollar, A.M., Hanaoke, M. and McClish, G.A. (1971) Dosimetry, tolerance and shelf-life extension related to disinfestation of tropical fruits by gamma irradiation. In: *Proceedings of a Panel on Disinfestation of Fruit by Irradiation*, International Atomic Energy Agency, Vienna, Austria, pp. 43–57.

O'Loughlin, G.T., East, R.A. and Meats, A. (1984) Survival, development rates and generation times of Queensland fruit fly, *Dacus tyroni* in a marginally favorable climate: experiments in Victoria. *Australian Journal of Zoology* 32, 353–361.

Ougham, H.J. and Stoddart, J.L. (1986) Synthesis of heat shock proteins and acquisition of thermotolerance in high temperature tolerant and high temperature susceptible lines of sorghum. *Plant Science* 44, 163–167.

Paull, R.E. (1990a) Chilling injury in tropical and subtropical crops. In: Wang, C.Y. (ed.), *Chilling Injury of Horticultural Crops*. CRC Press, Boca Raton, Florida, pp. 17–36.

Paull, R.E. (1990b) Postharvest heat treatments and fruit ripening. *Postharvest News and Information* 1, 355–363.

Paull, R.E. and Chen, N.J. (1990) Heat shock response in field grown ripening papaya fruit. *Journal of the American Society for Horticultural Science* 115, 623–631.

Rippon, L.E. and Smith, R.J. (1979) Postharvest treatment of Chinese gooseberries (kiwifruit) for the control of Queensland fruit fly. *Rural News* 73, 34–35.

Ruckelshaus, W.D. (1984) Ethylene dibromide, amendment of notice of intent to cancel registration of pesticide products containing ethylene dibromide. Congressional Federal Register, Washington, DC, US Government Printing Office 49, 14182–14185.

Rutherford, A. (1914) The mango weevil. *Tropical Agriculture Peradeniya* 13, 410–411.

Saeki, S., Katayama, M. and Okumura, M. (1980) Effect of temperature upon the development of the oriental fruit fly and its possible destruction in the mainland of Japan. *Research Bulletin of the Plant Protection Service Japan* 16, 73–76.

Saul, S. and Seifert, J. (1990) Methoprene on papaya: persistence and toxicity to different developmental stages of fruit flies (Diptera: Tephritidae). *Journal of Economic Entomology* 83, 901–904.

Saul, S.H., Mau, R.F.L. and Oi, D. (1985) Laboratory trials of methoprene-impregnated waxes for disinfesting papayas and peaches of the Mediterranean fruit fly (Diptera: Tephritidae). *Journal of Economic Entomology* 78, 652–655.

Saul, S.H., Mau, R.F. Kobayashi, R.M., Tsuda, D.M. and Nishina, M.S. (1987) Laboratory trials of methoprene-impregnated waxes for preventing survival of adult oriental fruit flies (Diptera: Tephritidae) from infested papayas. *Journal of Economic Entomology* 80, 494–496.

Saunders, G.W. and Elder, R.J. (1966) Sterilization of banana fruit infested with banana fruit fly. *Queensland Journal of Agricultural Science* 23, 81–85.

Schultz, H.G., Bruhn, C.M. and Diaz-Krauf, K.V. (1989) Consumer attitude towards irradiated foods: Effects of labeling and benefits information. *Food Technology* 43, 80–86.

Seo, S.T., Chambers, D.L., Kobayashi, R.M., Lee, C.Y.L. and Komura, M. (1970a) Infestation of oriental fruit flies in '53-116' and 'Smooth Cayenne' pineapple varieties grown near supporting guava or other host fruits. *Journal of Economic Entomology* 63, 1830–1831.

Seo, S.T., Chambers, D.L., Komura, M. and Lee, C.Y.L. (1970b) Mortality of mango weevils in mangoes treated by dielectric heating. *Journal of Economic Entomology* 63, 1977–1978.

Seo, S.T., Kobayashi, R.M., Chambers, D.L., Steiner, L.F., Balock, J.W., Komura, M. and Lee, C.Y.L. (1971) Fumigation with methyl bromide plus refrigeration to control infestations of fruit flies in agricultural commodities. *Journal of Economic Entomology* 64, 1270–1274.

Seo, S.T., Chambers, D.L., Akamine, E.K., Komura, M. and Lee, C.Y.L. (1972) Hot water–ethylene dibromide fumigation—refrigeration treatment for mangoes infested by oriental and Mediterranean fruit flies. *Journal of Economic Entomology* 65, 1372–1374.

Seo, S.T., Chambers, D.L., Lee, C.Y.L., Komura, M., Fujimoto, M. and Kamakahi, D. (1973a) Resistance of pineapple variety '59-443' to field populations of oriental fruit flies and melon flies. *Journal of Economic Entomology* 66, 522–523.

Seo, S.T., Kobayashi, R.M., Chambers, D.L., Dollar, A.M. and Hanoaka, M. (1973b) Hawaiian fruit flies in papaya, bell pepper and egg plant: Quarantine treatment with gamma irradiation. *Journal of Economic Entomology* 66, 937–939.
Seo, S.T., Hu, B.K.S., Komura, M., Lee, C.Y.L. and Harris, E.J. (1974a) *Dacus dorsalis*. Vapor heat treatment in papaya. *Journal of Economic Entomology* 67, 240–242.
Seo, S.T., Kobayashi, R., Chambers, D.L., Steiner, L.F., Lee, C.Y.L. and Komura, M. (1974b) Mango weevil: cobalt-60 gamma-irradiation of packaged mangoes. *Journal of Economic Entomology* 67, 504–505.
Seo, S.T., Akamine, E.K., Goo, T.T.S., Harris, E.J. and Lee, C.Y.L. (1979a) Oriental and Mediterranean fruit flies: fumigation of papaya, avocado, tomato, bell pepper, eggplant, and banana with phosphine. *Journal of Economic Entomology* 72, 354–359.
Seo. S.T., Chambers, D.L., Akamine, E.K., Harris, E.J., Lee, C.Y.L. and Komura. M. (1979b) Oriental fruit fly: factors affecting ethylene dibromide fumigation of infested papayas. *Journal of Economic Entomology* 72, 350–353.
Seo, S.T., Hu, B.K.S., Komura, M., Lee, C.Y.L. and Harris, E.J. (1979c) Oriental and Mediterranean fruit flies: fumigation of papaya, avocado, tomato, bell pepper, eggplant and banana with phosphine. *Journal of Economic Entomology* 72, 354–359.
Seo, S.T., Farias, G.J. and Harris, E.J. (1982) Oriental fruit fly: ripening of fruit and its effect on index infestation of Hawaiian papayas. *Journal of Economic Entomology* 75, 173–178.
Sharp, J.L. (1986) Hot-water treatment for control of *Anastrepha suspenda* (Diptera: Tephritidae) in mangoes. *Journal of Economic Entomology* 79, 706–708.
Sharp, J.L. (1989) Hot-water immersion appliance for quarantine research. *Journal of Economic Entomology* 82, 189–192.
Sharp, J.L. (1990) Immersion in heated water as a quarantine treatment for California stone fruit infested with the Caribbean fruit fly (Diptera: Tephritidae). *Journal of Economic Entomology* 83, 1468–1470.
Sharp, J.L. and Picho-Martinez, H. (1990) Hot-water quarantine treatment to control fruit flies in mangoes imported into the United States from Peru. *Journal of Economic Entomology* 83, 1940–1943.
Sharp, J.L. and Spalding, D.H. (1984) Hot water as a quarantine treatment for Florida mangoes infested with Caribbean fruit fly. *Proceedings of the Florida State Horiticultural Society* 97, 355–357.
Sharp, J.L., Ouye, M.T., Thalman, R., Hart, W., Ingle, S. and Chew, V. (1988) Submersion of 'Francis' mango in hot water as a quarantine treatment for the West Indian fruit fly and the Caribbean fruit fly (Diptera: Tephritidae). *Journal of Economic Entomology* 81, 1431–1436.
Sharp, J.L., Ouye, M.T., Hart, W., Ingle, S., Hallman, G., Gould, W. and Chew, V. (1989) Immersion of Florida mangoes in hot water as a quarantine treatment for Caribbean fruit fly (Diptera: Tephritidae). *Journal of Economic Entomology* 82, 186–188.
Shukla, R.P. and Tandon, P.L. (1985) Bioecology and management of the mango weevil *Sternochetus mangifera* (Fabricius) (Coleoptera: Curculionidae). *International Journal of Tropical Agriculture* 3, 293–303.

Sinclair, W.B. and Lindgren, D.L. (1955) Vapor heat sterilization of California citrus and avocado fruits against fruit-fly insect. *Journal of Economic Entomology* 48, 133–138.

Smilanick, J.L. and Fouse, D.C. (1989) Quality of nectarines stored in insecticidal low O_2 atmospheres at 5 and 15°C. *Journal of the American Society for Horticultural Science* 114, 431–436.

Smith, E.S.C. (1977) Studies on the biology and commodity control of the banana fruit fly, *Dacus musae* (Tryon) in Papua New Guinea. *Papua New Guinea Agricultural Journal* 28, 47–56.

Soderstrom, E.L. (1977) Research on the use of a low oxygen atmosphere for insect control in dried fruits and tree nuts. In: Dewey, D.H. (ed.), Horticultural Report No. 28, Michigan State University, East Lansing, Michigan, pp. 205–206.

Soderstrom, E.L. and Brandl, D.G. (1990) Controlled atmospheres for the preservation of tree nuts and dried fruits. In: Calderon, M. and Barkai-Golan, R. (eds), *Food Preservation by Modified Atmospheres*, CRC Press, Boca Raton, Florida, pp. 83–92.

Sommer, N.F. and Mitchell, F.G. (1986) Gamma irradiation – a quarantine treatment for fresh fruits and vegetables? *HortScience* 21, 356–360.

Spalding, D.H., King, J.R., Benschoter, C.A., von Windeguth, D.L., Reeder, W.F. and Burditt, A.K. (1977) Ethylene dibromide, methyl bromide and phosphine fumigation of tomatoes. *Proceedings of the Florida State Horticultural Society* 91, 156–158.

Spitler, G.H., Armstrong, J.W. and Couey, H.M. (1984) Mediterranean fruit fly host status of commercial lemons. *Journal of Economic Entomology* 77, 1441–1444.

Steward, J.K., Aharoni, Y., Hartsell, P.O. and Young, D.K. (1980) Acetaldehyde fumigation at reduced pressures to control the green peach aphid on wrapped and packed head lettuce. *Journal of Economic Entomology* 73, 149–152.

Subramanyam, C.K. (1925) A note on the life history of *Cryptorhyncus mangiferae*, Fab. *Madras Agricultural Department Yearbook* 29, 36.

Sugimoto, T., Furasawa, K. and Mizobuchi, M. (1983) The effectiveness of vapor heat treatment against the oriental fruit fly, *Dacus dorsalis* Hendel, in green pepper and fruit tolerance to the treatment. *Research Bulletin of the Plant Protection Service Japan* 19, 81–88.

Sun, Y.P. (1946) An analysis of some important factors affecting the results of fumigation on insects. *St Paul, Minnesota Agricultural Experimental Station Technical Bulletin* No. 177. St Paul, Minnesota, USA, 6 pp.

Swaine, G., Hargreaves, P.A., Jackson, D.E. and Corcoran, R.J. (1984a) Dimethoate dipping of tomatoes against Queensland fruit fly, *Dacus tryoni* (Froggatt). *Australian Journal of Experimental Agriculture and Animal Husbandry* 24, 447–449.

Swaine, G., Melksham, K.J. and Corcoran, R.J. (1984b) Dimethoate dipping of 'Kensington' mango against Queensland fruit fly. *Australian Journal of Experimental Agriculture and Animal Husbandry* 24, 620–623.

Tan, K.H. and Lee, S.L. (1982) Species diversity and abundance of *Dacus* (Diptera: Tephritidae) in five ecosystems of Penang, West Malaysia. *Bulletin of Entomological Research* 72, 709–716.

Thiem, H. (1937) Die Mittelmeerfruchtfliege. *Flugblatt* No. 551, 1–6.

United States Department of Agriculture, Bureau of Entomology and Plant Quarantine (1940) Administrative instructions amending authorization of the shipment of fruits and vegetables from Hawaii to the mainland subject to fumigation with methyl bromide under supervision. USDA, BEPQ Memorandum No. 510. Washington, DC, USA, 1 p.

United States Department of Agriculture, Bureau of Entomology and Plant Quarantine (1951) Ethylene dibromide fumigation approved as a condition for certification of certain fruits and vegetables under Hawaiian fruit and vegetable quarantine No. 13. USDA, BEPQ Memorandum No. 592. Washington, DC, USA, 1 p.

Vargas, R.I., Harris, E.J. and Nishida, T. (1983a) Distribution and season occurrence of *Certatitis capitata* (Wiedemann) (Diptera: Tephritidae) on the island of Kauai in the Hawaiian Islands. *Journal of Economic Entomology* 12, 303–310.

Vargas, R.I., Nishida, T. and Beardsley, J.W. (1983b) Distribution of abundance of *Dacus dorsalis* (Diptera: Tephritidae) in native and exotic forest areas on Kauai. *Journal of Economic Entomology* 12, 1185–1189.

Vargas, R.I., Miyashita, D. and Nishida, T. (1984) Life history and demographic parameters of three laboratory-reared tephritids (Diptera: Tephritidae). *Annals of the Entomological Society of America* 77, 651–656.

Vincent, L.E. and Lindgren, E.L. (1972) Hydrogen phosphide and ethyl formate: Fumigation of insects infesting dates and other dried fruits. *Journal of Economic Entomology* 65, 1667–1669.

Wade, N.L. and Rigney, C.J. (1979) Phytotoxicity of ethylene dibromide to cherry and banana fruit. *Journal of the American Society for Horticultural Science* 104, 900–903.

Waller, J.B. (1990) Insecticidal soaps for postharvest control of thrips in asparagus. *Proceedings of 43rd New Zealand Weed and Pest Control Conference*, 60–62.

Williamson, D.L., Summy, K.R., Hart, W.G., Sanchez-R., M., Wolfenbarger, D.A. and Burton, B.D. (1985) Efficacy and phytotoxicity of methyl bromide as a fumigant for the Mexican fruit fly in grapefruit. *Journal of Economic Entomology* 79, 172–175.

Wit, A.K.H. and van de Vrie, M. (1985a) Gamma radiation for postharvest control of insects and mites in cut flowers. *Mededelingen van de Faculteit Landbouwwetenschappen Rijksuniversiteit Gent* 50, 697–704.

Wit, A.K.H. and van de Vrie, M. (1985b) Fumigation of insects and mites in cut flowers for postharvest control. *Mededelingen van de Faculteit Landbouwwetenschappen Rijksuniversiteit Gent* 50, 705–712.

Yokoyama, V.Y. and Miller, G.T. (1987) High temperature for control of oriental fruit moth (Lepidoptera: Tortricidae) in stone fruit. *Journal of Economic Entomology* 80, 641–645.

Yokoyama, V.Y., Miller, G.T. and Hartsell, P.L. (1987) Methyl bromide fumigation for quarantine control of codling moth (Lepidoptera: Tortricidae) on nectarines. *Journal of Economic Entomology* 80, 840–842.

Yokoyama, V.Y., Miller, G.T. and Dowell, R.V. (1991) Responses of codling moth (Lepidoptera: Tortricidae) to high temperature, a potential treatment for export commodities. *Journal of Economic Entomology* 84, 528–531.

Zauberman, G., Schiffmann-Nadel, M. and Yanko, U. (1973) Susceptibility to chilling injury of three avocado cultivars at various stages of ripening. *HortScience* 8, 511–513.

Zee, F.T., Nishina, M.S., Chan, H.T. and Nishijima, K.A. (1989) Blossom end defects and fruit fly infestation in papayas following hot water quarantine treatment. *HortScience* 24, 323–325.

Quarantine Criteria I

Plant Host Status and Natural Resistance

2

PATRICK D. GREANY
*United States Department of Agriculture,
Agricultural Research Service,
1700 SW 23rd Drive, Gainesville, FL 32604, USA.*

Prior to attaining senescence, fruit often are resistant to infestation by tephritid fruit flies (Greany, 1989). Upon maturation of the seeds, however, many fruits lose their innate resistance to bacterial, fungal and insect infestation, when there is little need to sustain resistance beyond the point of seed maturity. In fact, fruits often advertise their readiness for infestation or consumption by emanating attractive volatiles (Prokopy *et al.*, 1973; Fein *et al.*, 1982; Reissig *et al.*, 1982; Prokopy and Roitberg 1984) and by colourful visual displays (Prokopy *et al.*, 1983), thereby increasing the chance that the seeds will be liberated and dispersed. Conversely, before ripening, fruit generally retains an unapparent green caste, which is unattractive to most fruit flies (Prokopy *et al.*, 1983).

Resistance of plants to insect infestation, including potential hosts for tephritid fruit flies, is due to biophysical and/or biochemical attributes that affect insect behaviour and/or physiological deterrents to insect growth and development (Greany, 1989). This chapter focuses on exploiting innate resistance in fruits to reduce the need for postharvest disinfestation measures. Disinfestation may be directly injurious to the commodities, or leave behind undesirable toxic residues. At least three possibilities exist to utilize fruit innate resistance.

1. Quantitatively assessing fruit susceptibility (plant host status) as a means of judging whether to expose the fruit to disinfestation procedures. Some fruits may be exempt from treatment if they display immunity in laboratory and field assessment studies. For fruits that are facultatively susceptible, research could be performed to discern the fruit characteristics that confer resistance or correlate with resistance. The impact of commodity senescence on these characteristics and resulting fruit susceptibility need to be assessed.

From: Paull, R.E. and Armstrong, J.W. (eds) (1994) *Insect Pests and Fresh Horticultural Products: Treatments and Responses.* CAB INTERNATIONAL, Wallingford, UK.

2. Delaying fruit senescence to avoid insect infestation. For example, plant growth regulators have proven useful in delaying fruit senescence.
3. Introducing genes that confer increased insect resistance or inhibit fruit senescence. For example, modern genetic engineering techniques have already been used to create culitivars that do not produce ethylene, allowing postharvest ripening of climacteric fruit (such as tomatoes) to be performed artificially after the threat of insect infestation is past.

These approaches have certain potential advantages over traditional postharvest disinfestation techniques. Advantages include: (i) potential reduction of insecticidal applications in the field; (ii) no fumigant residues; (iii) no special handling requirements (e.g. transport to special fumigation or other treatment facility); (iv) no weakening of fruit by quarantine treatments; (v) sustained postharvest resistance to bacterial and fungal pathogens. The major disadvantage is the potential lack of assured security.

Plant Host Status

Host lists for fruit flies frequently give little or no consideration to the physiological stage of fruit development. This may lead to the false conclusion that the commodity is a non-host. To determine host status properly, fruit should be exposed to fruit flies throughout the harvest season, under a variety of preharvest environmental conditions, and at different sites before it is concluded that they are non-hosts. For example, 'Sharwil' avocados were believed to be resistant at 'harvest maturity' to the Mediterranean fruit fly, *Ceratitis capitata* (Wiedemann), the melon fly, *Dacus cucurbitae* Coquillett, and the oriental fruit fly, *Dacus dorsalis* Hendel, in Hawaii (Armstrong *et al.*, 1983). However, it was recently discovered that under some conditions, 'Sharwil' avocados may be infested by these species (J. Armstrong, personal communication). Ideally, the physiological state of relatively susceptible fruit should be quantified and correlated with their susceptibility to fruit flies. Physiological properties of the fruit that are easily measured and which could be useful include peel colour and puncture resistance. By measuring factors such as these, incorrect judgements of susceptibility might be avoided.

Conversely, when infestation is forced in the laboratory (= artificial infestation), fruits that would not normally be attacked by fruit fly females may be successfully attacked, which could lead to the inclusion of that fruit on a list of suitable host plants. Other techniques that could also lead to erroneous listings as suitable hosts include the injection of fruit fly eggs into fruits, introduction of eggs into artificial wounds in the peel, or placing larvae on excised host plant tissues for development (Carey, 1984). These approaches bypass or disrupt regions of the peel that might otherwise afford resistance to egg and/or larval survival. These artificial infestation techniques do not provide a realistic assessment of the true host status of a fruit in an ecological context, and the unnecessary inclusion of a fruit cultivar on a host list could lead to regulatory disinfestation requirements before shipment to areas from which the fruit fly is quarantined.

On occasion, a successful infestation of a plant is accepted without qualification, leading to its inclusion on a host list. For example, Nguyen and Fraser (1989) found that the 'Persian' lime, *Citrus aurantifolia* (Christm.) Swingle, was included as a host of the Caribbean fruit fly, *Anastrepha suspensa* (Loew), by Swanson and Baranowski (1972) based upon the recovery of a single puparium from a lime during a 3-year study. Conversely, Nguyen and Fraser (1989) recovered no puparia from over 150 'Persian' limes that were each exposed in the laboratory to several hundred gravid Caribbean fruit fly females. Similarly, Hennessey *et al.* (1992) found no Caribbean fruit fly infestations in >100 000 'Tahiti' (= 'Persian') limes collected from 184 different groves on 60 harvest dates over a year-long period. These data led Hennessey *et al.* (1992) to conclude that this lime cultivar should be exempt from postharvest disinfestation requirements for the Caribbean fruit fly. In another example, Spitler *et al.* (1984) also found that 'Eureka' and 'Lisbon' lemons were virtually immune to attack by the Mediterranean fruit fly, *Ceratitis capitata*, (yielding <1 survivor per 33 333 eggs), and they concluded that lemons should not be considered a host for this fruit fly species. However, such examples may not be universal for all fruit flies. Eskafi (1988) found that lemons in Guatemala were occasionally infested by *Anastrepha* spp. (primarily *A. ludens*), especially when the fruit were fully mature. He attributed this to the long ovipositor (>4 mm) of *A. ludens*, and its ability to bypass resistance factors in the flavedo by depositing its eggs into the albedo (Eskafi, 1988).

Therefore, it is important to examine all the factors affecting the host status of a fruit. Of particular importance is the degree of fruit senescence when tested as a host. As a practical matter, an assessment of the maturity status of fruit that serve as suitable hosts when ripe may be needed in order to determine whether to apply prophylactic field applications of insecticides. For example, Salles (1993) showed that the South American fruit fly, *Anastrepha fraterculus* (Wiedemann), rarely attacks immature peaches. He found there was no need for bait spray applications prior to the onset of fruit maturity, which reduced the number of sprays needed from as many as five to one per season.

Quantitative assessment of plant host status

Resistance/susceptibility of a plant to insect attack can be assessed quantitatively by bioassay (exposing the plant to attack), or by measuring the level of the factor(s) conferring resistance. Of course, such measurements can be done only if the factors are known, and if measurement techniques are available. The bioassay method may not be feasible in all instances, mainly because of the time required for the flies to develop in the fruit, and because the bioassay causes destruction of the test fruit. If it is possible to measure specific biophysical or biochemical attributes, such as peel colour or puncture resistance, or the level of an allelochemical implicated in resistance, then the assessment of fruit resistance could be made much more easily and quickly. This would be important in deciding whether disinfestation of the fruit is required after harvest. Furthermore, some external biophysical properties, such as colour, lend themselves to rapid, non-destructive testing.

A number of studies have shown that the susceptibility of various fruits to tephritid fruit flies could be correlated with concentrations of specific compounds or levels of physical attributes (Greany, 1989). For example, Seo *et al*. (1982, 1983) showed that mature green papayas were highly resistant to oriental fruit fly infestation, and that resistance was correlated with fruit colour and the concentration of an allelochemical, benzyl isothiocyanate. To date, however, there have been no shipments of a commodity to quarantine-sensitive locations based solely upon quantitative measurement of putative resistance factors.

Studies of grapefruit susceptibility to infestation by the Caribbean fruit fly showed an association between resistance to infestation and peel colour and the resistance of the peel to mechanical puncture (Greany *et al*., 1991). High levels of oil in the peel and high concentrations of limonin (Shaw *et al*., 1991), which was shown to be insecticidal (Klocke and Kubo, 1982), also may play a role in resistance to infestation. These factors are related to fruit senescence, and they may be used by Florida regulatory officials to judge the susceptibility of Florida grapefruit to the Caribbean fruit fly throughout the harvest season. This is accomplished by use of rapid, non-destructive methods for measuring peel colour (Greany *et al*., 1991). Further development of rapid, non-destructive methods for peel softness measurement will increase the utility of this resistance-measurement strategy. For example, the sonic resonance method developed by Abbott *et al*. (1992) for non-destructively determining apple firmness may provide a viable means of evaluating citrus peel firmness. Through the measurement of fruit resistance to infestation, it may be possible to ship grapefruit safely without postharvest disinfestation treatments when the fruit meet or exceed Caribbean fruit fly resistance standards.

Manipulating Susceptibility by Delaying Senescence

Because fruit susceptibility to insect attack is often related to fruit senescence, delaying senescence without deleterious effects to fruit quality would be a desirable strategy. This possibility may be achieved through the use of plant growth regulators (PGRs). Campbell (1987) states that

> the use of selective PGRs to alter plant chemistry and induce resistance to insect attack has great applied potential. The concept of using PGRs for this purpose is, however, not new. It is somewhat perplexing that this strategy in pest control has been overlooked. Such a strategy would serve as a means of controlling pestiferous insects, without causing direct toxicity to beneficial insects. Furthermore, PGRs could be used to induce resistance in crop plants that are already agronomically suitable in terms of productivity, harvestability and taste without having to rely on the time-consuming process of breeding for insect resistance.

Numerous examples of the effects of PGRs on insects and mites were reviewed by Campbell (1987), Campbell *et al.* (1984), and Fischer *et al.* (1990), with discussions on the genetic induction of plant resistance to insects. The use of PGRs to manipulate host plant resistance specific to tephritids was discussed by Greany (1989).

Recent studies were done using the naturally-occurring plant growth regulator, gibberellic acid (GA_3), to delay senescence in citrus fruit and to sustain fruit resistance to fruit flies (Greany *et al.*, 1991). Grapefruit and oranges sprayed while still green (prior to colourbreak) with GA_3 exhibited sustained resistance to Caribbean and Mediterranean fruit fly infestation (Greany *et al.*, 1991; Rössler and Greany 1990, respectively). Similar reductions in susceptibility of GA_3-treated oranges and grapefruit to infestation by the Mexican fruit fly, *Anastrepha ludens* (Loew) were observed recently (Aluja *et al.*, unpublished data). Studies in Australia also showed that the Queensland fruit fly, *Bactrocera (Dacus) tryoni* (Froggatt) was less successful in infesting GA_3-treated oranges and grapefruit (Smith *et al.*, unpublished data). Recent studies in Brazil showed that GA_3-treated oranges were markedly less susceptible to attack by the South American fruit fly, as well as the Mediterranean fruit fly (Malavasi *et al.*, unpublished data).

Gibberellic acid (GA_3) which is commercially available and is approved for use on citrus by the US Environmental Protection Agency, does not affect internal ripening of citrus fruit (Coggins and Henning, 1988). In addition to reducing susceptibility of citrus fruit to fruit flies, GA_3 also reduces fruit susceptibility to postharvest decay (El-Otmani and Coggins, 1991). In Brazil, a marked reduction in spontaneous late-season fruit drop in oranges as a result of GA_3 treatment also was observed (Malavasi *et al.*, unpublished data), as noted to a lesser degree by El-Otmani *et al.* (1990). This is economically important because it may allow convenient scheduling of harvests without losses due to fruit drop and/or fruit fly infestation when fruit are left on the tree for extended periods. Finally, the use of non-toxic GA_3 for fruit fly prophylaxis should be more environmentally friendly than some other alternatives, such as malathion bait sprays, which have been shown to be capable of disrupting the biological control of various pests (Cohen *et al.*, 1987; Ehler and Kinsey, 1991; Hoelmer and Dahlsten, 1993).

Genetic Engineering of Insect Resistant Fruits

Campbell (1987) stated that 'the exploitation of plant biotechnology in recent years may eventually lead to the direct introduction of resistant genes into crops. However, to this day, the specific biochemical bases of resistance in crop plants toward insects and pathogens is still generally unknown.' While it is true that the specific bases for resistance are generally unknown, many plants do exhibit resistance to fruit flies, especially before becoming senescent, as noted above. For many climacteric fruit, which ripen upon ethylene stimulation (McGlasson *et al.*, 1978), it may be possible to avoid fruit fly attack by inhibiting ethylene production. Molecular methods might be used to inhibit senescence by interfering with endogenous ethylene production, as already accomplished in transgenic tobacco and tomatoes (Bestwick *et al.*, 1991; Langhoff *et al.*, 1992). Those climacteric fruit that show virtual immunity to fruit flies prior to ripening, such as mangoes and papayas (Joel, 1980; Seo *et al.*, 1982, 1983), may be ideal candidates for such genetic manipulation. After harvest, fruit could be held safe from the threat of fruit fly infestation, and then exposed to ethylene to complete ripening. Because of the recognition of the resistance of immature papayas to fruit fly infestation, the USDA Animal and Plant Health Inspection Service has agreed to allow shipment of 'Solo' papayas from selected regions of western Costa Rica to the United States without postharvest treatment if harvested less than half ripe (Anonymous, 1992). By inhibiting ethylene production in genetically-modified strains of papayas, it should be possible to increase the security of this protocol. Ethylene exposure near the point of marketing should also reduce losses from postharvest decay, an increasingly important prospect as the availability of approved fungicides for fresh fruit diminishes rapidly, coincident with the loss of postharvest fumigants such as ethylene dibromide.

Citrus fruits present another opportunity for genetic manipulation to achieve increased resistance to fruit flies. As noted above, lemons and limes are nearly immune to most tephritid fruit flies, and oranges are intermediate in resistance between lemons and grapefruit. If the factor(s) responsible for the resistance of lemons and limes could be expressed equally in oranges and grapefruit, citrus fruit would rarely be infested by fruit flies. This would be most easily accomplished if lemon/lime resistance to fruit flies is due to the action of a single enzyme, i.e. a single primary gene product. A specific difference in the flavonoid composition of lemons, oranges, and grapefruit correlates with their differential susceptibility to fruit flies and may account for differences in susceptibility among these fruit (Greany and Shapiro, 1993). The flavonoid eriocitrin, which occurs in lemons, is converted from a biochemical precursor, naringenin, which occurs in grapefruit. Conversion of naringenin to eriocitrin is accomplished through the action of the enzyme flavanone oxidase. Oranges possess an ester of eriocitrin, hesperitin. Elliger *et al.* (1980) showed eriocitrin to be deleterious to *Heliocoverpa zea* (Boddie). Naringenin showed little activity against *H. zea*, and hesperitin was intermediate in activity (Elliger *et al.*, 1980), corresponding with observed differences in fruit fly resistance among lemons, oranges, and grapefruit. Bioassays to determine whether eriocitrin is a key factor in the resistance of lemons to fruit flies are being undertaken. If so, it may be possible to impart heightened resistance to grapefruit by developing a transgenic strain of grapefruit with a propensity to convert naringenin to eriocitrin, as occurs naturally in lemons.

Similarly, Reissig *et al.* (1990) demonstrated that several crab apple (*Malus toringoides* Hughes) clones and species are resistant to development of apple maggot larvae. Incorporation of factors accounting for crab apple resistance could be very helpful in developing apple cultivars that are resistant to the apple maggot, *Rhagoletis pomonella* (Walsh). While the basis for resistance in the crab apples has not been identified, a clue may lie in the phenolic compounds they possess (Pree, 1977).

Conclusion

With the loss of traditional fumigants for fruit and vegetable disinfestation, the decreasing availability of approved insecticides and fungicides, and increasing concern for environmental safety and public health, it is imperative that new approaches be developed to ensure the continued supply of fresh fruits and vegetables to domestic and foreign markets. Many benefits may be derived from greater recognition and utilization of natural plant resistance mechanisms, and the use of new biotechnological tools including PGRs and recombinant techniques may help us achieve these goals.

References

Abbott, J.A., Affeldt, H.A. and Liljedahl, L.A. (1992) Firmness measurement of stored 'Delicious' apples by sensory methods, Magness-Taylor, and sonic transmission. *Journal of the American Society for Horticultural Science* 117, 590–595.

Anonymous (1992) Importation of papayas from Costa Rica. *Federal Register* 57 (121), 27896–27898.
Armstrong, J.W., Mitchell, W.C. and Farias, G.J. (1983) Resistance of 'Sharwil' avocados at harvest maturity to infestation by three fruit fly species (Diptera: Tephritidae) in Hawaii. *Journal of Economic Entomology* 76, 119–121.
Bestwick, R.K., Wagoner, W., Kellogg, J.A., Pengelly, W., Brown, L. and Ferro, A. (1991) Reduced ethylene biosynthesis in transgenic tobacco plants expressing S-adenosylmethionine hydrolase. *Journal of Cellular Biochemistry, The Genetic Dissection of Plant Cell Processes*, Supplement 15A, p. 110.
Campbell, B.C. (1987) The effects of plant growth regulation and herbicides on host plant quality to insects. In: Heinrichs, B. (ed.), *Plant Stress Interactions*. John Wiley & Sons, New York, pp. 205–247.
Campbell, B.C., Chan, B.G., Creasy, L.L., Dreyer, D.L., Rabin, L.B. and Waiss, A.C., Jr (1984) Bioregulation of host plant resistance to insects. In: Ory, R.L. and Rittig, F.R. (eds), *Bioregulators: Chemistry and Uses*. ACS Symposium Series 257, American Chemical Society, Washington DC, pp. 193–203.
Carey, J.R. (1984) Host-specific demographic studies of the Mediterranean fruit fly *Ceratitis capitata*. *Ecological Entomology* 9, 261–270.
Coggins, C.W., Jr and Henning, G.L. (1988) A comprehensive California field study of the influence of preharvest applications of gibberellic acid on the rind quality of Valencia oranges. *Israel Journal of Botany* 37, 145–154.
Cohen, E., Podoler, H. and El-Hamlauwi, M. (1987) Effects of the malathion-bait mixture used on citrus to control *Ceratitis capitata* (Wiedemann) (Diptera: Tephritidae) on the Florida red scale, *Chrysomphalus aonidum* (L.) (Hemiptera: Diaspididae), and its parasitoid *Aphytis holoxanthus* DeBach (Hymenoptera: Aphelinidae). *Bulletin of Entomological Research* 77, 303–307.
Ehler, L.E. and Kinsey, M.G. (1991) Ecological recovery of a gall midge and its parasite guild following disturbance. *Environmental Entomology* 20, 1295–1300.
Elliger, C.A., Chan, B.C. and Waiss, A.C., Jr (1980) Flavonoids as larval growth inhibitors. *Naturwissenschaften* 67, 338–339.
El-Otmani, M. and Coggins, C.W., Jr (1991) Growth regulator effects on retention of quality of stored citrus fruits. *Scientia Horticulturae* 45, 261–272.
El-Otmani, M., M'Barek, A.A. and Coggins, C.W., Jr (1990) GA_3 and 2,4-D prolong on-tree storage of citrus in Morocco. *Scientia Horticulturae* 44, 241–249.
Eskafi, F. (1988) Infestation of citrus by *Anastrepha* spp. and *Ceratitis capitata* (Diptera: Tephritidae) in high coastal plains of Guatemala. *Environmental Entomology* 17, 52–58.
Fein, B.L., Reissig, W.H. and Roelofs, W.L. (1982) Identification of apple volatiles attractive to the apple maggot, *Rhagoletis pomonella*. *Journal of Chemical Ecology* 8, 1473–1487.
Fischer, D.C., Kogan, M. and Greany, P. (1990) Inducers of plant resistance to insects. In: Hodgson, E., and Kuhr, R.J. (eds), *Safer Insecticides: Development and Use*. Marcel Dekker, New York, pp. 257–280.
Greany, P.D. (1989) Host plant resistance to tephritids: an under-exploited control strategy. In: Robinson, A.S. and Hooper, G. (eds), *World Crop Pests, Vol. 3A, Fruit Flies, Their Biology, Natural Enemies and Control*. Elsevier, Amsterdam, pp. 353–362.

Greany, P.D. and Shapiro, J.P. (1993) Manipulating and enhancing citrus fruit resistance to the Caribbean fruit fly. *Florida Entomologist* 76, 258–263.

Greany, P.D., McDonald, R.E., Schroeder, W.J. and Shaw, P.E. (1991) Improvement in efficacy of gibberellic acid treatments in reducing susceptibility of grapefruit to attack by Caribbean fruit fly. *Florida Entomologist* 74, 570–580.

Hennessey, M.K., Baranowski, R.M. and Sharp, J.L. (1992) Absence of natural infestation of Caribbean fruit fly (Diptera: Tephritidae) in commercial Florida 'Tahiti' lime fruits. *Journal of Economic Entomology* 85, 1843–1845.

Hoelmer, K.A. and Dahlsten, D.L. (1993) Effects of malathion bait spray on *Aleyrodes spiraeoides* (Homoptera: Aleyrodidae) and its parasitoids in Northern California. *Environmental Entomology* 22, 49–56.

Joel, D.M. (1980) Resin ducts in the mango fruit: a defence system. *Journal of Experimental Botany* 31, 1707–1718.

Klocke, J.A. and Kubo, I. (1982) Citrus limonoid by-products as insect control agents. *Entomologia Experimentalis et Applicata* 32, 299–301.

Langhoff, D.P., Wagoner, W., Kellogg, J.A., Stamp, J.A. and Bestwick, R.K. (1992) Effect of *S*-adenosylmethionine hydrolase expression on ethylene biosynthesis in transgenic tomatoes. *Journal of Cellular Biochemistry, Crop Improvement via Biotechnology*, Supplement 16F, p. 225.

McGlasson, W.B., Wade, N.I. and Adato, I. (1978) Phytohormones and fruit ripening. In: Letham, D.S., Goodman, P.B. and Higgins, T.J.V. (eds), *Phytohormones and Related Compounds: a Comprehensive Treatise*, Vol. 2. Elsevier, Amsterdam, pp. 447–493.

Nguyen, R. and Fraser, S. (1989) Lack of suitability of commercial limes and lemons as hosts of *Anastrepha suspensa* (Diptera: Tephritidae). *Florida Entomologist* 72, 718–720.

Pree, D.J. (1977) Resistance to development of larvae of the apple maggot in crab apples. *Journal of Economic Entomology* 70, 611–614.

Prokopy, R.J. and Roitberg, B.D. (1984) Foraging behavior of true fruit flies. *American Scientist* 72, 41–49.

Prokopy, R.J., Moericke, V. and Bush, G.L. (1973) Attraction of apple maggot flies to odor of apples. *Environmental Entomology* 2, 744–749.

Prokopy, R.J., Owens, E.D. and Prestwich, G.D. (1983) Visual detection of plants by herbivorous insects. *Annual Review of Entomology* 28, 337–364.

Reissig, W.H., Fein, B.L. and Roelofs, W.L. (1982) Field tests of synthetic apple volatiles as apple maggot attractants. *Environmental Entomology* 11, 1294–1298.

Reissig, W.H., Brown, S.K., Lamb, R.C. and Cummins, J.N. (1990) Laboratory and field studies of resistance of crab apple clones to *Rhagoletis pomonella* (Diptera: Tephritidae). *Environmental Entomology* 19, 565–572.

Rössler, Y., and Greany, P.D. (1990) Enhancement of citrus resistance to the Mediterranean fruit fly. *Entomologia Experimentalis et Applicata* 54, 89–96.

Salles, L.A. (1993) Periodos de ataque e de controle da mosca da fruta em pessegueiro (Period of attack and control of *Anastrepha fraterculus* (Wied.) in peach). *Hortisul.* 3 (in press).

Seo, S.T., Farias, G. and Harris, E.J. (1982) Oriental fruit fly: ripening of fruit and its effect on index of infestation of Hawaiian papayas. *Journal of Economic Entomology* 75, 173–178.

Seo, S.T., Tang, C.S., Sanidad, S. and Takenaka, T.H. (1983) Hawaiian fruit flies (Diptera: Tephritidae): Variation of index of infestation with benzyl isothiocyanate concentration and color of maturing papaya. *Journal of Economic Entomology* 76, 535–538.

Shaw, P.E., Calkins, C.O., McDonald, R.E., Greany, P.D., Webb, J.C., Nisperos-Carriedo, M.O. and Barros, S.M. (1991) Changes in limonin and naringin levels in grapefruit albedo with maturity and the effects of gibberellic acid acid on these changes. *Phytochemistry* 30, 3215–3219.

Spitler, G.H., Armstrong, J.W. and Couey, H.M. (1984) Mediterranean fruit fly (Diptera: Tephritidae) host status of commercial lemon. *Journal of Economic Entomology* 77, 1441–1444.

Swanson, R.W. and Baranowski, R.M. (1972) Host range and infestation by the Caribbean fruit fly, *Anastrepha suspensa* (Diptera: Tephritidae) in South Florida. *Proceedings of the Florida State Horticultural Society* 85, 271–274.

Statistical Concept and Minimum Threshold Concept

3

Jacqueline L. Robertson[1], Haiganoush K. Preisler[1] and E. Ruth Frampton[2]
[1]*United States Department of Agriculture, Forest Service, PO Box 245, Berkeley, CA 94701, USA;*
[2]*Ministry of Agriculture and Fisheries, Lincoln, New Zealand.*

Introduction

For insect pests of fresh horticultural products, as with any other problem in quarantine entomology, statistical analyses are used to estimate the probability that a treatment will successfully prevent the pest from becoming established in the area into which the product is imported. To evaluate the probable success of a treatment, laboratory bioassays are done first. Once a treatment has been selected, a large-scale confirmatory test follows. This sequence of testing resembles the protocol generally used to evaluate the probable success of a treatment to control any pest species on any crop, but is more complex because control also includes the problem of prevention of successful establishment.

In this chapter, we review the statistical methods used in commodity treatment bioassays that are done in the laboratory and that serve as the basis for confirmatory tests. In particular, we address problems of experimental design and data interpretation inherent in such bioassays. Finally, we describe the statistical basis of confirmatory tests based strictly on pest mortality, versus the concept of the maximum pest limit.

Quantal Response Bioassays

A *quantal response bioassay* is any experiment in which the response of the test organisms varies in relation to a measurable characteristic of a stimulus (Robertson and Preisler, 1992). In quarantine entomology, the stimulus may be a fumigant, temperature, an atmosphere with altered gaseous composition, or any other treatment that has a deleterious effect

From: Paull, R.E. and Armstrong, J.W. (eds) (1994) *Insect Pests and Fresh Horticultural Products: Treatments and Responses.* CAB INTERNATIONAL, Wallingford, UK.

on the pest (e.g. Armstrong and Couey, 1989; Moffitt and Burditt, 1989; Sharp and Picho-Martinez, 1990; Yokoyama *et al.*, 1990).

All quantal response bioassays differ in terms of the *response variables* (dependent variables) and the *explanatory variables* (independent variables) involved. Explanatory variables are the measurable characteristics of a stimulus or stimuli (e.g. concentration of a chemical, temperature) that cause responses of the target arthropod species to vary. Response variables are the random outcomes (e.g. death or survival) of the exposure to the stimulus: they vary in relation to its intensity.

Robertson and Preisler (1992) described various types of quantal response bioassays that can be done. The simplest type (and the type most often used in quarantine entomology) is a *binary response experiment with one explanatory variable*. In the rare instances in which more than one variable has been tested in a commodity treatment bioassay, the variables have been reduced to a single variable by multiplication of one variable by the other (e.g. the concentration times time product; Monro, 1969).

As long as estimates of the probability of death continue to be required by regulatory agencies of the area of importation, the advantages of bioassays with multiple relevant explanatory variables will probably continue to be disregarded despite the fact that such multivariates provide valuable information about variables (other than the probability of death) that affect the probability that a pest might become successfully established in a previously pest-free area when the pest species is imported on a fresh horticultural product. In particular, the probability of survival, which is related to many factors, is more important than a simple dose–response relationship with just one explanatory variable (dose) and death as the only response variable.

Among the germane multiple variables are level of infestation of the pest when the commodity is harvested, suitability of a particular species or cultivar as a host for the quarantine pest, infestation rate, and size of the shipment (Armstrong, 1985; Armstrong and Couey, 1989; Vail *et al.*, 1992). As we describe below, these types of variables must be investigated to establish treatment efficacy based on survival rather than death.

Death and treatment efficacy: the probit 9

History of the probit 9 requirement

Until adoption of the maximum pest limit concept by New Zealand (R.T. Baker *et al.*, 1990), successful treatment in quarantine entomology was the most rigidly defined of any specialized type of arthropod control. In 1939, A.G. Baker stated that 'The security demanded as a basis for recommendation is determined by reading on the regression line the exposure coordinated with a probit of 9. In percentages this probit represents a mortality of 99.996 83 percent, or a survival of approximately 32 out of 1 000 000.'

The basis for A.G. Baker's (1939) recommendation seems quite simple: like his contemporaries, he probably considered probit analysis to be the standard method for the analysis of dose–response data. Baker also concluded that 99.996 83% mortality would ensure that the risk of accidental introduction of a tephritid fruit fly would be virtually non-existent. However, knowledge of exotic pests has increased substantially since 1939 (e.g. Landolt *et al.*, 1984; Armstrong and Couey, 1989; Yokoyama *et al.*, 1990; Vale *et al.*, 1992) and a growing body of evidence suggests that mortality (particularly 99.996 83% mortality) is far too narrow a criterion upon which to evaluate treatment efficacy (e.g. Landolt *et al.*, 1984; Robertson *et al.*, 1984; R.T.Baker *et al.*, 1990; Vail *et al.*, 1992).

A.G. Baker's (1939) recommendation, like any other, should have been subject to debate and revision as knowledge about particular insect pests increased, and as entomologists became more familiar with biostatistics. Instead, the 'probit of 9' was codified as the requirement for 'probit 9 security' by the US Department of Agriculture and comparable regulatory agencies of other nations. In the United States and most other countries that import fresh horticultural products that might be contaminated with exotic pests, the requirement for probit 9 security has not changed despite 50 years of additional research in entomology and biostatistics. Scientifically, this situation is untenable.

Assumptions of the probit 9 requirement

Three interrelated assumptions are inherent in the probit 9 requirement: (i) 99.996 8% effectiveness is the minimum level necessary for quarantine security; (ii) the probit model is always suitable for the analyses of data from commodity treatment bioassays; (iii) only the probability of death is relevant to the future establishment of the pest species in a new environment. As convenient as these assumptions may be for regulatory agencies, they have limited scientific bases.

Even if we assume that assumptions (i) and (iii) are correct, and that death should be used as the only criterion for treatment efficacy, a more general statistical definition of quarantine security is necessary to avoid assumption (ii). As described in the next section, laboratory bioassays for commodity treatments are merely quantal response bioassays, and all quantal response bioassays have the same statistical characteristics.

Q: a more general term to replace probit 9

The statistical statement of binary response with a single explanatory variable is

$$p_i = F(\alpha + \beta x_i) \qquad [3.1]$$

p_i is the probability of response (e.g. death), x_i is the ith treatment level (e.g. dose) or a function of that level (e.g. logarithm of dose), α is the intercept of the regression line, β is the slope of the regression line, and F is a distribution function.

Transformation of the units of both the response variable and the explanatory variable is not always necessary for the relative frequency of responses to fit a linear model, despite Baker's 1939 statement that 'the use of probits and logarithms or a logarithmic scale, as here used, converts the relationship between mortality and length of exposure to a linear one . . .'. Although units of the explanatory variable are usually converted to logarithmic values, this transformation is not always necessary for data to fit a line. For example, units of temperature or radiation tested in commodity treatments often need not be transformed at all for the data to fit the linear model (e.g. Burditt and Hungate, 1988).

In binary quantal response bioassays, the points along the fitted regression line are such that, for a given level of the explanatory variable, there is a corresponding probability of response. Examples of the level of the corresponding probability of the response variable are the LD (lethal dose), LC (lethal concentration), or LT (lethal time). For example, the LD_{50} is the dose expected to be lethal to 50% of a population tested. In terms of lethal doses, the probit 9 criterion corresponds with the 99.9968% mortality (e.g. an $LD_{99.9968}$).

We suggest that a more general term, Q, would be appropriate for quarantine treatments. A Q to cause mortality equivalent to an $LD_{99.9968}$ can be called a $Q_{99.9968}$. When the distribution function, F, in equation [3.1] is assumed to be the normal distribution, $Q_{99.9968}$ is equivalent to the probit 9.

The way a particular bioassay should be done (i.e. the experimental design) to estimate any lethal level depends on the response level of interest. The level necessary for 99.9968% mortality is especially difficult to estimate with any degree of certainty (regardless of the experimental design) because of its location in the upper extreme of the probability distribution (Copenhaver and Mielke, 1977). In addition, the numerical value that results from the estimation procedure depends on the statistical model used: the probit model is only one of several available.

Probit or logit analysis

Of the numerous types available (e.g. Robertson and Preisler, 1992), the two distribution functions most commonly used in equation [3.1] are the normal and the logistic. The normal distribution is assumed in probit analysis; the model specified by Equation [3.1] is $p_i = \Phi(\alpha + \beta x_i)$ where Φ is the standard normal (Gaussian) distribution function. For the logit model

$$p_i = \frac{1}{1 + e^{-(\alpha+\beta x_i)}} = \frac{e^{\alpha+\beta x_i}}{1 + e^{\alpha+\beta x_i}} \qquad [3.2]$$

where e ($\simeq$ 2.718 28) is the base of the natural logarithm. As mentioned previously, convention was the source of use of the probit model by entomologists in 1939.

Although a debate about use of one model versus the other has continued among statisticians (Berkson, 1951; Finney, 1964; Copenhaver and Mielke, 1977), no scientific evidence has ever been presented to show that data from a binary quantal response bioassay are always distributed as either a normal or a logistic function. Whether one model or the other should be used is difficult to determine because similar results are usually obtained with either one except at the extreme ends of the probability distribution. The difficulties involved in this problem provide an ample basis for our recommendation that a generalized term, Q, be used in quarantine entomology.

Expectations that one model will always fit biological data are totally unfounded. The most sensible scientific approach is to examine the goodness-of-fit of each data set and then select the model that best fits the data. In many instances, either the probit or logit model seems to fit bioassay data adequately. However, an adequate fit does not necessarily mean the best fit in the sense that another model that was not examined may be more appropriate for a particular data set. Estimates of treatment efficacy reflect both the statistical model and the assumptions inherent in the model.

EXPERIMENTAL DESIGN: DOSE SELECTION AND SAMPLE SIZE

Dose selection and sample size are the two most crucial aspects of experimental design in any binary response bioassay with one explanatory variable. Before the investigation described by Robertson *et al.* (1984), many researchers routinely used the guidelines for experimental design of dose–response bioassays described by Finney (1971) without realizing that these recommendations for dose selection pertain to estimation of the LD_{50} and not necessarily to other LDs.

The applicability of the methods described by Brown (1966) and Freeman (1970) also are limited to the 50% response level or to responses in that vicinity. Tsutakawa (1980) presented another approach to the problem of dose selection for efficient estimation of an arbitrary lethal dose, but his method is not general and its computational difficulty precludes routine use. Thus, guidelines for the numbers of subjects to test at each dose and for the experiment as a whole were needed.

Empirical evidence (Haverty and Robertson, 1982) suggested that different designs were required for precise estimation of the LD_{50} and the LD_{90}. A Monte Carlo computer simulation was used to test the effects of sample size and dose placement on the precision of LD_{50} and LD_{90} estimates and to find the optimal percentage of mortality that should be observed for three to eight doses in experiments with total sample sizes of 60 to 720 (Robertson *et al.*, 1984). The criterion used for precision was the width of the 95% confidence limits around an LD_{50} or LD_{90} estimated with the logit model.

Minimum sample size

The investigation of Robertson *et al.* (1984) showed that at least 120 test subjects are required for a reliable dose–response experiment. Use of 60–64 test subjects (the smallest sample size that was tested) is not desirable because of the poor precision of the LD_{50} and LD_{90} estimates, and the frequent occurrence of infinite 95% confidence limits (that is, confidence limits that go from minus infinity to plus infinity and therefore, are equivalent to no confidence limits at all). Because infinite 95% confidence limits occur occasionally even with a sample size of 120, Robertson *et al.* (1984) concluded that at least 240 test subjects should be used to estimate a dose–response relationship with acceptable precision.

Although optimal sample size requirements for bioassays done to estimate a $Q_{99.9968}$ have not be determined, we suspect that far more than 30 000 test subjects may be required based on the experience of the US National Center for Toxicological Research in a study ('megamouse') done to estimate the ED_{01} (effective dose to kill 1%) for a carcinogen, 2-acetylaminofluorene, to laboratory mice (Ottoboni, 1984). In the megamouse study, a sample size of 24 000 was used and the result was still not definitive. Among other problems, no way exists to select the model that would be most appropriate for the data analysis.

The megamouse problem is particularly relevant to quarantine entomology and is one which investigators must recognize from the outset of any experiment with a possible treatment of horticultural crops. Particularly bothersome is the fact that, although some species of exotic pests such as tephritid fruit flies might infest fruits with extremely large numbers of larvae, a single larva of other species such as codling moth (*Cydia pomonella* L.) can be found in one fruit. Tests with very large numbers of lepidopterous pests such as codling moth are neither necessary (Yokoyama *et al.*, 1990; Hartsell *et al.*, 1991) nor practical.

Dose placement

Optimal placement of doses varies depending on the lethal dose of interest and the number of doses to be tested (Robertson *et al.*, 1984). For example, Robertson *et al.* (1984) found that precisely estimated LD_{50}s are obtained when responses are evenly distributed between 25 and 75%. Precise LD_{90} estimates, however, require that one or two doses cause at most 10% mortality and that most doses cause between 75 and 95%

mortality. The placement of treatment levels necessary to estimate $Q_{99.9968}$ has not been determined because no study has shown that the probit model can always be assumed to be the best and because placement depends so heavily on the model assumed to be appropriate (see Table 2.1 in the Examples section).

NUMBER OF DOSES

Müller and Schmitt (1990) addressed the problem of how to choose the number of doses to provide the best estimate of the LD_{50} based on the criterion of minimizing the asymptotic variance at the 50% response level. They concluded that use of as many doses as possible is preferable and that use of only three doses cannot be recommended. For any routine bioassay, use of at least five doses has been recommended. Far more are probably required to obtain a reliable estimate of 99.9968% mortality; most would have to cause 100% mortality. Regardless of the number of doses tested, a response curve cannot be estimated if all but one cause 100% mortality (Robertson and Preisler, 1992).

GOODNESS-OF-FIT

The measure of how well data fit the assumptions of the model (whether probit, logit, or some other related model; whether or not error is binomial) is the goodness-of-fit. The usual way to test fit is with a χ^2 test. In a χ^2 test, values (responses) predicted by the model are compared with values actually observed in the bioassay. If the values differ significantly at $P = 0.05$, the model does not fit the data and a more appropriate model should be sought. However, alternatives to the probit or logit models are not easy to use (Robertson and Preisler, 1992).

Various methods to compensate for lack-of-fit have been used. Finney (1971) multiplied the variances by a heterogeneity factor to account for extra variation that causes poor fit. A similar method is used in various probit or logit programs [e.g. POLO-PC (LeOra Software Inc., 1987); SAS PROBIT (SAS Institute, Inc., 1982)]. However, this method does not identify the reasons why the data do not fit the model. A more meaningful approach is to examine possible causes of lack-of-fit by means of residual plots (Preisler, 1988).

RESIDUAL PLOTS

A residual is the difference between an observed response and an expected response. Because each binomial response has a variance related to the size of the response, the response must be standardized by division by its standard error before residuals are plotted. The standard error is

$$\sqrt{np(1-p)}$$

where n is the number of subjects tested at the specific dose and p is the estimated response probability. For good fit, residuals plotted against predicted values or against dose usually lie within a horizontal band around zero (mostly between –2 and 2).

Causes of Poor Fit

Preisler (1988) listed five causes of significant departure from the probit-binomial model that might be the cause for lack-of-fit. Some of these causes may be identified by residual plots, some may not. First, outliers that do not fit the model might be the cause of lack-of-fit, especially if they occur at a response level near 0 or 100%. An outlier should be discarded from the data set only if strong evidence suggests that an error in recording has occurred. Otherwise, the outlier could indicate interaction with or the existence of an important independent variable.

The second cause of lack-of-fit may be omission of a significant explanatory variable (e.g. body weight, temperature) from the model. For example, a univariate model has been used but a multiple probit or logit model is appropriate. (The need for studies of this type in quarantine entomology will be discussed in more detail later in this chapter.)

Third, the probit curve might not fit the data. Fourth, responses of subjects with a factor in common might be correlated; as a result, error terms are not independent. For example, insects from the same generation might be more alike in their response to a pesticide than when compared with insects in another generation. Finally, Preisler (1988) identified a situation in which error terms are not binomial.

For lack-of-fit caused by a lack of independence of error term and by extrabinomial variability (departure from the assumptions of a binomial distribution), Preisler (1988) described a method to obtain maximum likelihood estimates for parameters in a compound–binomial– probit model. This method is based on addition of a random effect factor to the model.

Experimental Design: Time as a Variable

In bioassays with pests in horticultural products, time–dose–response data have frequently been analysed by fitting probit lines (with time replacing dose) to mortality data for a fixed dose over time. However, unless different groups are used for each observation period, this method is not appropriate because responses at different time points will be correlated (Preisler and Robertson, 1989).

Independent Versus Serial Designs

Biossays that include both time and dose as variables can have one of two possible designs. With the *independent sampling design*, separate groups of test subjects are treated with a fixed dose. Each group is then observed for a different period of time, and numbers of responses are recorded at each observation period. For probit or logit models to be appropriate, responses that occur in a given group of test subjects must be recorded only once.

The other alternative, the *serial sampling design*, involves treatment of each group of subjects with a given dose, inclusion of several doses in the experiment as a whole, and recording of responses for each dose group at a series of times after treatment. A number of observation intervals are necessary because responses differ with each dose (i.e. a high dose might kill more test subjects in a short period of time, but a low dose might kill fewer subjects during the same interval).

GENERAL STATISTICAL MODELS FOR STUDIES WITH TIME

For the independent design, statistical models of time–response data for a fixed dose are straightforward. The probability of response by a given time t can be modelled by the probit, logit, or other curves with time (or some function thereof such as the logarithm of time) as the single independent variable. The general statistical model for the binary response with time as the explanatory variable is

$$p_i = F(\alpha + \beta t_i) \qquad [3.3]$$

where p_i is the probability of response, t_i is the ith time (or some function of time), α is the intercept of the regression line, β is the slope of the regression line, and F is a distribution function [e.g. probit, logit, complementary log–log (CLL)].

The general statement of the CLL relationship in the serial time–mortality design is as follows. For a fixed exposure time t_j ($j = 1,\ldots,J$), to a treatment at a concentration d_i ($i = 1,\ldots,I$), probability of mortality of a test subject by time t_j is

$$1 - e^{-e^{\gamma_j + \beta \log_{10}(d_i)}} \qquad [3.4]$$

where β is an unknown parameter and γ_j are unknown categorical variables corresponding with the times t_j. The linear part of Equation [3.4], $\gamma_j + \beta\log_{10}(d_i)$, is the CLL line. The model assumes that $\log_e[-\log_e(1 - p_{ij})]$ is linear in the covariates.

For the independent time–mortality design, γ_j is replaced by the logarithm of t_j and $\log_{10}(d_i)$ is replaced by the constant $\log_{10}(D)$ where D is the fixed dose used in the experiment. This statement of the CLL relationship is actually the Weibull function, which has been used to model responses to some chemicals over time (e.g. Su *et al.*, 1987). Further details of analyses of studies with time as a variable are provided by Preisler and Robertson (1989) and Robertson and Preisler (1992).

Examples

Data from three bioassays – two commodity treatment experiments done to estimate a $Q_{99.9968}$ [original data for *Bactrocera dorsalis* Hendel (= *Dacus dorsalis* Hendel) data provided in original form by A.K. Burditt, Jr, USDA, ARS, retired; results for *Ceratitis capitata* (Wiedemann) on mangoes published by Sharp and Picho-Martinez, 1990] – and a sample of bioassay data from a pesticide on a forest insect [mexacarbate on the western spruce budworm, *Choristoneura occidentalis* Freeman, done to estimate an LD_{50} and LD_{90} (Savin *et al.*, 1977)] demonstrate the special problems inherent in use of the probit analysis, especially at the probit 9 level.

Table 3.1 shows that the choice of model (in these examples, probit or logit) for each of the three data sets had a negligible effect at LD_{90} or LT_{90},

Table 3.1. Comparison of estimates from probit and logit models with experimental data sets for *B. dorsalis* (C30, C31),[a] *Ceratis capitata* (Mango),[b] and *Choristoneura occidentalis* (CO)[c]

Data	*n*[d]	Model[e]	Max[f]	$Q_{99.9968}$[g] (95% CL)	LD_{90} (95% CL)	χ^2 (df)
C30	67 473	P	9	8.94 (8.12–9.76)	3.27 (3.04–3.51)	129 (7)
		L		11.31 (9.91–12.71)	3.07 (2.80–3.34)	180 (7)
C31	28 047	P	8	7.79 (5.62–9.95)	2.82 (2.33–3.30)	463 (6)
		L		8.85 (6.77–10.93)	2.81 (2.50–3.11)	229 (6)
Mango	24 451	P	70	76.1 (66.0–86.1)	37.1 (33.7–40.5)	246 (6)
		L		101.9 (85.5–118.3)	36.2 (32.5–40.0)	248 (6)
CO	364	P	0.5	1.71 (0.97–3.00)	0.18 (0.13–0.25)	2.9 (4)
		L		7.94 (3.44–18.30)	0.18 (0.15–0.21)	2.2 (4)

[a]Data for generations 30 (C30) and 31 (C31) of *B. dorsalis* eggs exposed to 2.8°C. Units are days. Data provided by A.K. Burditt, Jr.
[b]Data (Mango) from time–temperature experiments on wild and laboratory strains of *C. capitata* in mangoes immersed in water at 46.1 ± 0.25°C (Sharp and Picho-Martinez, 1990). Units are minutes.
[c]Data for mexacarbate typically applied to western spruce budworm (CO).
[d]*n* is the total number of test subjects tested in the experiment.
[e]*P* is the probit model, L is the logit model.
[f]Max is the largest dose or time tested.
[g]$Q_{99.9968} = LD_{99.9968}$.

but it had a large effect on estimates for 99.9968% mortality. For the studies with *B. dorsalis* and *C. capitata*, large values of the χ^2 goodness-of-fit statistic versus degrees of freedom indicated that neither model adequately fitted the data. The goodness-of-fit for the mexacarbate data is very good for both models. However, estimates of the $Q_{99.9968}$ for the two models were dramatically (and significantly) different. There is no way to choose between the probit or logit model. Estimation of the $Q_{99.9968}$ required extrapolation of the curve to a dose that was far outside the range tested in the experiment. Extrapolation is not a sound practice; data such as those for *C. occidentalis* should only be used to estimate a maximum of an LD_{95} because this was the highest kill actually observed.

As mentioned previously, large χ^2 values such as those for the *B. dorsalis* and *C. capitata* data sets are usually managed by multiplication of the variances and covariances by the heterogeneity factor (e.g. SAS Institute Inc., 1982; Sharp and Picho-Martinez, 1990). Besides ignoring the fundamental problem, this practice assumes that all causes of poor fit affect calculations of the standard error but not the point estimates. As mentioned previously, causes of poor fit can be studied by examining plots of standardized residuals for the dose levels tested (Preisler, 1988). When we examined plots of the temperature data for two generations of *B. dorsalis*, no systematic patterns (Fig. 3.1) that might indicate departure from the probit model were apparent.

Both plots showed some evidence of extra-binomial variation (i.e. variation about the mean is greater than expected under the binomial model); an outlier was apparent in the data for the F_{31} *B. dorsalis* exposed to cold. Thus, another model that incorporates extra-binomial variation within the probit framework might give a better fit. Table 3.2 shows estimates of $LT_{99.9968}$ (with 95% confidence limits) for seven generations of *B. dorsalis* produced by fitting such a model.

For generation 31, the outlier was not included. The values of χ^2 thus obtained seem to indicate that the data fit this model reasonably well. All estimates of $Q_{99.9968}$ are larger than those produced by the simple probit model (see corresponding estimates in Table 3.1). Estimates of $QT_{99.9968}$ for the various generations also show differences in response (Table 3.2). For example, the 95% confidence limits for generation 30 and generation 36 do not overlap. The average $Q_{99.9968}$ over the seven generations in this experiment is 9.855 days (SD = 0.56). Assuming that the seven generations included in this experiment provide a random sample of the responses of present and future generations of *B. dorsalis*, 95% of the generations would have a $Q_{99.9968}$ that is less than $9.855 + (1.65 \times 0.56) = 10.78$°C. The treatment level that could be used in a confirmatory test to verify that the treatment meets quarantine security requirements is thus 10.78 days when eggs are exposed to a temperature of 2.8°C.

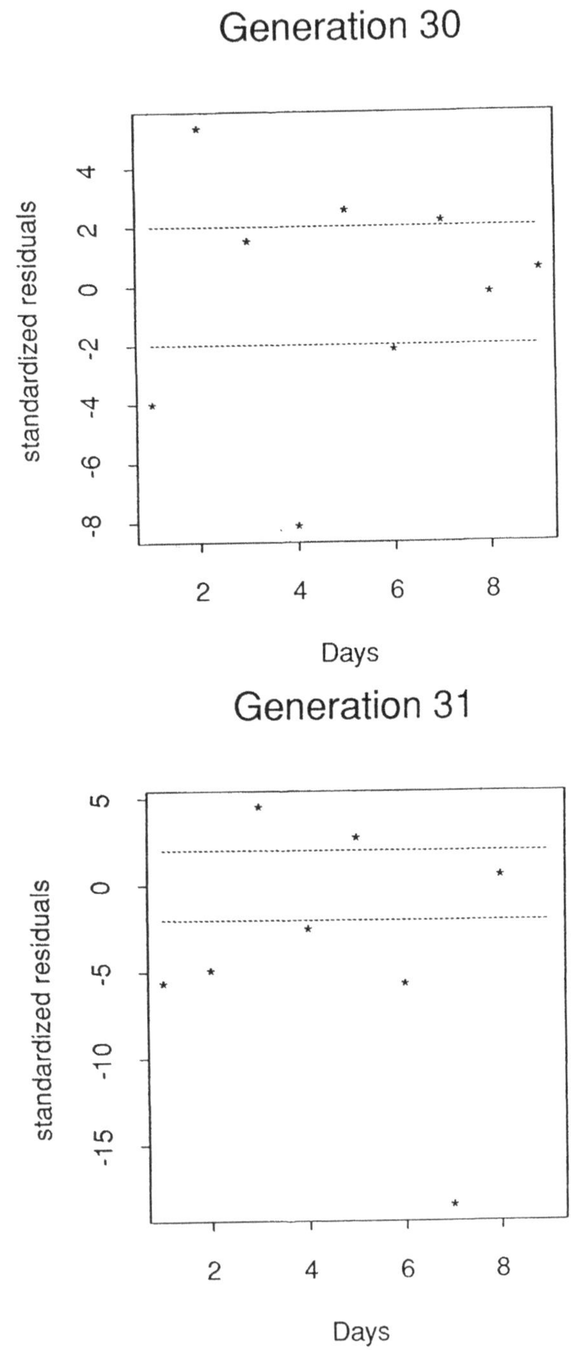

Fig. 3.1 Plots of standardized residuals developed by fitting probit curves to responses of *B. dorsalis* eggs from generation 30 (upper) and 31 (lower) exposed to 2.8°C for different numbers of days. Most of the residuals are outside the horizontal band of ±2 units about zero.

Table 3.2. Results for seven generations of *B. dorsalis* treated with cold, with estimates obtained by fitting a probit model with a random effect factor in the probit line to account for extra-binomial variation.

Generation	Max	$Q_{99.9968}$ (95% CL)	χ^2	df
30	9	9.52 (8.79–10.24)	2.68	6
31	7	10.64 (8.47–12.82)	1.52	4
32	10	9.46 (8.71–10.21)	8.96	7
33	8	9.38 (8.46–10.30)	2.54	5
34	9	9.30 (7.89–10.70)	1.69	6
35	9	10.10 (9.17–11.04)	3.12	6
37	12	10.54 (10.34–10.75)	7.25	9

Source: A.K. Burditt, Jr Units are days at 2.8°C.

Estimation of treatment efficacy when numbers of test subjects are unknown

One of the reasons that the probit model does not fit the data for *C. capitata* in mangoes is that mortalities were estimated rather than observed (Sharp and Picho-Martinez, 1990). In these experiments, the number of larvae treated at each level was estimated from a control group not immersed in water. Both controls and treated mangoes had been exposed to natural oviposition by adult fruit flies. Unfortunately, infestation by natural oviposition can be variable (e.g. Armstrong, 1983; Armstrong and Couey, 1984).

The problem of *n* being unknown causes special problems in the analysis when mortality is calculated to be negative. Usually, data points with negative mortality are disregarded (e.g. Sharp and Picho-Martinez, 1990). A more efficient way to deal with experiments of this type is to model the observed survival rather than death by using the Poisson or Poisson-lognormal models (Robertson and Preisler, 1992). Consideration of survival not only avoids the problem of apparent negative mortality, but it includes the extra source of variation caused by the fact that exact numbers of larvae treated at each level are unknown.

Confirmatory tests

In confirmatory tests, large numbers of the target species are tested with a fixed treatment level (usually the $Q_{99.9968}$) and an upper 95% boundary for the probability of at least one survivor is calculated. For example, if n larvae are treated and no survivors are observed, then the 95% upper boundary is p_u $1 - (0.05)^{1/n}$. If $n = 95\,000$, for example, then $p_u = 31.53$ per million. To meet the probit 9 criterion, < 32 survivors per million are required. Therefore, the treatment would meet the probit 9. If one or more survivors is observed, then the tables given by Couey and Chew (1986) can be used to calculate p_u. At the stage of confirmatory testing, the assumption that the data follow a probit curve is unnecessary.

The Maximum Pest Limits

Introduction

In 1990, R.T. Baker *et al.* described the concept of setting maximum pest limits for produce being imported into New Zealand. This method consists of limits on the maximum number of tephritid fruit flies imported into the port of Auckland at a given time, assuming that a single male and a single female do not emerge, mate, and found a new population.

The maximum pest limit concept is derived form the suggestion of Landolt *et al.* (1984) that quarantine efficacy should be evaluated on the basis of the probability of successful reproduction. These authors suggested that factors such as mating, dispersal, feeding, and finding hosts in a new location should be considered. These factors, which Landolt *et al.* (1984) collectively considered to constitute 'risk', are explanatory variables that should be defined. Treatment efficacy could then be estimated based on infestation levels and lot size of shipments of fresh horticultural products.

Infestation level and lot size

In our description of methods relevant simply to mortality as a variable, we outlined experimental and statistical procedures used to determine a treatment level for use in quarantine treatments, and to estimate the efficacy of a given treatment by a confirmatory test. The efficacy of a treatment is the percentage of eggs, larvae or adults that a given treatment is expected to kill. Because efficacy is a percentage, it does not provide an estimate of the actual number of survivors (or those killed) in a given lot (or shipment). The actual number depends on three additional factors, i.e. the proportion of infested fruit, the mean infestation level of fruit from a given area, and the size of the lot. With these factors in mind, Landolt *et al.* (1984) proposed 'that the level of security recommended for quarantine treatments of tephritid-infested fruit be set at a particular probability of an introduction occurring, specified as the probability of a potential mating pair arriving per shipment'.

Probability of survivors per lot

By making some assumptions about the distribution of the number of larvae per fruit, we derived a formula for the probability of the number of live larvae in a shipment of a particular size. First, we assume that two adults of the opposite sex are the minimum number required to establish a population. Then, using a Poisson approximation (R.T. Baker *et al.* 1990), we calculate *Pr(r* survivors in a lot size *N)* as

$$\frac{e^{-\lambda}\lambda^{r}}{r!} \quad [3.5]$$

where $\lambda = N\phi\mu p$, N = lot size, ϕ = treatment efficacy, μ=mean number of larvae per fruit and p = proportion of fruit infested. Therefore, the probability of having ≥two larvae in a lot size *N* is

$$q = Pr[r \geq 2] = (1 - e^{-\lambda/2})^2 \quad [3.6]$$

For example, with a lot size of $N =$ 60 000 fruit, a proportion of infested fruit $p =$ 0.001 (1 infested fruit per 1000 fruits), a mean number of larvae per infested fruit μ = 10 and a treatment efficacy of $\phi = 32/10^6$, λ= 0.0192, and q=0.000 091 28. Therefore, the probability that any lot has ≥two survivors is 1 in 10 955. If the proportion of infested fruit is $p =$ 0.0001 (i.e. 1 infested fruit per 10 000 fruit), then the expected probability of ≥two survivors is 1 in 1 085 776.

These calculations clearly indicate that the lot size will influence the risk of introducing a mating pair from a shipment. For practical purposes, a maximum lot size arriving at a given point of entry at a given time could thus be set to minimize the probability of introducing a mating pair into a country. Vail *et al.* (1992) reached the same conclusion in their investigation of horticultural commodities infested with codling moth, *Cydia pomonella* (L.). They also proposed use of a model, based on survival, that would reduce the chance of successful reproduction from imported walnuts, nectarines, and cherries.

Confidence bounds for probability of survivors

The probability q in Equation [3.6] is only an estimate because the efficacy of treatment, the proportion of infested fruit, and the mean infestation rate are estimates that include their own error terms. Given the standard errors of the estimates $\hat{\phi}$ and $\hat{\mu}$, and p, the delta method (Robertson and Preisler, 1992) for calculating approximate standard errors of functions of random variables could be used to calculate a standard error and 95% confidence limits for q. Alternatively, an upper confidence boundary for q could be calculated given similar bounds for the efficacy of a treatment, the proportion of infested fruit, and the infestation level of fruits from a given area.

Maximum pest limits and acceptance sampling

Fruit consignments that arrive in Auckland, New Zealand at a given time are sampled to estimate the number of immature fruit flies that may be present. The maximum number allowable is determined based on environmental conditions and the likelihood of the fruit flies to reproduce.

R.T. Baker *et al.* (1990) described methods for calculating the number of fruits to be tested in order to detect, with some accuracy, the maximum allowable infestation level. Their method relies on four assumptions: (i) the treatment acts independently on different individuals; (ii) the mean number of fruit fly larvae within a fruit is known; (iii) the efficacy of the treatment used in the country of origin is known; (iv) the maximum lot size per day is known.

Further research is needed to refine the concept of the maximum pest limit. For example, another determinant to pest establishment is the climatic suitability of the new environment into which the arthropod might be carried on or in an infested commodity (Worner, 1988). In the future, the concepts of maximum pest limits and ecoclimatic indices might be integrated to provide even more refined pest exclusion criteria.

Conclusions

The transition from exclusive use of mortality (especially the probit 9) as the criterion for treatment efficacy to use of probability of introduction and establishment, with concomitant development of maximum pest limits for important species of pests on horticultural commodities, will require use of interim methods to ensure that dose–mortality bioassays are as efficient and well designed as possible. Results for the three data analyses described in our examples suggest that three criteria should be used by regulatory agencies to ensure the statistical validity of bioassays of possible treatments for commodity protection.

First, the experimental design must be such that it covers the lethal dose of likely interest. Estimates obtained by extrapolation outside the range of doses actually tested are always suspect. Implicit in the experimental design are use of adequate numbers of test subjects and adequate replications of the experiment. Second, the model used must adequately fit the data. If fit is not adequate, causes for poor fit should be sought. If the model cannot be adapted, another model should be identified. Finally, commodity treatment bioassays should be done on more than one generation of the target species to account for natural variation among generations and experiments.

To help speed the demise of the probit 9 requirement, editors of scientific publications can help to ensure that the statistical analyses used in commodity treatment bioassays are scientifically based. The requirement of a regulatory agency for a probit 9 estimate should not mean that meaningless values be presented in the scientific literature. An author attempting to publish probit 9 values should be required to present evidence that the first two (and preferably, all three) considerations described in the preceding paragraph have been taken into account. In summary, the interpretation of bioassay data must be statistically valid regardless of bureaucratic requirements.

We also recommend use of the more general (and less restrictive) term Q to replace the probit 9 requirement. When the distribution function used in the binomial model is the normal distribution, a $Q_{99.9968}$ would be equal to the probit 9. However, different distribution functions could be used as well, permitting investigators to choose the model that best fits their data.

Finally, as quickly as possible, scientists should begin to define the variables that affect establishment of quarantine pests of horticultural products and any other commodity. These variables include infestation rates for specific commodities (Vail *et al.*, 1992) and cultivars (Yokoyama *et al.*, 1990); natural survival rates in specific commodities; dispersal; host-finding; likelihood of infestation based on host and pest phenology; and likelihood of infestation based on climatological factors (Worner, 1988).

References

Armstrong, J.W. (1983) Infestation biology of three fruit fly (Diptera: Tephritidae) species on 'Brazilian', 'Valery', and 'William's' cultivars of banana in Hawaii. *Journal of Economic Entomology* 76, 539–543.

Armstrong, J.W. (1985) Pest organism response to potential quarantine treatments. In: *Proceedings, Regional Conference on Plant Quarantine Support for Agricultural Development*. Asian Plant Quarantine Centre and Training Institute, Seroang, Selangor, Malaysia, pp. 25–31.

Armstrong, J.W. and Couey, H.M. (1984) Methyl bromide treatments at 30°C for California stonefruits infested with Mediterranean fruit fly (Diptera: Tephritidae). *Journal of Economic Entomology* 77, 1229–1232.

Armstrong, J.W. and Couey, H.M. (1989) Fumigation, heat and cold. In: Robinson, A.S. and Hooper, G. (eds), *Fruit Flies, Their Biology, Natural Enemies and Control*. Elsevier, Amersterdam, The Netherlands, pp. 411–424.

Baker, A.G. (1939) The basis for treatment of products where fruitflies are involved as a condition for entry into the United States, *USDA Circular 551*, Washington, DC.

Baker, R.T., Cowley, J.W., Harte, D.W. and Frampton, E.R. (1990) Development of a maximum pest limit for fruit flies (Diptera: Tephritidae) in produce imported from New Zealand. *Journal of Economic Entomology* 83, 13–17.

Berkson, J. (1951) Why I prefer logits to probits. *Biometrics* 33, 327–339.

Brown, B.W., Jr (1966) Planning a quantal assay of potency. *Biometrics* 22, 322–329.

Burditt, A.K., Jr and Hungate, F.P. (1988) Gamma irradiation as a quarantine treatment for cherries infested with western cherry fruit fly (Diptera: Tephritidae). *Journal of Economic Entomology* 81, 859–862.

Copenhaver, T.W. and Mielke, P.W. (1977) Quantit analysis: a quantal assay refinement. *Biometrics* 33, 175–186.

Couey, H.M. and Chew, V. (1986) Confidence limits and sample size in quarantine entomology. *Journal of Economic Entomology* 79, 887–890.

Finney, D.J. (1964) *Statistical Method in Biological Assay*. Griffin, London.

Finney, D.J. (1971) *Probit Analysis*, 3rd edn. Cambridge University Press, Cambridge, England.

Freeman, P.B. (1970) Optimal Bayesian sequential estimation of the median effective dose. *Biometrika* 57, 79–89.

Hartsell, P.L., Vail, P.V., Tebbetts, J.S. and Nelson, H.D. (1991) Methyl bromide quarantine treatment for codling moth (Lepidoptera: Tortricidae) in unshelled walnuts. *Journal of Economic Entomology* 84, 1289–1293.

Haverty, M.I. and Robertson, J.L. (1982) Laboratory bioassays for selecting candidate insecticides and application rates for field tests on the western spruce budworm. *Journal of Economic Entomology* 75, 183–187.

Landolt, P.J., Chambers, D.L. and Chew, V. (1984) Alternative to the use of probit 9 as a criterion for quarantine treatments of fruit fly (Diptera: Tephritidae)-infested fruit. *Journal of Economic Entomology* 77, 285–287.

LeOra Software Inc. (1987) POLO-PC: A user's guide to Probit *Or LO*git analysis. 1119 Shattuck Ave, Berkeley, California.

Moffitt, H.R. and Burditt, A.K., Jr (1989) Low-temperature storage as a postharvest treatment for codling moth. *Journal of Economic Entomology* 82, 1679–1681.

Monro, H.A.U. (1969) Manual of fumigation for insect control. *FAO Agriculture Studies* No. 79. Rome, Italy.

Müller, H-G. and Schmitt, T. (1990) Choice of number of doses for maximum likelihood estimation of the ED50 for quantal dose– response data. *Biometrics* 46, 117–126.

Ottoboni, M.A. (1984). *The Dose Makes the Poison*. Vincente Books, Berkeley, California.

Preisler, H.K. 1988. Assessing insecticide bioassay data with extra-binomial variation. *Journal of Economic Entomology* 81, 759–765.

Preisler, H.K. and Robertson, J.L. (1989) Analysis of time–dose–mortality data. *Journal of Economic Entomology* 82, 1534–1542.

Robertson, J.L. and Preisler, H.K. (1992) *Pesticide Bioassays with Arthropods*. CRC Press, Boca Raton, Florida.

Robertson, J.L., Smith, K.C., Savin, N.E. and Lavigne, R.J. (1984) Effects of dose selection and sample size on the precision of lethal dose estimates in dose–mortality regression. *Journal of Economic Entomology* 77, 833–837.

SAS Institute Inc. (1982) *User's Guide: Statistics*. SAS Institute, Cary, North Carolina.

Savin, N.E., Robertson, J.L. and Russell, R.M. (1977) A critical evaluation of bioassay in insecticide research: likelihood ratio test of dose–mortality regression. *Bulletin of the Entomological Society of America* 23, 257–266.

Sharp, J.L. and Picho-Martinez, H. (1990) Hot-water quarantine treatment to control fruit flies in mangoes imported into the United States from Peru. *Journal of Economic Entomology* 83, 1941–1943.

Su, N-Y., Tamashiro, M. and Haverty, M.I. (1987) Characterization of slow-acting insecticides for remedial control of the Formosan subterranean termite (Isoptera: Rhinotermitidae). *Journal of Economic Entomology* 80, 1–4.

Tsutakawa, R.K. (1980) Selection of dose levels for estimating a percentage point of a logistic quantal response curve. *Applied Statistics* 29, 25–33.

Vail, P.V., Tebbets, J.S., Mackey, B.E. and Curtis, C.E. (1992) Quarantine treatments: A biological approach to decision-making for selected hosts of codling moth (Lepidoptera: Tortricidae). *Journal of Economic Entomology* 86, 70–75.

Worner, S.P. (1988). Ecoclimatic assessment of potential establishment of exotic pests. *Journal of Economic Entomology* 81, 973–983.

Yokoyama, V.Y., Miller, G.T. and Hartsell, P.L. (1990) A methyl bromide quarantine treatment to control codling moth (Lepidoptera: Tortricidae) on nectarines packed in shipping containers for export to Japan and effect on fruit attributes. *Journal of Economic Entomology* 83, 2335–2339.

II

Physiological and Biochemical Response of Insects to Possible Disinfestation Procedures

Chemical Fumigants 4

JOHN D. STARK
Washington State University,
Puyallup Research and Extension Center, 7612 Pioneer Way East,
Puyallup, WA 98371-4998, USA.

Introduction

Fumigation continues to be a major weapon in mankind's continual battle against pests of food and other commodities. Fumigants have been used for a long time. In fact, reports of sulfur fumes being used to disinfest homes are recorded as far back as 1000 BC. The practice of modern fumigation began approximately 140 years ago with the discovery of the fumigant properties of carbon disulfide and hydrocyanic acid. Today, fumigants are commonly used to disinfest pest species from dwellings, food stores and fresh produce prior to marketing. Chemical fumigation is still the preferred method of pest control when the swift and complete removal of a pest species from a structure, container, or commodity is required. However, research into alternative methods of pest disinfestation is progressing rapidly (see Chapters 5–7).

The fumigation process involves creating a lethal concentration of toxic gas for a time sufficient to destroy a targeted pest. Fumigation is usually carried out in a structure such as a grain storage container, ship, warehouse, food processing plant, or specially designed chamber.

Many chemical compounds exhibit sufficient vapour pressure and toxicity to act as fumigants, but few can be used as such because they are corrosive to containers, destructive to plant tissue, or leave an unwanted residue in the commodity (Monro, 1974). The toxicity of some insecticides can be attributed to vapour action, but these are not true fumigants and are not considered here. A 'true fumigant' is a toxic chemical that exerts its effects while in a gaseous state (Price, 1985). They are generally low molecular weight compounds with simple structures (Fig. 4.1). Only true fumigants used to fumigate fresh horticultural products will be discussed in this chapter.

From: Paull, R.E. and Armstrong, J.W. (eds) (1994) *Insect Pests and Fresh Horticultural Products: Treatments and Responses.* CAB INTERNATIONAL, Wallingford, UK.

CH_3Br Methyl bromide	PH_3 Phosphine	$BrH_2C{-}CH_2Br$ Ethylene dibromide
$HC{\equiv}N$ Hydrogen cyanide	$CH_2{=}CHC{\equiv}N$ Acrylonitrile	CS_2 Carbon disulfide

Fig. 4.1 Chemical structures of some commonly used fumigants.

Some fumigants are gases at room temperature while others are solids or liquids. Fumigants are categorized as physical poisons if they follow Ferguson's principle and are toxic at thermodynamic activities between 0.1 and 1.0 (Ferguson and Pirie, 1948). Most of the commonly used fumigants are toxic at much lower thermodynamic activities and are therefore just volatile chemical poisons.

Insect species differ greatly in their susceptibility to fumigants (Monro, 1974). In general, larvae of holometabolous insects are the most vulnerable stage, while pupae are the most tolerant (Monro *et al.*, 1952).

The most important factor influencing the efficacy of a fumigant as an insect control agent is temperature. This is because as temperature increases, so does an insect's respiration rate. Therefore, insects will take in more active ingredient at higher temperatures (Sun, 1946). Also, the actual material that a pest species infests is important because some fumigants sorb to certain materials more than to others, resulting in less available toxicant. Humidity is not as important as temperature, although very water-soluble compounds adhere to water molecules in the atmosphere which also reduces the amount of available toxicant.

Compared to other insecticides, little work has been published on the mode of action of fumigants or their physiological and toxicological effects on insects. Because fumigants pose a health risk to humans, primarily through accidental poisoning, a fairly large amount of research has been conducted on the effects of fumigants on various species of mammals.

As mentioned above, fumigants exert their toxic effect in the gaseous state and as such, the primary route of entry into insects is through the respiratory system (Monro, 1974). However, depending upon the fumigant, absorption through the cuticle can also play an important role in uptake (Monro, 1974). Entry of fumigants into insect eggs takes place through the chorion or through respiratory channels (Monro, 1974). A chemical is a poison when, at a particular dose, it disrupts an important physiological function(s). The degree of toxicity exhibited by a compound is dependent upon the innate susceptibility of a particular species which, in turn, depends upon many factors such as type of exposure (route of entry into the body), vigour of the organism, quantity and quality of detoxification enzymes, excretion rates, etc. Basically, the toxicity of a compound depends upon the amount of active ingredient that reaches and interacts with target sites (Stark *et al.*, 1992).

The properties and effects of several of the more commonly used fumigants are discussed below.

Methyl Bromide

Methyl bromide is a colourless, odourless gas with a boiling point of –3.56°C and vapour pressure of 1420 mmHg at 20°C (see Fig. 4.1 for structure). It is non-flammable in air and is stable and not corrosive. The insecticidal properties of methyl bromide were first discovered by the French entomologist, Le Goupil (1932), when he added this compound to another fumigant as a fire retardant. The mixture was more effective than the original product alone and led to the discovery that methyl bromide was a potent fumigant in its own right. Methyl bromide is the fumigant of choice for many commodities because it is easy to use, inexpensive, and does not usually leave a harmful residue. It is also widely used because many plants, vegetables, and fruits are quite tolerant to it when it is used at the concentrations necessary for effective insect control. Methyl bromide has the added benefit of being effective at low temperatures. This product is commonly used to fumigate various products stored in mills, ships, warehouses, and railway cars (Monro, 1974) and can be used to control insects, mites, rodents, fungi, bacteria, and viruses (Majumder, 1954; Bruch, 1961; Richardson and Monro, 1962; Price, 1985). Methyl bromide is not as toxic to many insect species as some of the other fumigants such as hydrogen cyanide, ethylene oxide, and acrylonitrile (Monro, 1974), but it possesses other properties that make it an excellent fumigant. Methyl bromide exhibits rapid and deep penetration, its vapours dissipate rapidly, it is toxic to many types of pests, and as mentioned above, it does not damage many types of plant material. Methyl bromide is a hazardous material to work with because it is colourless and odourless. A substantial number of cases of accidental poisonings have occurred in man and, therefore, many studies have been conducted on the effects of methyl bromide on various mammalian species. In man and other mammals, exposure to sublethal concentrations results in neurological symptoms, while exposure to higher concentrations results in circulatory damage. Dermal exposure to this product can result in severe blistering of skin (Watrous, 1942).

The mode of action of methyl bromide is not presently known, but has been the subject of much study. The biochemical effects induced by methyl bromide in insects are varied. Unlike respiratory poisons, such as hydrogen cyanide, exposure to lethal concentrations of methyl bromide does not result in a rapid paralysis in insects (von Oettingen, 1955).

Instead, a delay in the onset of symptoms is seen. Mammals appear to react to methyl bromide in a similar manner in that they also exhibit delays in the manifestation of symptoms and death (von Oettingen, 1955). The symptoms of methyl bromide poisoning resemble that of a narcotic, but no anaesthetic effect is observed in insects and respiratory inhibition is not the primary mode of action (Bond, 1965). Methyl bromide is not a physical poison because its thermodynamic activity at concentrations lethal to insects is much lower that 0.1 (Ferguson and Pirie, 1948). The symptoms exhibited by insects exposed to lethal concentrations of methyl bromide are a condition of irritation, followed by uncoordinated movement, spasms, and eventually, death (Bond, 1956).

Many studies have been conducted with various animal species and animal products to determine the mode of action of methyl bromide. Miller and Haggard (1943) suggested that the toxicity of methyl bromide is attributable to the release of free bromide ions after finding high intracellular levels of ions in rats after lethal exposure. This theory is not widely accepted today. The early suggestion that the toxicity of methyl bromide was due to the formation of methyl alcohol was also rejected because of the differences in the toxicity and symptomatology of these two compounds (Irish *et al.*, 1940). Another early hypothesis was that methyl bromide disrupted cellular membranes, particularly those of nerve cells.

A series of studies have indicated that methyl bromide interferes with SH enzymes and the SH groups in proteins. Blackburn *et al.* (1944) discovered that methyl bromide methylated the thiol groups in wool protein. Lewis (1948) observed that methyl bromide reacted with a number of sulfhydryl (SH) enzymes *in vitro* and proposed that the toxicity of this compound was due to disruption of enzyme function. Loveday and Winteringham (1951) also found that methyl bromide methylated free SH groups in insects while Winteringham and Barnes (1955) claimed that toxicity was due to methylation of proteins.

Definitive proof that methyl bromide has an effect on respiratory enzymes was first reported by Dixon and Needham (1946) who found that this compound inhibits hexokinase, *in vitro*. Further work by Winteringham *et al.* (1958) with the house-fly suggested that death is caused by a reversible breakdown of adenosine triphosphate (ATP).

More recently, a study by Nishimura *et al.* (1980) revealed that glutathione was methylated and sulfhydryl groups were alkylated by methyl bromide. Also, cytotoxicity declined when reduced glutathione was added. Thus, the study by Nishimura *et al.* (1980) indicates that the toxic effects of methyl bromide can be reversed and that the mode of action is due to reaction with the SH groups of respiratory enzymes.

Shivanandappa and Rajendran (1987) found that glutathione levels in the Khapra beetle, *Trogoderma granarium*, are depleted following exposure to methyl bromide. Glutathione-*S*-transferase activity is also induced by methyl bromide.

Other effects that have been seen in insects are a cessation of cell division in codling moth eggs (Cheetham, 1990). Cheetham claimed that methyl bromide is a general cellular toxin with little substrate specificity.

The evidence presented above indicates that the mode of action of methyl bromide is far from clear. Methyl bromide certainly appears to exert at least part of its toxic effect by methylating SH groups in various enzyme systems that are important in cellular respiration. Interference with SH groups will cause harmful effects on many systems within an organism.

Phosphine

Phosphine or hydrogen phosphide (PH_3) is a colourless, flammable gas with the distinctive odour of decaying fish. This fumigant has wide application and is used against pests infesting grain and flour as well as those infesting plant products (Monro, 1974). Phosphine gas is released when aluminum or magnesium phosphide comes into contact with water and was first tested against insects and rodents in 1933. It is very toxic to man and other mammals as well as to insects (Van den Bruel and Bollaerts, 1956; Lindgren *et al.*, 1958; Bond and Monro, 1961; Lindgren and Vincent, 1966).

Phosphine and hydrogen cyanide appear to have a similar mode of action, namely respiratory inhibition. In mammals, exposure to high concentrations leads to an extreme drop in blood pressure followed by death (Hayes and Laws, 1991). Klimmer (1969) felt that phosphine was a blood poison, and while it does react with mammalian haemoglobin, this hypothesis is not readily accepted today.

Insects exposed to phosphine are rapidly paralysed. Also, like hydrogen cyanide, exposure of insects to oxygen after phosphine exposure, greatly enhances the toxicity of this fumigant (Bond, 1963b; Hobbs and Bond, 1989). In fact, phosphine is virtually non-toxic to insects when oxygen is not present (Bond *et al.*, 1969). The primary route of entry into the body of insects is through the spiracles. However, exposure to high concentrations of phosphine causes narcosis which in turn reduces absorption of the compound (Winks, 1985).

Results of the majority of studies with insects suggest that phosphine, like hydrogen cyanide, acts as a respiratory poison primarily by inhibiting cytochrome-*c* oxidase (Nakakita *et al.*, 1974; Chefurka *et al.*, 1976; Kashi and Chefurka, 1976; Nakakita, 1976; Price, 1980a,b). However, in other studies with three insect species (*Rhyzopertha dominica*, *Oryzaephilus surinamensis*, *Crytolestes ferrugineus*) exposure to phosphine resulted in little *in vivo* inhibition of cytochrome-*c* oxidase (Price, 1980b; Price and Dance, 1983). Phosphine was, however, found to inhibit catalase (Price *et al.*, 1982). The resistant strains of the above insects are found to absorb less phosphine than sensitive strains. The effects of phosphine in five grain beetle species appear to be similar to that of anoxia (Kashi, 1981a,b). However, Price and Bell (1981) found that *Ephestia cautella* embryos were affected differently by anoxia and phosphine. Anoxia produced delays in development and reduced cell differentiation, while phosphine exposure did not cause developmental delays. Also, Price and Walter (1987) showed that exposure of the lesser grain borer, *Rhyzopertha dominica* (F.) to phosphine does not affect levels of ATP, pyruvate and lactate as does anoxia or HCN.

The study by Price and Bell (1981) revealed that exposure to phosphine produced an accumulation of pyruvate in embryos of *E. cautella*. Even though it appears that phosphine acts like cyanide, Price (1985) states that cytochrome-*c* oxidase may not be the primary biochemical lesion of this fumigant because its biochemical effects do not completely mimic those of anoxia.

More recently, Bolter and Chefurka (1990a) discovered that phosphine stimulates the release of hydrogen peroxide from insect mitochondria. Further work indicated that inhibition of cytochrome-*c* oxidase results in the production of superoxide radicals which form hydrogen peroxide (Bolter and Chefurka, 1990b).

Phosphine has also been shown to inhibit acetylcholinesterase in *Ephestia cautella* (Al-Hakkak *et al.*, 1989).

Therefore, it appears that phosphine's mode of action is not as simple as originally believed and that enzyme systems other than cytochrome-*c* oxidase are inhibited by this compound.

Ethylene Dibromide

Ethylene dibromide (EDB), or 1,2-dibromoethane (Fig. 4.1), is a colourless non-flammmable liquid that has a chloroform-like odour. Its fumigant properties were first reported by Neifert *et al.* (1925). Use of EDB as a fumigant increased after it was used to eliminate oriental fruit fly from fruits in Hawaii (Balock, 1951; Balock and Lindgren, 1951). EDB is very toxic to insects as well as humans and other mammals and is an excellent fumigant. Until recently, this product was used extensively to fumigate fresh fruits, vegetables, and stored grain. However, the discovery that EDB is carcinogenic (Nachtomi and Sharma, 1977; Ames *et al.*, 1980) has resulted in a ban on its use in the United States. However, it is still used in some parts of the world.

One of the advantages of EDB over other fumigants is that many fruits are tolerant to concentrations required to kill insect pests (Monro,

1974). A disadvantage to using this product is that it is strongly sorbed by fruits and grains. Entry of EDB into insects is primarily through the spiracles, but it also enters through the integument. After a lethal exposure, insects show no signs of poisoning and, like methyl bromide, death is delayed and may not occur for several days (Bond and Monro, 1961). Death in mammals is also delayed and is associated with general respiratory failure and cardiac arrest (Hayes and Laws, 1991). EDB is known to exert sublethal effects on certain insects. Confused flour beetle females exposed to sublethal concentrations displayed reduced fecundity and fertility (Loschiavo, 1960).

The mode of action of EDB has not been clearly elucidated. Heppel *et al.* (1947) discovered that the toxic effects of EDB were reduced upon addition of cysteine, methionine, and other SH-containing compounds indicating that this product interfered with SH enzymes. In 1948, Lewis found that EDB reduced the activity of the SH-containing enzyme, papain, *in vitro*. However, in *Calliphora* larvae, *in vivo* inhibition of SH groups could not be detected (Loveday and Winteringham, 1949). Hirade and Ninomiya (1950) were able to correlate the toxicity of EDB with *in vitro* inhibition of SH groups of reduced glutathione, while Morikawa (1964) correlated EDB toxicity with *in vivo* inhibition of triosephosphate dehydrogenase. In another study with the house fly, EDB was found to inhibit succinic dehydrogenase and glycolysis in general, but did not reduce ATP levels (Pant, 1958). Thus, Pant (1958) suggested that EDB blocked SH groups of phosphoglyceraldehyde dehydrogenase. Therefore, it appears that EDB works primarily by interfering with the SH enzymes associated with respiration, but its mode of action is still unclear.

Hydrogen Cyanide

Hydrogen cyanide (HCN) is a colourless gas or liquid that has a characteristic almond-like odour. The liquid form is also known as prussic acid. HCN is flammable and boils at 25.67°C. It was first used to kill insects by Buckton (1854). By 1886 it was used to kill scale insects infesting citrus trees in California by tenting trees (Quayle, 1938). HCN was one of the first widely used fumigants (Monro, 1974) and today is primarily used to fumigate stored products such as grain. It is very water soluble and sorbed by many materials so is not suitable for use with many fresh products. It is mentioned here because it is still used to fumigate citrus fruits in certain parts of the world and has been shown to be an effective fumigant for a wide range of insect pests on fresh flowers (Hansen *et al.*, 1991). HCN is also used for dormant nursery stock and some living plant material. It does not affect the viability of seeds at concentrations that kill insect pests, however, actively growing plants can be damaged after exposure (Monro, 1974). It is a particularly good fumigant for dry products.

HCN is one of the most toxic fumigants to insects and to man. Symptoms of poisoning include the rapid paralysis indicative of respiratory inhibition. Sublethal exposure of insects can cause what appears to be death, but insects may recover. Lindgren (1938) termed this recovery 'protective stupefaction' and it is apparently due to reduction or cessation of breathing. The entry of HCN into insects is primarily through the spiracles and tracheae (Lindgren, 1938). Keilin (1929) determined that HCN inhibited the respiratory enzyme cytochrome-*c* oxidase. The cyanide ion has since been shown to inhibit many enzymes, particularly those that contain copper and iron (Dixon and Webb, 1958). Cytochrome-*c* oxidase, however, is the primary enzyme inhibited by this compound. Cytochrome oxidase is essential for tissue respiration and the generation of ATP. Inhibition of this enzyme results in a shutting down of cellular respiration leading to respiratory failure and death. The mode of action in mammals is fairly straightforward, but insects can withstand long periods without oxygen. The symptoms of HCN poisoning in insects resemble those of anoxia, but differences exist. Bond (1963a) found that administration of oxygen to insects exposed to lethal concentrations of HCN was actually deleterious, which was the reverse of the situation seen in mammals.

Acrylonitrile

Acrylonitrile (Fig. 4.1) is a colourless, explosive and flammable compound that boils at 77.3–77.5°C and has a vapour pressure of 105 Torr at 25°C. This product was first used as an insecticide in 1939 and is toxic to many insect species as well as to man (Lindgren *et al.*, 1954; Krohne and Lindgren, 1958; Bond and Monro, 1961). Acrylonitrile is used primarily for fumigation of stored grains (Ruppel *et al.*, 1960) and control of dry wood termites (Monro, 1974). Acrylonitrile has limited use as a fumigant for fresh horticultural produce because it is highly toxic to nursery stock and growing plants and damages many fresh fruits (Claypool and Vines, 1956). Although its use as a fumigant for fresh horticultural produce is limited, it is used to disinfest nuts and dates (Monro, 1974) and tobacco (Tenhet, 1957). Mixed with carbon tetrachloride it can be used to control potato tuber moth in stored potatoes (Pradham *et al.*, 1960). Owing to its similarity to cyanide, its mode of action has been attributed to release of the cyanide ion (see above). However, other studies indicate that toxicity is dependent upon the acrylonitrile molecule itself as well as release of cyanide (Magos, 1961). Acrylonitrile has been shown to inhibit trehalase and phosphorylases in *Tribolium castaneum* (Rajendran and Muthu, 1981) and to deplete glutathione in the Khapra beetle, *Trogoderma granarium* (Shivanandappa and Rajendran 1987).

Carbon Disulfide

Carbon disulfide (CS_2) is a clear to yellowish liquid at room temperature and was one of the first widely used fumigants. It was first used against the granary weevil in 1854. By 1869 it was used as a soil fumigant against grape phylloxera in France. Because CS_2 is an extremely hazardous product due to its flammability and explosiveness, it is usually mixed with other non-flammable products. CS_2 is used primarily to fumigate stored grain, nematodes in soil, and nursery stock. It has limited use for fresh products, but has been used to fumigate some fruits (Osburn and Lipp, 1935). CS_2 has good penetrating properties and it does not injure dry seeds (Hinds, 1917). However, growing plants can be injured or killed after exposure (Fleming and Baker, 1935). Dried beans and peas have also been successfully fumigated with CS_2 (Monro, 1974).

Pant (1958) found that glycolysis was inhibited in house-flies exposed to CS_2. These flies had reduced levels of ATP and ADP which may have been a result of inhibition of cytochrome-*c* oxidase (Price, 1985). CS_2 combines with amino acids and the thiol groups of proteins to form dithiocarbamates and thiazolidone compounds causing disruption in the normal metabolism of nerve cells (Cohen *et al.*, 1958). Dithiocarbamates chelate copper and zinc within cells and inhibit important enzymes such as dopamine hydroxylase. CS_2 can also react with the free amine group, pyridoamine and produce a condition that resembles a vitamin B_6 deficiency in mammals. The toxicity of CS_2 to insects, therefore, appears to be a result of an overall inhibition of glycolysis, probably due to inhibition of cytochrome-*c* oxidase (Pant, 1958).

Resistance

There are approximately 400 recorded cases of insecticide resistance in insect species today. Most cases of resistance involve contact insecticides, although resistance to fumigants has also been recorded.

As early as 1965, *Tribolium confusum* Duv. and *T. castaneum* (Herbst.) developed resistance to ethylene dibromide and HCN, but not to methyl bromide (Lindgren and Vincent, 1965). More recent studies with methyl bromide indicate that resistance to this compound is primarily due to differential metabolism. Insects metabolize methyl bromide by conjugation with glutathione which produces *S*-methyl glutathione and *S*-methyl cysteine (Starratt and Bond, 1981). Resistant strains of the granary weevil *Sitophilus granarius* (L.) had twice as much glutathione-*S*-transferase activity as a susceptible strain and the resistant strain produced a third metabolite, *S*-methyl glutathione sulfoxide (Starratt and Bond, 1981). Resistant strains of *Rhyzopertha dominica* and *T. castaneum* exhibited lower uptake of phosphine than susceptible strains (Chaudhry and Price, 1989). Phosphine was primarily converted to hypophosphite in these insects and the rate of phosphine oxidation was greater in resistant than in susceptible strains.

Surveys indicate that resistance to fumigants is fairly widespread. Hole (1981) studied resistance in seven species of stored grain beetles to methyl bromide and phosphine and found that many of the species were tolerant to both of these products. Zettler (1991) found that 18% of strains of *T. confusum* and *T. castaneum* collected throughout the United States were resistant to phosphine. White and Lambkin (1990) found that resistance to phosphine was common in field-collected grain beetle species in Australia.

Resistance to the true fumigants will probably increase in the future. Therefore, development and implementation of alternative methods of disinfestation are essential.

Future Outlook

The ban on the use of ethylene dibromide should serve as a warning for the continued use of true fumigants in the future. Acrylonitrile is known to be carcinogenic (Guengerich *et al.*, 1981), teratogenic (Willhite *et al.*, 1981), and mutagenic (Duverger-Van Bogaert *et al.*, 1982). Carbon disulfide has been shown to produce abnormalities in frog embryos (Ghate, 1985). A recent study has indicated that methyl bromide is weakly mutagenic in bacterial and mammalian cells, in addition to insects (Kramers *et al.*, 1985). Work by Al-Hakkak (1988) with *Drosophila melanogaster* indicated that phosphine induced sex-linked recessive mutations. That these products may be carcinogenic, mutagenic, or teratogenic is disturbing in light of the ethylene dibromide situation. Also, halocarbons, including methyl bromide, have been implicated in the destruction of the ozone layer (Prather and Watson, 1990). Although methyl bromide and phosphine are two of the most widely used fumigants, results of this nature could put their future use in jeopardy.

Due to the occurrence of resistance in some species, ethylene dibromide is no longer used in some countries, and a renewed interest has arisen in the use of mixtures of true fumigants with other materials such as CO_2 to control pests of stored products and fresh produce. Bond and Morse (1982) tested mixtures of methyl bromide and phosphine to control *T. castaneum* and found that the mixture was more effective than either product alone. Combinations of methyl bromide and 20% CO_2 were found to be more effective in the control of *T. castaneum* than was methyl bromide alone (Calderon and Leesch, 1983). Mixtures of fumigants with gamma radiation have also been explored (Mehta *et al.*, 1984). Additive effects of methyl bromide and gamma irradiation were evident against *T. confusum* (Abdel-Salam and Nakhlah, 1987). Rajendran and Muthu (1989) found that mixtures of certain fumigants resulted in reduced efficacy against the eggs of *T. castaneum*, however, a phosphine–CO_2 mixture was more effective than either alone. The toxicity of carbon disulfide to *Bruchophagus melipes* was greatly increased by carrying out fumigation under reduced atmospheric pressure (Verma, 1986).

The difficulty in developing new fumigants is that they must be relatively safe to use, effective against pest species, and must not damage plant material or leave a harmful residue. The stringent requirements for the registration of new pesticides makes it unlikely that many new fumigants will be commercialized in the near future.

Because many fumigants are genotoxic and resistance to these products is becoming more widespread, fumigation practices will have to change in the coming years. We will have to rely less on the true fumigants and more on alternative methods of control, such as controlled atmospheres, radiation, and heat treatments.

References

Abdel-Salam, K.A. and Nakhlah, J.M. (1987) Effect of a combination of treatments of gamma radiation and methyl bromide fumigation on the confused flour beetles *Tribolium confusum* Duv. *Insect Science and its Application* 8, 103–106.

Al-Hakkak, Z.S. (1988) Mutagencity of phosphine gas in *Drosophila melanogaster*. *Journal of Biological and Sciences Research* 19, 739–745.

Al-Hakkak, Z.S., Al-Azzawi, M.J., Al-Adhamy, B.W. and Khalil, S.A. (1989) Inhibitory action of phosphine on acetylcholinesterase of *Ephestia cautella* (Lepidoptera: Pyralidae). *Journal of Stored Products Research* 25, 171–174.

Ames, B., Infante, P. and Reitz, R. (1980) Ethylene dichloride: a potential health risk? Banbury Report No. 5. Cold Spring Harbor Laboratory.

Balock, J.W. (1951) Ethylene dibromide for destroying fruit fly infestations in fruits and vegetables. *Science*, 114, 122.

Balock, J.W. and Lindgren, D.L. (1951) Toxicity of various compounds as fumigants for eggs and larvae of the oriental fruit fly. *Journal of Economic Entomology* 44, 657–659.

Blackburn S.E., Consden, R. and Phillips, H. (1944) The action of sulphites on cysteine disulphide linkages of wool. 4. Methylation of the thiol groups of bisulphited wools. *Biochemical Journal* 38, 25–29.

Bolter, C.J. and Chefurka, W. (1990a) The effect of phosphine treatment on superoxide dismutase, catalase, and peroxidase in the granary weevil,

Sitophilus granarius. Pesticide Biochemistry and Physiology 36, 52–60.

Bolter, C.J. and Chefurka, W. (1990b) Extramitochondrial release of hydrogen peroxide from the insect and mouse liver mitochondria using respiratory inhibitors phosphine, myxothiazol, and antimycin and spectral analysis of inhibited cytochromes. *Archives of Biochemistry and Biophysics* 278, 65–72.

Bond, E.J. (1956) The effect of methyl bromide on the respiration of the cadelle *Tenebroides mauritanicus* (L.). (Coleoptera: Astomidae). *Canadian Journal of Zoology* 34, 405–415.

Bond, E.J. (1963a) The action of fumigants on insects. IV. The effects of oxygen on the toxicity of fumigants to insects. *Canadian Journal of Biochemistry and Physiology* 41, 993–1004.

Bond, E.J. (1963b) Effect of oxygen on cyanide poisoning in insects. *Nature* 193, 1002–1003.

Bond, E.J. (1965) The influence of oxygen on the metabolism of *Sitophilus granarius* (L.) during cyanide poisoning. *Journal of Insect Physiology* 11, 765–777.

Bond, E.J. and Monro, H.A.U. (1961) The toxicity of various fumigants to the cadelle *Tenebroides mauritanicus. Journal of Economic Entomology* 54, 451–454.

Bond, E.J. and Morse, P.M. (1982) Joint action of methyl bromide and phosphine on *Tribolium castaneum* (Herbst) (Coleoptera: Tenebrionidae). *Journal of Stored Products Research* 18, 83–94.

Bond, E.J., Robinson, J.R. and Buckland, C.T. (1969) The toxic action of phosphine. Absorption and symptoms of poisoning in insects. *Journal of Stored Products Research* 5, 289–298.

Bruch, C.W. (1961) Gaseous sterilization. *Annual Review of Microbiology* 15, 245–262.

Buckton, G.B. (1854) On the application of cyanide of potassium to killing insects for the cabinet. *Zoologist* 12, 4503–4506.

Calderon, M. and Leesch, J.G. (1983) Effect of reduced pressure and CO_2 on the toxicity of methyl bromide to two species of stored-product insects. *Journal of Economic Entomology* 76, 1125–1128.

Chaudhry, M.Q. and Price, N.R. (1989) Biochemistry of phosphine uptake in susceptible and resistant strains of two species of stored product beetles. *Comparative Biochemistry and Physiology* 94C, 425–429.

Cheetham, T. (1990) Pathological alterations in embryos of the codling moth (Lepidoptera: Tortricidae) induced by methyl bromide. *Annals of the Entomological Society of America* 83, 59–67.

Chefurka W.A., Kashi, K.P. and Bond, E.J. (1976) The effects of phosphine on electron transport in mitochondria. *Pesticide Biochemistry and Physiology* 6, 65–84.

Claypool, L.L. and Vines, H.M. (1956) Commodity tolerance studies of deciduous fruits to moist heat and fumigants. *Hilgardia* 24, 297–355.

Cohen, E.E., Paulus, H.J., Kennan, R.G. and Scheel, L.D. (1958) Skin adsorption of carbon disulfide vapor in rabbits. I. Associated changes in blood protein and zinc. *AMA Archives of Industrial Health* 17, 164–169.

Dixon, M. and Needham, D.M. (1946) Biochemical research on chemical warfare agents. *Nature* 158, 432–438.

Dixon, M. and Webb, E.C. (1958) *The Enzymes*. Longman, London.

Duverger-Van Bogaert, M., Lambotte-Vandepaer, M., de Meester, C., Mercier,

M. and Poncelet, F. (1982) Role of glutathione in liver-mediated mutagencity of acrylonitrile. *Toxicological Letters* 11, 3–4.

Ferguson, J. and Pirie, H. (1948) The toxicity of vapours to the grain weevil. *Annals of Applied Biology* 35, 532–550.

Fleming, W.E. and Baker, F.E. (1935) The use of carbon disulphide against Japanese beetle. US Department of Agriculture, Washington, DC, *Technical Bulletin* No. 478.

Ghate, H.V. (1985) Toxicity and teratogenic effects of carbon disulphide in frog embryo. *River Biology* 78, 129–132.

Guengerich, F.P., Geiger, L.E., Hogy, L.L. and Wright, P.L. (1981) *In vitro* metabolism of acrylonitrile to 2-cyanoethylene oxide, reaction with glutathione, and irreversible binding to proteins and nucleic acids. *Cancer Research* 41, 4925–4933.

Hansen, J.D., Hara, A.H., Chan, H.T., Jr and Tenbrink, V.L. (1991) Efficacy of hydrogen cyanide fumigation as a treatment for pests of Hawaiian cut flowers and foliage after harvest. *Journal of Economic Entomology* 84, 532–536.

Hayes, W.J., Jr and Laws, E.R. (1991) Solvents, fumigants, and related compounds. In: *Handbook of Pesticide Toxicology*, Vol. 2. Academic Press, San Diego, pp. 637–730.

Heppel, L.A., Porterfield, V.T. and Sharples, N.E. (1947) Toxicology of 1,2-dichloroethane (ethylene dichloride) IV, its detoxication by L-cysteine, D,L-methionine and certain other sulphur containing compounds. *Journal of Pharmacology and Experimental Therapeutics* 91, 385–394.

Hinds, W.E. (1917) Carbon bisulphide as an insecticide. US Department of Agriculture, Washington, DC, *Farmers' Bulletin* No. 799.

Hirade, J. and Ninomiya, A. (1950) The mechanism of toxic action of organic halogen compounds. *Journal of Biochemistry* 37, 19–34.

Hobbs, S.K. and Bond, E.J. (1989) Response of *Tribolium castaneum* (Herbst) (Coleoptera: Tenebrionidae) to sublethal treatments with phosphine. *Journal of Stored Products Research* 25, 137–146.

Hole, B.D. (1981) Variation in tolerance of seven species of stored product Coleoptera to methyl bromide and phosphine in strains from twenty-nine countries. *Bulletin of Entomological Research* 71, 299–306.

Irish, D.D., Adams, E.M., Spencer, H.C. and Rowe, V.K. (1940) Response attending exposure of laboratory animals to vapors of methyl bromide. *Journal of Industrial Hygiene and Toxicology* 22, 218–230.

Kashi, K.P. (1981a) Response of five species of stored product insects to phosphine in oxygen-deficient atmospheres. *Pesticide Science* 12, 111–115.

Kashi, K.P. (1981b) Toxicity of phosphine to five species of stored product insects in atmospheres of air and nitrogen. *Pesticide Science* 12, 116–122.

Kashi, K.P. and Chefurka, W.A. (1976) The effect of phosphine on absorption and circular dichroic spectra of cytochrome-c and cytochrome-c oxidase. *Pesticide Biochemistry and Physiology* 6, 350–362.

Keilin, D. (1929) Cytochrome and respiratory enzymes. *Proceedings of the Royal Society*, B 104, 206–252.

Klimmer, O.R. (1969) Beitrag zur Wirkung des Phosphorwasserstoffes (PH_3). Frage de sog chronischen Phosphorwasserstoffvergiftung. *Archives of Toxicology* 24, 164–187.

Kramers, P.G.N., Voodg, C.E., Knaap, A.G.A.C. and der Heijden, C.A. (1985) Mutagenicity of methyl bromide in a series of short-term tests. *Mutation Research* 155, 41–47.

Krohne, H.E. and Lindgren, D.L. (1958) Susceptibility of life stages of *Sitophilus oryza* to various fumigants. *Journal of Economic Entomology* 51, 157–158.

Le Goupil, M. (1932) Les proprietés insecticide du bromure de methyle. *Revue de Pathologie Végétale et d'Entomologie Agricole de France* 19, 169–172.

Lewis, S.E. (1948). Inhibition of SH enzymes by methyl bromide. *Nature* 161, 692–693.

Lindgren, D.L. (1938) The stupefaction of red scale *Aonidiella aurantii* by hydrocyanic acid. *Hilgardia* 11, 213–255.

Lindgren, D.L. and Vincent, L.E. (1965) The susceptibility of laboratory and field cultures of *Tribolium confusum* and *T. castaneum* to ethylene dibromide, HCN and methyl bromide. *Journal of Economic Entomology* 58, 551–556.

Lindgren, D.L. and Vincent, L.E. (1966) Relative toxicity of hydrogen phosphide to various stored-product insects. *Journal of Stored Products Research* 2, 141–146.

Lindgren, D.L., Vincent, L.E. and Krohne, H.E. (1954) Relative effectiveness of ten fumigants to adults of eight species of stored product insects. *Journal of Economic Entomology* 47, 923–926.

Lindgren, D.L., Vincent, L.E. and Strong, R.G. (1958) Studies on hydrogen phosphide as a fumigant. *Journal of Economic Entomology* 51, 900–903.

Loschiavo, S.R. (1960) Effects of low doses of ethylene dibromide on stages of confused flour beetle. *Journal of Economic Entomology* 53, 762–767.

Loveday, P.N. and Winteringham, F.P.W. (1949) *Pest Infestation Research*. HMSO, London, p. 29.

Loveday P.N. and Winteringham F.P.W. (1951) Fate of insecticides in insects. In: *Pest Infestation Research 1948*, HMSO, London, pp. 29–31.

Magos, L. (1961) A study of acrylonitrile poisoning in relation to methaemoglobin-CN complex formation. *British Journal of Industrial Medicine* 19, 283–286.

Majumder, S.K. (1954) Control of microbial spoilage in food-stuffs with methyl bromide. *Mysore Centre Food Techical Research Institute Bulletin* 3, 269–271.

Mehta, V.K., Sethi, G.R. and Garg, A.K. (1984) Effect of gamma radiation and fumigants on *Tribolium castaneum* (Herbst). *Journal of Nuclear Agriculture and Biology* 13, 109–111.

Miller, D.P. and Haggard, H.W. (1943) Intracellular penetration of bromide as a feature in the toxicity of alkyl bromides. *Journal of Industrial Hygiene and Toxicology* 25, 423.

Monro, H.A.U. (1974) *Manual of Fumigation for Insect Control*, 2nd edn. Food and Agriculture Organization of the United Nations, Rome, Italy.

Monro, H.A.U., Cunningham, C.R. and King, J.E. (1952) Hydrogen cyanide and methyl bromide as fumigants for insect control in empty cargo ships. *Scientific Agriculture* 32, 241–265.

Morikawa, O. (1964) Effects of nematocides on the glycolysis of the American cockroach. *Japanese Journal of Applied Entomology and Zoology* 8, 227–285.

Nachtomi, E. abnd Sharma, D.S.R. (1977) *In vivo* repair of rat liver DNA damaged by ethylene dibromide. *Biochemical Pharmacology* 26, 1941–1945.

Nakakita, H. (1976) The inhibitory site of phosphine. *Journal of Pesticide Science* 1, 235–238.

Nakakita, H., Saito, T. and Iyatomi, K. (1974) Effect of phosphine on the respiration of adult *Sitophilus zeamais*. *Journal of Stored Products Research* 10, 87–92.

Neifert, I.E., Cook, F.C., Roark, R.C., Tonkin, W.H., Back, E.A. and Cotton, R.T. (1925) Fumigation against grain weevils with various volatile organic compounds. US Department of Agriculture, Washington, DC, *Departmental Bulletin* No. 1313, 40 pp.

Nishimura, M.M., Umeda, S., Ishizo, S. and Sato, M. (1980) Cytotoxicity of methyl bromide: effect of methyl bromide on cultured mammalian cells. *Journal of Toxicological Science* 5, 321–330.

Osburn, M.R. and Lipp, J.W. (1935) Fumigation of fresh fruit to destroy the adult Japanese beetle. US Department of Agriculture, Washington, DC, Circular No. 373.

Pant, C.P. (1958) Studies on the mode of action of insecticides. PhD Thesis, Iowa State College.

Pradhan, S., Bhambhani, H.J. and Wadhi, S.R. (1960) Fumigation of infested potatoes. *Indian Journal of Entomology* 22, 181–189.

Prather, M.J. and Watson, R.T. (1990) Stratospheric ozone depletion and future levels of atmospheric chlorine and bromine. *Nature* 344, 730–734.

Price, N.R. (1980a) The effect of phosphine on respiration and mitochondrial oxidation in susceptible and resistant strains of *Rhyzopertha dominica*. *Insect Biochemistry* 10, 147–150.

Price, N.R. (1980b) Some aspects of the inhibition of cytochrome-c oxidase by phosphine in susceptible and resistant strains of *R. dominica*. *Insect Biochemistry* 10, 65–71.

Price, N.R. (1985) The mode of action of fumigants. *Journal of Stored Products Research* 21, 157–164.

Price, N.R. and Bell, C.H. (1981) Structure and development of embryos of *E. cautella* (Walker) during anoxia and phosphine treatment. *International Journal of Invertebrate Reproduction* 3, 17–25.

Price, N.R. and Dance, S.J. (1983) Some biochemical aspects of phosphine action and resistance in three species of stored product beetles. *Comparative Biochemistry and Physiology* 76C, 277–281.

Price, N.R. and Walter, C.M. (1987) A comparison of some effects of phosphine, hydrogen cyanide and anoxia in the lesser grain borer, *Rhyzopertha dominica* (F.) (Coleoptera: Bostrychidae). *Comparative Biochemistry and Physiology* 86C, 33–36.

Price, N.R., Mills, K.A. and Humphries, L.A. (1982) Phosphine toxicity and catalase activity in susceptible and resistant strains of the lesser grain borer. *Comparative Biochemistry and Physiology* 73C, 411–413.

Quayle, H.J. (1938) *Insects of Citrus and Other Subtropical Fruits*. New York, Comstock. 583 p.

Rajendran, S. and Muthu, M. (1981) Effects of acrylonitrile on trehalase, phosphorylase and acetylcholinesterase activities in *Tribolium castaneum* Herbst and *Trogoderma granarium* Everts. *Experientia* 37, 886–887.

Rajendran, S. and Muthu, M. (1989) The toxic action of phosphine in combination with some alkyl halide fumigants and carbon dioxide against the eggs of *Tribolium castaneum* Herbst (Coleoptera: Tenebrionidae). *Journal of Stored Products Research* 25, 225–230.

Richardson, L.T. and Monro, H.A.U. (1962) Fumigation of jute bags with ethylene oxide and methyl bromide to eradicate potato ring rot bacteria. *Applied Microbiology* 10, 448–451.

Ruppel, R.F., Bravo, G. and Hatheway, W.H. (1960) Effectiveness of mixtures of acrylonitrile and carbon tetrachloride against three pests of stored corn. *Journal of Economic Entomology* 53, 238–242.

Shivanandappa, T. and Rajendran, S. (1987) Induction of glutathione *S*-transferase by fumigants in larvae of the Khapra beetle, *Trogoderma granarium* (E.). *Pesticide Biochemistry and Physiology* 28, 121–126.

Stark, J.D., Wong, T.T.Y., Vargas, R.I. and Thalman, R.K. (1992) Survival, longevity, and reproduction of tephritid fruit fly parasitoids (Hymenoptera: Braconidae) reared from fruit flies exposed to azadirachtin. *Journal of Economic Entomology* 85, 1125–1129.

Starratt, A.N. and Bond, F.J. (1981) Metabolism of methyl bromide by susceptible and resistant strains of the granary weevil *Sitophilus granarius* (L.). *Pesticide Biochemistry and Physiology* 15, 275–282.

Sun, Y.P. (1946) An analysis of some important factors affecting the results of fumigation on insects. St Paul, Minnesota Agricultural Experiment Station. *Technical Bulletin* No. 177.

Tenhet, J.N. (1957) Tobacco fumigants and fumigation. *Tobacco* 144, 22–26.

Van den Bruel, W.E. and Bollaerts, D. (1956) Hydrogen phosphide fumigation. *Parasitica* 12, 32–52.

Verma, B.R. (1986) Susceptibility of *Bruchophagus mellipes* Gahan in seeds to reduced pressure fumigation. *Journal of Entomological Research* 10, 44–50.

von Oettingen, N.F. (1955) The halogenated hydrocarbons, their toxicity and potential dangers. US Public Health Service, Washington, DC, Publication No. 414.

Watrous, R.M. (1942). Methyl bromide: local and mild systemic effects. *Industrial Medecine* 11, 575.

White, G.G. and Lambkin, T.A. (1990) Baseline responses to phosphine and resistance status of stored-grain beetle pests in Queensland, Australia. *Journal of Economic Entomology* 83, 1738–1744.

Willhite, C.C., Ferm, V.H. and Smith, R.P. (1981) Teratogenic effects of aliphatic nitriles. *Teratology* 23, 317–323.

Winks, R.G. (1985) The toxicity of phosphine to adults of *Tribolium castaneum* (Herbst): phosphine induced narcosis. *Journal of Stored Products Research* 21, 25–29.

Winteringham, F.P.W. and Barnes, J.M. (1955) Comparative response of insects and mammals to certain halogenated hydrocarbons used as insecticides. *Physiology Review* 35, 701–739.

Winteringham, F.P.W., Hellyer, G.C. and McKay, M.A. (1958) Effects of methyl bromide on phosphorus metabolism in the adult housefly, *Musca domestica* L. *Biochemical Journal* 69, 640–648.

Zettler, J.L. (1991) Pesticide resistance in *Tribolium castaneum* and *T. confusum* (Coleoptera: Tenebrionidae) from flour mills in the United States. *Journal of Economic Entomology* 84, 763–767.

Irradiation 5

James L. Nation[1] and Arthur K. Burditt, Jr[2]
[1]*Department of Entomology, University of Florida, Gainesville, FL 32611, USA;*
[2]*9068-C SW 82 Terrace, Ocala, FL 34481-9438, USA.*

Irradiation as a Quarantine Treatment

The term irradiation includes radio waves, radar, infrared, microwaves, visible, ultraviolet, electrons, X-rays and gamma rays. However, for use as treatments to eliminate infestations of pests of quarantine importance the Food and Drug Administration in the United States has only approved use of gamma radiation from ^{60}Co and ^{137}Cs, electrons from an accelerator of $\leqslant 10$ MeV, and X-rays with an energy of $\leqslant 5$ MeV for food irradiation up to a dose of 1000 Gy (Food and Drug Administration, 1984). The USDA Animal and Plant Health Inspection Service has authorized irradiation as a quarantine treatment for Hawaiian papayas infested with larvae of three species of fruit flies: the oriental fruit fly, *Bactrocera (Dacus) dorsalis* (Hendel), Mediterranean fruit fly, *Ceratitis capitata* (Wiedemann), and melon fly, *B. cucurbitae* Coquillett.

History

Koidsumi (1930) first suggested that X-rays could be used as a quarantine treatment to kill oriental fruit fly and melon fly larvae within fruits and vegetables in Formosa. Subsequently (Balock *et al.*, 1956, 1966) proposed that gamma rays from cobalt-60 be used as a commodity treatment for the three species of Hawaiian fruit flies. Macfarlane (1966), Shipp and Osborn (1968) and Eric *et al.* (1970) recommended irradiation as a treatment for

Paper No. R-02980 of the Research Article Series, University of Florida Agricultural Experiment Station.

From: Paull, R.E. and Armstrong, J.W. (eds) (1994) *Insect Pests and Fresh Horticultural Products: Treatments and Responses.* CAB INTERNATIONAL, Wallingford, UK.

fruit infested with Queensland fruit fly, *B. tryoni* (Froggatt), larvae in Australia, and Benschoter and Telich (1964) reported the use of X-rays for the Mexican fruit fly, *Anastrepha ludens* (Loew). Since 1970 there have been a number of other studies on the use of irradiation as a quarantine treatment (Burditt 1982).

The joint Food and Agriculture Organization–International Atomic Energy Agency (FAO/IAEA), Division of Atomic Energy in Food and Agriculture has coordinated research and sponsored meetings to review results of research being conducted on the use of irradiation to solve quarantine problems. These have included a panel in 1970 on the use of irradiation to solve quarantine problems in the international fruit trade in Honolulu (Anonymous, 1971), coordinated research programmes on insect disinfestation of food and agricultural products by irradiation (Anonymous, 1984b, 1985) and on the use of irradiation as a quarantine treatment of food and agricultural commodities (Anonymous 1986b, 1989, 1991b), a consultants' group on use of irradiation as a quarantine treatment of agricultural commodities (Anonymous, 1984a), and task force meetings on irradiation as a quarantine treatment (Anonymous, 1986a) and on irradiation as a quarantine treatment of fresh fruits and vegetables (Anonymous, 1991a). Recently (Anonymous, 1992) a research coordination meeting on the research programme on irradiation as a quarantine treatment against nematodes and insects other than fruit fly was held in Gainesville, Florida.

The 1970 meeting identified the potential of irradiation as a quarantine disinfestation treatment for fruit, primarily against fruit flies and the mango weevil, *Cryptorhynchus mangiferae* (F.) (Anonymous, 1971). Doses required for quarantine treatment of infested fruit, criteria required for quarantine security, effects of irradiation on fruits, and methods to implement the technique for disinfestation by radiation were discussed.

Later meetings (Anonymous, 1985) recommended that research should be conducted on whether a general quarantine dose could be developed for a particular pest species regardless of host commodity. A minimum dose for the most studied case of the tephritid fruit flies was 150 Gy to treat eggs and larvae to prevent emergence of normal adults (Anonymous, 1986a). A generic 300 Gy minimum dose to treat other species was also adopted at the same meeting. Treatment schedules were proposed for future incorporation in the *FAO International Plant Quarantine Treatment Manual* (Karpati *et al.*, 1983).

The focus changed at later meetings and it was recommended that a bibliography of key papers be compiled, a test be developed to verify that a pest has been irradiated and is incapable of reproduction, studies be initiated to determine effectiveness of quarantine protocol profiles for pests and tolerance of host plants, training programmes be developed for research personnel and for plant protection and quarantine regulatory personnel, and that research include electron beam and X-ray studies (Anonymous, 1986b, 1989, 1991b). Markers to verify that any surviving irradiated pests cannot reproduce and become established in a non-infested area were given added emphasis and the programme was expanded to include mites, nematodes, thrips and other species of insects (Anonymous, 1992).

Efficacy of Irradiation as a Quarantine Treatment

Historically the criterion for evaluating the effectiveness of fumigation, cold storage or other quarantine treatments to eliminate infestations of fruit flies and other pests of quarantine importance has been based on preventing development of live larvae in the fruit and their successful pupation (Baker, 1939). In order to kill eggs or larvae outright or to prevent pupation of mature larvae requires very high doses (in the range of 1000 Gy) of irradiation (Balock *et al.*, 1963; Lippold *et al.*, 1968). Balock *et al.* (1963) demonstrated that emergence of tephritid fruit fly adults could be reduced or prevented by exposure of eggs, larvae, or 1–3-day-old pupae to less than 150 Gy of gamma radiation from a cobalt-60 source. It has been proposed that the criterion for efficacy of irradiation as a quarantine treatment should be based on preventing adult emergence (Balock *et al.*, 1963) or on the inability of the insect to perpetuate the species (Ouye and Gilmore, 1985). These guidelines for quarantine security have been accepted by FAO/IAEA and it has been recommended to regulatory agencies that a dose which will prevent emergence of adults capable of flight, mating or production of fertile offspring should be approved for quarantine security.

The dose required to reduce egg hatch, larval pupation or adult emergence has been used to compare radiation effects on different species (Balock *et al.*, 1963). Doses of <36 to >1000 Gy applied to eggs, reduced hatch, depending upon age of the eggs, <50 Gy reduced pupation and <35 Gy reduced adult emergence. Doses <700 and <50 Gy, applied to immature larvae, reduced pupation and adult emergence, respectively. When mature larvae were reared in medium and irradiated doses $\leqslant 1600$ Gy reduced pupation and $\leqslant 50$ Gy reduced adult emergence. However, doses up to 80 Gy were required to reduce adult emergence when mature larvae were treated in fresh medium or in medium with water added.

Although a dose of 50 to 300 Gy had been found to be adequate to prevent development, pupation and adult emergence or reproduction when insect eggs or larvae are irradiated, exposure of fruit fly larvae in a liquid medium such as fruit with a high juice content, or an environment with a low oxygen level, reduced the efficacy of an irradiation treatment and required an increase in dose in order to achieve quarantine security (Balock *et al.*, 1963; Kaneshiro *et al.*, 1983). Also, research by Burditt *et al.* (1989) has shown that a significant decrease in adult emergence of mature, cocooned, non-diapausing codling moth, *Cydia pomonella* L., larvae resulted when the dose rate was increased from 1 to 200 Gy min^{-1} using a cobalt-60 source.

Although a minimal dose may prevent adult emergence, increasing the dose to a maximum of 1000 Gy could result in a reduction of feeding or an increase in mortality of larvae and thus prevent further injury from their activity. Small fruit, such as cherries, would be severely damaged if irradiated larvae were able to survive, continue to feed, reach maturity, and pupate. Therefore, Jona and Arzone (1979) suggested that the irradiation dose used for the European cherry fruit fly, *Rhagoletis cerasi* L., should 'kill the larvae expeditiously and preclude any growth before death'. Thus, they proposed treatment at the maximum dose of 1000 Gy in order to reduce feeding damage to the host fruit.

Gamma irradiation as a potential quarantine treatment for insects other than fruit flies has been reviewed by Heather (1992). He found that research has been reported on Coleoptera and Lepidoptera as pests of stored products as well as mites, thrips, and aphids that infest fruits, vegetables, and ornamentals. Such pests may be present as immatures and adults. Therefore, it is essential that a quarantine treatment should sterilize any adults that may be present. In general, a dose of 300 Gy has been found to be effective as a quarantine treatment for such species (Anonymous, 1986a).

Physiological and Biochemical Effects of Radiation in Insects

Insect development

Some insects and certain stages of development are very resistant to radiation (O'Brien and Wolfe, 1964; Ducoff, 1972). As little as 4 Gy radiation given to *Drosophila melanogaster* Meigen eggs during cleavage causes 50% mortality. The pupal stage requires 400 Gy to cause the same mortality (Nothel, 1968). Cogburn *et al.* (1966) found that 132 Gy, the lowest dose reported, killed eggs of the Indian meal moth, *Plodia interpunctella* (Hübner), and Angoumois grain moth, *Sitotroga cerealella* (Olivier), and larvae of Indian meal moth, but only killed some larvae of Angoumois grain moth. Similar stage-dependent and species-dependent sensitivities have been found for many insects.

Radiation-induced injury in an early stage of development may be expressed later in development, when a particular tissue becomes critical to the function or survival of the insect. Mortality of *D. melanogaster* can be divided into the three categories of 'immediate death', 'early death' and 'late death' when embryos in the early blastoderm stage (1.75 h old) were irradiated with X-rays (Nothel, 1968). Doses up to 4 Gy cause no immediate deaths, but do cause 'early death' occurring during gastrulation, and some 'late death' due to damage to developing larval organs. Exposure of the blastoderm stage to 80 Gy, however, causes 50% 'immediate death' due to complete destruction of dividing cells, and the remaining 50% die during gastrulation (Nothel, 1968). Once hatch has occurred, radiation exposure affects mainly the imaginal discs. These discs are the non-differentiated tissues that give rise to legs, wings, compound eyes, and other organs in the adult during pupation. Radiation at certain stages can mimic the action of genetic mutations that occur in *D. melanogaster*. Reduction in dorsocentral bristles and extension of the wings at 45 degrees to the body, a condition virtually identical to the mutant *Dichaete* in *D. melanogaster*, can be induced in up to 100% of surviving flies by irradiating 5-h-old prepupae. Similar effects could not be produced by irradiating younger or older stages (Fritz-Niggli, 1951; cited in Nothel, 1968).

An explanation for the variability in insect sensitivity has usually been attributed to the principles elaborated by Bergonie and Tribondeau (1959), who concluded that damage is proportional to the degree of reproductive activity in cells and to the level of differentiation of the cells. The early stages of embryogenesis in the egg and the period when adult tissues are being developed during the early part of the pupal stage are especially sensitive stages in insects. This is probably because cells are actively dividing and differentiating.

Cultured insect cells, even during division, are more resistant to irradiation damage than are cultured mammalian cells. Among insect cell lines there is considerable variation in resistance (Koval, 1983). The most extensively studied insect cell line has been TN-368 cells derived from minced ovaries of adult cabbage looper moths, *Trichoplusia ni* (Hübner) (Hink, 1970). The cells are primarily fibroblast-like cells. The cells are substantially more resistant to death by X- and γ-irradiation (Koval, 1983) and to 254 nm ultraviolet (UV) light (Koval, 1986a) than mammalian cells. The mechanism(s) of resistance cannot be completely explained at present. There is evidence for both DNA-repair processes and inducible cellular recovery capabilities in TN-368 cells exposed to radiation (Koval *et al.*, 1978; Koval 1986a,b, 1988). The cells also were more resistant than mammalian cell lines to several DNA-damaging chemical agents that cross-link DNA–DNA and DNA–protein. Studies of the recovery from these chemical agents suggests that the cross-linking action may be very important lesions that cause death in the cultured cells.

Midgut tissue

The midgut is the principal site for digestion and absorption of food. Midgut secretory and absorptive cells are frequently replaced by the growth of new cells from small, undifferentiated regenerative cells located near the base of the epithelial cells. In many insects the regenerative cells are clumped together in nidi, or 'nests', at frequent intervals along the length of the midgut. Large bands of circular muscles wrap around the gut and promote squeezing and mixing action, while smaller bands of longitudinal muscles usually lie on the outside of the circular muscles. Changes in all the cell types and in the morphology of the midgut have been observed after high radiation doses (Jafri and Ismail, 1977), but the regenerative cells are the most sensitive and the first to show damage (Riemann and Flint, 1967; Jafri and Ismail, 1977).

Ducoff's (1972) review of midgut irradiation damage studies tentatively concluded that such damage was one of the important causes of the immediate death. Lee (1964) showed that lead shielding of the body region containing the midgut of the queen honey bee, *Apis mellifera* L., protected the queens for up to 3 weeks from the otherwise lethal effects of 100 Gy of X-rays. Boll weevil, *Anthonomus grandis grandis* Boheman, adults were sterilized with 90–100 Gy of radiation, but became sluggish, fed poorly, and showed rapid mortality (Riemann and Flint, 1967). Even lower doses of 20–40 Gy caused extensive damage to the regenerative midgut cells of the weevils, with subsequent loss of the secretory epithelium of the midgut. Hence, replacement of the adult weevil's midgut secretory epithelium continued throughout the several months of life span. Both cell divisions and differentiation are occurring as evidenced by mitotic figures in various portions of the midgut. By using partial body shielding techniques, Riemann and Flint (1967) showed conclusively that irradiation of the midgut was the critical factor. Moreover, growth and invasion of gut fungi were more extensive in irradiated insects than in normal insects. Twenty Gy often prevented spermatogonia in testes 10 days after irradiation, and 60 Gy were completely effective in stopping spermatogenesis in these tests.

The importance of the boll weevil and the potential for using irradiated weevils for control led to attempts to fractionate the radiation dose and produce sterilization with less mortality, with varying degrees of success (Haynes *et al.*, 1977, 1979). Leopold (1986) found that feeding newly emerged boll weevil adults with the standard laboratory diet for 2 days, and then irradiating them (100 Gy) induced more mortality than occurred in adults fed only water, 5% sucrose in water solution, or 1% urea. Very low levels of mitotic activity ($<0.05\%$ of nidi had dividing cells) were found in weevils fed the liquid diets with 90% for weevils on the standard laboratory diet. Apparently the liquid diets and/or lack of protein did not stimulate nidi into mitotic activity, and hence protected the cells from radiation damage. Abnormalities in the midgut of irradiated weevils occur, including hyperplastic growths of non-differentiating cells. Occasionally overgrowths pushed into the gut lumen and nearly filled the lumen with tissue.

The midgut is also very sensitive tissue in the larvae of confused flour beetle, *Tribolium confusum* Jacquelin du Val (Jafri and Ismail, 1977). The nidi of regenerative cells were still numerous 24 h after irradiation with 53 Gy from a ^{60}Co source, but gradually disappeared with increasing dose (up to 700 Gy) and with increasing time following treatment. The 53 Gy dose caused columnar epithelial cells to begin to separate from each other and from the basement membrane within 24 h post-treatment. Some cells are vacuolated, and the brush border is not always continuous. Twelve days after irradiation the circular gut muscles appeared to be only loosely connected to the underlying tissue. In some cases the epithelial cells shed into the lumen. Epithelial cell swelling may be the first indicator of radiation damage. Similar, accelerated changes in epithelial and regenerative cells occur when larvae are given doses of 100, 500, and 700 Gy of irradiation. Adult and larval yellow mealworm, *Tenebrio molitor* L., receiving 50 to 1000 Gy, have changes mainly in the regenerative cells (Brower and Ashraf, 1972), and damage is positively correlated with dose and time following exposure.

No changes in midgut epithelium or any reduction in protein intake were detected in female *Bactrocera cucumis* fruit flies sterilized by 100 or 400 Gy doses applied 8 days after puparium formation. The peritrophic membrane was present in both non-irradiated and irradiated (400 Gy) up to 20 days after irradiation. Furthermore, there was no difference in oxygen consumption between controls and flies irradiated with up to 800 Gy. Flies that were irradiated with the higher doses did die sooner than control flies (Bailey, 1973). Little (1967) found no adverse effects of irradiation upon the midgut of Mediterranean fruit flies.

In the oriental fruit fly, *B. dorsalis*, Prasad *et al.* (1980) found no midgut damage from 100 Gy, but 150 Gy and 200 Gy did considerable damage to the regenerative cells and the mature epithelium. The appearance of the epithelial cells was altered at 150 Gy with cell membrane breakdown and appearance of vacuoles. Existing epithelial and regenerative cells were difficult to find 10 days post-treatment. The muscle layer of the gut was the only remaining tissue. The midgut of fruit flies is less sensitive to radiation than that of other insects, including boll weevils and honey bees. The ingestion of liquid or semi-liquid food by adult fruit flies, with possibly lower protein content may mean there is less need to replace midgut cells. A more detailed study of the midgut in tephritids, with attention to how much replacement of epithelial cells occurs normally might be worth while.

Nervous tissue

Irradiation can damage nervous tissue, especially during its growth. In some insects it appears that nearly all the nerve cells are present at hatching, with growth occurring primarily by an increase in cell size thereafter, whereas in other insects there is evidence of extensive post-embryonic cell divisions in the nervous system. In first instar larvae of the Japanese beetle, *Popillia japonica* Newman, there is little cell division in the brain. Brain cell number increases sufficiently in the second instar to account for one cycle of cell division, but in the third instar there is little cell division (Abercrombie, 1936). The growth of the central nervous system in the hemipteran, *Oncopeltus fasciatus* (Dallas), the large milkweed bug, occurs by both cell multiplication and increase in size. Nerve cell divisions were observed also in adults up to one month of age (Johansson, 1957). The terminal abdominal ganglion in immature *Acheta domesticus* (L.), the house cricket, grew mainly by increase in cell size (about 70 times) rather than in cell number during postembryonic development (Edwards, 1968). In contrast, glial cells, the protective non-conducting cells associated with neurons throughout the nervous system, increased in number from about 1000 to 17 000 cells.

The thoracic ganglia in larval yellow fever mosquito, *Aedes aegypti* (L.), grow by cell multiplication (Trager, 1937). Mitotic figures occur in the supraoesophageal ganglion of the Mediterranean fruit fly, but were almost eliminated by irradiation at 80 Gy (Rahman *et al.*, 1990).

Irradiation (5 to 150 Gy) of any of the immature stages of the Mediterranean fruit fly results in a highly significant reduction in size increase of the supraoesophageal ganglion relative to the proventriculus. The minimum size of the supraoesophageal ganglion (only about 30% of that of the controls) was found in individuals treated with the highest dose, and there seemed to be a predictable relationship of size reduction with radiation dose. The major reduction in size occurred in the optic discs of the ganglionic mass (Rahman *et al.*, 1990).

A similar reduction in size (about 72–76%) of the supraoesophageal ganglion in third instar larvae of the melon and the oriental fruit flies occurred when larvae were treated at several stages of development (Rahman *et al.*, 1992). Although the size of the proventriculus was related to the developmental stage of the insects, its size was not altered by irradiation. Jessup *et al.* (1992) also found similar size reduction in the supra-oesophageal ganglion, relative to the proventriculus, in irradiated larvae of the Queensland fruit fly. This size reduction in the supraoesophageal ganglion after irradiation in several tephritid fruit flies may provide the basis for an assay of irradiation in a quarantine situation. This would require some skill in dissecting fruit fly larvae. Similar changes in the central nervous system of other economically important insects have not been reported.

Behavioural Changes

Bailey (1975) suggested that behavioural changes in insects may be a more sensitive indicator of radiation damage than morphological or cytological changes. Behavioral changes are undoubtedly mediated through the nervous and possibly endocrine systems, and may indicate very subtle changes in nerve centres, nerve function or hormone secretion.

Carpenter ants exposed to ionizing radiations (X-rays, less than 0.5 R s^{-1}) wave their antennae as the earliest indicator of exposure, and a slightly higher level leads to a burst of rapid running activity (Martinsen and Kimeldorf, 1972). The compound eyes and antennal receptors (i.e. olfactory receptors) may be sensitive detectors of ionizing radiations.

The mating propensity of several insect species, e.g. codling moth, Indian meal moth, and a fruit fly (Bailey, 1975; Brower, 1975; Hutt and White, 1975), increases following irradiation. Possibly the radiation acts upon the higher brain centres that regulate the mating response in female insects. Brower (1975) suggested, however, that inadequate sperm transfer during mating or failure of eupyrene sperm to reach the spermatheca may be factors in the Indian meal moth. Adult eclosion behaviour is changed in *Bactrocera cucumis* by a single large dose of radiation given 12 h out of phase with the normal light-induced rhythm (Bailey, 1975). The radiation caused a bimodal emergence rhythm of adults, with the extra peak of emergence shifted towards the time of exposure to the radiation. Irradiation of secondary screwworm flies, *Cochliomyia macellaria* (F.), with 250 Gy gamma rays reduced sustained flight, flight time and distance flown, and resulted in lower wing-beat frequency than control flies (Noblet *et al.*, 1968). Similar effects were found in fall armyworm moths, *Spodoptera frugiperda* (J.E. Smith), receiving 400 Gy of gamma radiation (Shepard *et al.*, 1968). These results may be physiological, structural, and/or behavioural.

Changes in dispersal and locomotor behaviour may reflect changes in nervous and/or muscle systems. Locomotor activity declines progressively as *B. cucumis* is irradiated with 400, 600, and 800 Gy doses, but not if irradiated with only 100 Gy (Bailey, 1975). Hamada (1980) found that the distribution and dispersal of irradiated (100 Gy) melon flies is not the same as that of control flies. The low recapture of irradiated flies is partly caused by the reduced percentage of flies escaping from the release cage (up to 10% fewer), and many others were unable to fly. Possible damage to wing muscle function, nervous control centres, and/or the metabolic machinery that would supply rapid energy to the muscles may have occurred. In contrast Fletcher and Economopoulos (1976) found no differences in the dispersal rates of control and gamma irradiated (110 Gy) laboratory-cultured olive fruit flies. Nair and Bhakthan (1969) and

Bhakthan and Nair (1972) showed that irradiation of adult house flies less than 24 h old with 100 or 200 Gy results in damage to the sarcoplasmic reticulum of the flight muscles and disintegration of mitochondria. They suggested that the radiation-induced changes in the sarcoplasmic reticulum critical to muscle function were due to altered membrane permeability.

Feeding behaviour of insects after irradiation may change and reflect damage to the nervous system, especially the sensory system. Irradiation of first instars of the cutworm, *Spodoptera litura* (F.), resulted in delayed action expressed in the sixth instar as a reduction in nitrogen ingested and in nitrogen balance (Seth and Sehgal, 1985b). There is a lengthening (100% increase) of the active feeding period of sixth instar larvae that had received a dose of 70 Gy in the third instar (Seth and Sehgal, 1985a). The duration of the wandering stage (cessation of feeding and searching for a pupation site) was lengthened. No physiological explanation was offered, but it seems possible that the radiation treatments may have altered the hormonal balance in the larvae. Normally the juvenile hormone level falls in larval insects prior to pupation, and if no decline occurs, or it is kept high by the addition of exogenous hormone or hormone analogue (methoprene), lepidopterous insects especially are likely to delay pupation and continue to feed.

Pharate adult Mediterranean and Caribbean fruit flies, *Anastrepha suspensa* (Loew), irradiated 2 days prior to emergence with 100 Gy had reduced response to a food bait and reduced total intake of food by both sexes (Galun *et al.*, 1985). Mediterranean fruit fly males, which normally consume more sucrose than females, showed reduced consumption of sugar, and both sexes showed reduced aggregation at food sources. Possible physiological or biochemical damage to sensory structures, midgut or other tissues of the flies may have caused the reduced food intake and aggregation, or changes in physical activity. Galun and Warburg (1968) found that 40 Gy slowed the rate of digestion of a blood meal in male *Ornithodoros tholozani*, with digestion in female ticks being slowed at 10 Gy. Males required 160 Gy doses to cause 99% dominant lethals in their reproductive cells, while females only required 40 Gy doses to prevent egg development. The digestion of a blood meal in ticks is normally accompanied by an increase in oxygen consumption, but in these ticks there was no reduction in the metabolic rate caused by up to 160 Gy irradiation. Apparently the oxygen consumption in these freshly fed ticks is not the simple result of growth of the ovary and eggs, which could be stopped by radiation doses that did not alter respiration.

Enzyme Activity and Other Biochemical Constituents

Changes in enzymatic activity of irradiated insects have been reported (Batchelor, 1989). Irradiation (420 Gy in air) of the navel orangeworm, *Amyelois transitella* (Walker), caused a nearly two-fold increase in superoxide dismutase (SOD) 3 h post-treatment, with return to the control levels 6 h post-treatment. A slight increase in SOD activity also occurred in irradiated diapausing codling moth larvae. The rapid increase in SOD suggests that irradiation-induced production of free radicals stimulated the increase in the level of the scavenger, SOD. Glutathione transferase, another protective enzyme, showed no changes that could be correlated with radiation treatment in codling moth and navel orangeworm (Batchelor, 1989). Catalase activity shows considerable variability in both the codling moth and navel orangeworm, but it could not be demonstrated that radiation had any effect. The ATP level of these insects also could not be correlated with radiation treatments.

Glycogen levels are higher in diapausing codling moth larvae irradiated in 0.25% oxygen than in larvae irradiated in 3% oxygen or in air (about 21% oxygen) (Batchelor, 1989). The very low oxygen level apparently protected the larvae from radiation damage and conserved glycogen. This protection agrees with other data showing that codling moth larval mortality is reduced when irradiated in 0.25% oxygen compared with air or 3% oxygen. Similar results were found by Chandler *et al.* (1979) with house-fly larvae irradiated with 50 Gy; irradiated larvae showed 50% increase in glycogen 72 h post-treatment. Some vertebrates have shown similar elevations of glycogen levels in liver tissue after irradiation (Ahlersova *et al.*, 1980, cited in Batchelor, 1989; Kahn *et al.*, 1986).

No changes in the cuticular hydrocarbon components of the screwworm, *Cochliomyia hominivorax* (Coquerel), irradiated as pharate adults 4 days prior to emergence were detected (Pomonis, 1989). These lipids, which are secreted by the epidermal cells that lie immediately beneath the cuticle, serve a great variety of functions in insects, including semiochemical communication (Howard and Blomquist, 1982; Bartelt and Jackson, 1984; Peschke, 1987), possible protective mimicry (Howard *et al.*, 1980; Vander Meer and Wojcik, 1982; Vander Meer *et al.*, 1989), and water proofing (Jackson and Baker, 1970; Nelson, 1978; Lockey, 1988). Disruption of cuticular lipids has profound effects upon insects, and irradiation-induced damage to epidermal cells may be reflected in changes in cuticular lipid composition or quantity.

Longevity and Ageing

Control and irradiated (100 Gy) olive fruit flies, *Dacus oleae*, were equally able to survive and reproduce on a series of carbohydrate sources, with one exception (Tsiropoulos, 1980). The irradiated flies could not utilize the trisaccharide melezitose as well as non-irradiated flies, and irradiated males lived only half as long as the control males. Irradiated females also had a shortened life span on melezitose. Longevity of the secondary screwworm fly also decreased with increasing irradiation dosage at 20 Gy, 45 Gy, and higher dosages (Noblet *et al.*, 1968).

Several studies reported that radiation can increase the life span of some insects. Hunter and Krithayakiern (1971) found that low doses (5, 10, or 20 Gy of gamma radiation) increased the life expectancy of female house crickets, but not of males. Higher doses (40 to 100 Gy), however, reduced life expectancy of both sexes. Even the low dose of 5 Gy substantially reduced egg production by females (44.9 eggs vs. 304 eggs). The authors suggested that the longer life of irradiated females was in part due to the reduced nutrient and physiological demands of egg production, though other physiological processes may also be slowed down.

Tilton and Brower (1983) extensively reviewed the effects of irradiation upon the reproductive tissues of insects. Since that review a few additional insects have been studied, but no new principles or major changes in our knowledge have occurred. In general, the reproductive tissues of insects are very sensitive to irradiation damage, and insects can be sterilized with relatively low dosages of irradiation. Numerous reports have shown, however, that male insects irradiated for sterility are often not as competitive in seeking mates as untreated insects, though this varies with radiation dose and species (Tilton and Brower, 1983, and references cited therein). Reduced competitiveness of irradiated males could be due to changes in behaviour, physiology, or a combination of factors.

Damage to pheromone producing capability might be an important factor in loss of competitiveness of pheromone producing male insects, such as many bark beetles, the boll weevil, and many tephritid fruit flies. Haynes *et al.* (1977) extracted and measured the quantity of pheromone in the frass from male boll weevils sterilized by 25 fractionated doses of gamma radiation totaling 62.5 Gy. Irradiated males produced only one-third as much pheromone as control insects for the first 4 days after emergence, but for the next week they produced as much as control insects. Fractionated doses should allow sterile males to be produced for use in a sterile release programme with satisfactory pheromone production.

Ponce (1991) showed that male Caribbean fruit flies irradiated with a sterilizing dose (30 Gy) (Walder and Calkins, 1992) as pharate adults a few hours prior to emergence, produced normal quantities and a typical blend of pheromone components. Pheromone production is reduced significantly at 50, 70, and 100 Gy doses. The mating system of tephritid fruit flies is complex and it is not known if other behavioural factors not measured in this study also might contribute to reduced competitiveness.

References

Abercrombie, W.F. (1936) Studies on cell number and the progression factors in the growth of Japanese beetle larvae (*Popillia japonica* Newman). *Journal of Morphology* 59, 91–112.

Ahlersova, E., Ahlers, I. and Praslicka, M. (1980) Tissue glycogen and blood glucose in irradiated rats II. Effect of non-lethal doses of continuous gamma-radiation. *Folia Biologica* (Prague) 26, 423–431.

Anonymous (1971) Disinfestation of fruit by irradiation, *Proceedings of a Panel on the Use of Irradiation to Solve Quarantine in the International Fruit Trade.* Honolulu 1970, IAEA, Vienna.

Anonymous (1984a) Report on FAO/IAEA consultant group meeting on the use of irradiation as a quarantine treatment of agricultural commodities. *Food Irradiation Newsletter* 8, 6–13.

Anonymous (1984b) Report of the FAO/IAEA research coordination meeting on insect disinfestation of food and agricultural products by irradiation. *Food Irradiation Newsletter* 8, 14–28.

Anonymous (1985) Report of the second FAO/IAEA research co-ordination meeting on insect disinfestation of food and agricultural products by irradiation. *Food Irradiation Newsletter* 9, 12–28.

Anonymous (1986a) Excerpt of the task force meeting on irradiation as a quarantine treatment. Suriwongse Hotel, Chiang Mai, Thailand. 17–21 February 1986. *Food Irradiation Newsletter* 10, 5–10.

Anonymous (1986b) Report of the FAO/IAEA research coordination meeting on the use irradiation as a quarantine treatment of food and agricultural commodities. Chiang Mai, Thailand. 24–28 February 1986. *Food Irradiation Newsletter* 10, 17–28.

Anonymous (1989) Report of the second FAO/IAEA research coordination meeting on the use of irradiation as a quarantine treatment of food and agricultural commodities. Orlando, Florida, 23–27 May 1988. *Food Irradiation Newsletter* 13, 34–48.

Anonymous (1991a) Summary of task force meeting on irradiation as a quarantine treatment of fresh fruits and vegetables. Bethesda, MD. 7–11 January 1991. *Food Irradiation Newsletter* 15, 9–10.

Anonymous (1991b) Report of the final FAO/IAEA research coordination meeting on the use irradiation as a quarantine treatment of food and agricultural commodities. Malaysian Agricultural Research and Development Institute (MARDI), Kuala Lumpur, Malaysia. 27–31 August 1990. *Food Irradiation Newsletter* 15, 42–58.

Anonymous (1992) Report of the first FAO/IAEA research coordination meeting on irradiation as a quarantine treatment of mites, nematodes, and insects other than fruit fly. Division of Plant Industry, Florida Department of

Agriculture and Consumer Services, Gainesville, Florida. 27 April– May 1992. *Food Irradiation Newsletter* 16 (in press).

Bailey, P. (1973) Respiration and nutrition of radiation-sterilized female *Dacus cucumis* (Diptera, Tephritidae) *Entomologia Experimentalis et Applicata* 16, 433–444.

Bailey, P. (1975) Behavioral changes induced by irradiating *Dacus cucumis* with gamma rays. *Journal of Insect Physiology* 21, 1247–1250.

Baker, A.C. (1939) The basis for treatment of products where fruitflies are involved as a condition for entry into the United States. *US Department of Agriculture Circular* 551, 8pp.

Balock, J.W., Christenson, L.D. and Burr, G.O. (1956) Effect of gamma rays from cobalt 60 on immature stages of the oriental fruit fly (*Dacus dorsalis* Hendel) and possible application to commodity treatment problems. *Proceedings of the Annual Meeting of the Hawaii Academy of Sciences*, Honolulu 31, 18.

Balock, J.W., Burditt, A.K., Jr and Christenson, L.D. (1963) Effects of gamma radiation on various stages of three fruit fly species. *Journal of Economic Entomology* 56, 42–46.

Balock, J.W., Burditt, A.K., Jr, Seo, S.T. and Akamine, E.K. (1966) Gamma radiation as a quarantine treatment for Hawaiian fruit flies. *Journal of Economic Entomology* 59, 202–204.

Bartelt, R.J. and Jackson, L.L., (1984) Hydrocarbon component of the *Drosophila virilis* (Diptera, Drosophilidae) aggregation pheromone, (Z)-10-heneicosene. *Annals of the Entomological Society of America* 77, 364–371.

Batchelor, T.A. (1989) Potential use of biochemical data in the development of radiation-based insect disinfestation treatments for fresh commodities. PhD Dissertation, University of California, Davis, CA.

Benschoter, C.A. and Telich, C.J. (1964) Effect of gamma rays on immature stages of the Mexican fruit fly. *Journal of Economic Entomology* 57, 690–691.

Bergonie, J. and Tribondeau, L. (1959) Interpretation of some results of radiotherapy and an attempt at determining a logical technique of treatment. *Radiation Research* 11, 587–588.

Bhakthan, N.M.G. and Nair, K.K. (1972) Fine structural damage in the somatic tissues of gamma-irradiated house fly. I. Flight muscles. *Annals of the Entomological Society of America* 65, 504–508.

Brower, J.H. (1975) Gamma irradiation of adult *Plodia interpunctella*, Effects on mating, sterility, and number of progeny. *Annals of the Entomological Society of America* 68, 1086–1090.

Brower, J.H. and Ashraf, M. (1972) Histopathology of irradiated yellow mealworm adult and larval midguts. *Annals of the Entomological Society of America* 65, 950–954.

Burditt, A.K., Jr (1982) Food irradiation as a quarantine treatment of fruits. *Food Technology* (11) 51–54, 58–60, 62.

Burditt, A.K., Jr, Hungate, F.P. and Toba, H.H. (1989) Gamma irradiation, effect of dose rate on development of mature codling moth larvae and adult eclosion. *Radiation Physics and Chemistry* 34, 979–984.

Chandler, L., Roye, R.K. and Srivastava, P.N. (1979) Radiation-induced biochemical changes in the larvae of housefly, *Musca domestica*. *Journal of Radiation Research* 20, 137–145.

Cogburn, R.R., Tilton, E.W., and Burkholder, W.E. (1966) Gross effects of gamma radiation on the Indian meal moth and the Angoumois grain moth. *Journal of Economic Entomology* 59, 682–685.

Ducoff, H.S. (1972) Causes of death in irradiated adult insects. *Biological Reviews* 47, 211–240.

Edwards, J.S. (1968) Some questions for the insect nervous system. In: Beament, J.W.L. and Treherne, J.E. (eds), *Insects and Physiology*. American Elsevier, New York, pp. 163–173.

Eric, B., LeCompte, J., Klein, S. and Kricker, W. (1970) Study of disinfestation of bananas by gamma irradiation. *Food Technology in Australia* 22, 664–667.

Food and Drug Administration (1984) Irradiation in the production processing and handling of food. *Federal Register*, US Government Printing Office, Washington, DC, 49, 5714–5722.

Fletcher, B.S. and Economopoulos, A.P. (1976) Dispersal of normal and irradiated laboratory strains and wild strains of the olive fly *Dacus oleae* in an olive grove. *Entomologia Experimentalis et Applicata* 20, 183–194.

Fritz-Niggli, H. (1951) Erste biologische Versuche mit dem 31-MeV-Betatron. *Archiv Klaus-Stift. Vererb. Forsch.* 24, 454–459.

Galun, R. and Warburg, M. (1968) Irradiation effects on respiration and blood digestion in the tick *Ornithodoros tholozani* and their importance for the sterile-male technique In: *Proceedings of International Symposium on Isotopes and Radiation in Entomology*. IAEA, Vienna, pp. 249–258.

Galun, R., Gothilf, S.,Blondheim, S., Sharp, J.L., Mazor, M. and Lachman, A. (1985) Comparison of aggregation and feeding responses by normal and irradiated fruit flies, *Ceratitis capitata* and *Anastrepha suspensa* (Diptera, Tephritidae), *Environmental Entomology* 14, 726–732.

Hamada, R. (1980) Studies on the dispersal behavior of melon flies, *Dacus cucurbitae* Coquillett (Diptera, Tephritidae), and the influence of gamma-irradiation on dispersal. *Applied Entomology and Zoology* 15, 363–371.

Haynes, J.W., Mitlin, N., Davich, T., Dawson, J., McGovern, W. and McKibben, G. (1977) Sterilization of boll weevil pupae with fractionated doses of gamma irradiation. *Entomologia Experimentalis et Applicata* 21, 57–62.

Haynes, J.W., Wright, J. and Mattix, E. (1979) Fractionated vs. acute irradiation, The effect of treating adult boll weevils at different ages. *Entomolgia Experimentalis et Applicata* 26, 61–66.

Heather, N.W. (1992) Review of irradiation as a quarantine treatment for insects other than fruit flies. In: *Use of Irradiation as a Quarantine Treatment of Food and Agricultural Commodities. Proceedings of the Final Research Coordination Meeting, Kuala Lumpur, Malaysia*. 27–31 August 1990. IAEA, Vienna, pp. 203–218.

Hink, W.F. (1970) Established insect cell line from the cabbage looper, *Trichoplusia ni*. *Nature (London)* 226, 466–467.

Howard, R.W. and Blomquist, G.J. (1982) Chemical ecology and bio-chemistry of insect hydrocarbons. *Annual Review of Entomology* 27, 149–172.

Howard, R.W., McDaniel, C.A. and Blomquist, G.J. (1980) Chemical mimicry as an integrating mechanism, cuticular hydrocarbons of a termitophile and its host. *Science*, 210, 431–433.

Hunter, P.E. and Krithayakiern, V. (1971) Effect of gamma radiation upon life expectancy and reproduction in the house cricket, *Acheta domesticus* (Orthoptera, Gryllidae) *Annals of the Entomological Society America* 64, 119–123.

Hutt, R.B. and White, L.D. (1975) Codling moth. Effects of gamma radiation on mating propensity of each sex, and the limitation of mating by females. *Environmental Entomology* 4, 774–776.

Jackson, L.T. and Baker, G.L. (1970) Cuticular lipids of insects. *Lipids* 5, 239–246.

Jafri, R.H. and Ismail, M. (1977) Cytological lesions in the midgut of *Tribolium confusum* larvae exposed to gamma radiation. *Journal of Invertebrate Pathology* 29, 10–17.

Jessup, A.J., Rigney, C.J., Millar, A., Sloggett, R.F. and Quinn, N.M. (1992) Gamma irradiation as a commodity treatment against Queensland fruit fly in fresh fruit. In: *Use of Irradiation as a Quarantine Treatment of Food and Agricultural Commodities. Proceedings of the Final Research Coordination Meeting, Kuala Lumpur, Malaysia.* 27–31 August 1990. IAEA, Vienna, pp. 13–42.

Johansson, A.S. (1957) The nervous system of the milkweed bug, *Oncopeltus fasciatus* (Dallas) (Heteroptera Lygaeidae) *Transactions of the American Entomological Society* 83, 119–183.

Jona, R. and Arzone, A. (1979) Control of *Rhagoletis cerasi* in cherries by gamma irradiation. *Journal of Horticultural Science* 54, 167–170.

Kaneshiro, K.Y., Ohta, A.T., Kurihara, J.S., Kanegawa, K.M. and Nagamine, L.R. (1983) Gamma radiation treatment for disinfestation of the Mediterranean fruit fly in California grown fruits. I. Stone fruits. *Proceedings of the Hawaiian Entomological Society* 24, 245–259.

Karpati, J.F., Schotman, C.Y. and Zammarano, K.A. (eds) (1983) *International Plant Quarantine Treatment Manual*. Rome. FAO Plant Production and Protection Paper 50. FAO Rome.

Khan, A.S., Bhartiya, H.C. and Bhartiya, K.M. (1986) Effect of gamma irradiation and radioprotector MPG on liver glycogen in mice. *Radiobiologia Radiotherapia* 27 (H.4), 473–477.

Koidsumi, K. (1930) Quantitative studies on the lethal action of X-rays upon certain insects. *Journal of the Society of Tropical Agriculture* 2, 243–263.

Koval, T.M. (1983) Intrinsic resistance to the lethal effects of X-irradiation in insect and arachnid cells. *Proceedings of the National Academy of Sciences USA* 80, 4752–4755.

Koval, T.M. (1986a) Enhanced survival by photoreactivation and liquid-holding following UV damage of TN-386 insect cells. *Mutation Research* 166, 149–156.

Koval, T.M. (1986b) Inducible repair of ionizing radiation damage in higher eukaryotic cells. *Mutation Research* 173, 291–293.

Koval, T.M. (1988) Enhanced recovery from ionizing radiation damage in a lepidopteran insect cell line. *Radiation Research* 115, 413–420.

Koval, T.M., Myser, W.C., Hart, R.W. and Hink, W.F. (1978) Comparison of survival and unscheduled DNA synthesis between an insect and a mammalian cell line following X-ray treatments. *Mutation Research* 49, 431–435.

Lee, W.R. (1964) Partial body radiation of queen honeybees. *Journal of Apicultural Research* 3, 113–116.

Leopold, R.A. (1986) Enhancement of postirradiation longevity and mating of the boll weevil (Coleoptera, Curculionidae) by regulation of preirradiation feeding. *Journal of Economic Entomology* 79, 1484–1491.

Lippold, P.C., Gambrell, F.L. and Massey, L.M., Jr (1968) Effects of ionizing radiation on the European chafer, the plum curculio, and the large milkweed bug. *Annals of the Entomological Society America* 61, 151–158.

Little, H.F. (1967) Midgut epithelium of adult *Chelisoches morio* (Dermaptera, Chelisochidae) and *Ceratitis capitata* following ionizing radiation. *Annals of the Entomological Society America* 60, 412–414.

Lockey, K.H. (1988) Lipids of the insect cuticle, origin, composition and function. *Comparative Biochemistry and Physiology* 89B, 595–645.

Macfarlane, J.J. (1966) Control of the Queensland fruit fly by gamma irradiation. *Journal of Economic Entomology* 59, 884–889.

Martinsen, D.L. and Kimeldorf, D.J. (1972) The prompt detection of ionizing radiations by carpenter ants. *Biological Bulletin* 143, 403–419.

Nair, K.K. and Bhakthan, N.M.G. (1969) Preliminary studies on the ultrastructural damage in the flight muscles of gamma-irradiated housefly. *International Journal of Radiation Biology* 16, 397–399.

Nelson, D.R. (1978) Long-chain methyl-branched hydrocarbons: occurrence, biosynthesis, and function. *Advances in Insect Physiology* 13, 1–33.

Noblet, R., Shepard, M. and Hunter, P.E. (1968) Reproduction and flight in the secondary screw-worm fly, *Cochliomyia macellaria* Fabr., following gamma radiation of the pupae (Diptera, Calliphoridae), *Journal of the Georgia Entomological Society* 3, 173–180.

Nothel, H. (1968) Correlations, interactions and differences between radiation effects on longevity and natural aging. In: *Isotopes and Radiation in Entomology*. IAEA, Vienna, pp. 87–102.

O'Brien, R.D. and Wolfe, L.S. (1964) *Radiation, Radioactivity, and Insects*. Academic Press, New York, pp. 24–25.

Ouye, M.T. and Gilmore, J.E. (1985) The philosophy of quarantine treatment as related to low-dose radiation. In: Moy, J.H. (ed.), *Radiation Disinfestation of Food and Agricultural Products*. University of Hawaii, Honolulu, pp. 67–69.

Peschke, K. (1987) Cuticular hydrocarbons regulate mate recognition, male aggression, and female choice of the rove beetle, *Aleochara curtula*. *Journal of Chemical Ecology* 13, 1993–2008.

Pomonis, J.G. (1989) Cuticular hydrocarbons of the screwworm, *Cochliomyia hominivorax* (Diptera, Calliphoridae). Isolation, identification, and quantification as a function of age, sex, and irradiation. *Journal of Chemical Ecology* 15, 2301–2317.

Ponce (1991) Quantitative analysis of pheromone production in irradiated Caribbean fruit fly males, *Anastrepha suspensa* (Loew). PhD Dissertation, University of Florida, Gainesville, FL, 101pp.

Prasad, H.H., Sethi, G.R. and Singh, K.M. (1980) Effect of gamma radiation on the midgut of oriental fruitfly, *Dacus dorsalis* Hendel. *Indian Journal of Entomology* 42, 498–504.

Rahman, R., Rigney, C. and Busch-Petersen, E. (1990) Irradiation as a quarantine treatment against *Ceratitis capitata* (Diptera, Tephritidae). Anatomical and cytogenetic changes in mature larvae after gamma irradiation. *Journal of Economic Entomology* 83, 1449–1454.

Rahman, R., Bhuiya, A.D., Huda, S.M.S., Shahjahan, R.M., Nahar, G. and Wadud, M.A. (1992) Anatomical changes in mature larvae of two *Dacus* spp.

following irradiation. In: *Use of Irradiation as a Quarantine Treatment of Food and Agricultural Commodities, Proceedings of the Final Research Coordination Meeting, Kuala Lumpur, Malaysia.* 27–31 August 1990. IAEA, Vienna, pp. 133–139.

Riemann, J.G. and Flint, H.M. (1967) Irradiation effects on midguts and testes of the adult boll weevil, *Anthonomus grandis*, determined by histological and shielding studies. *Annals of the Entomological Society America* 60, 298–308.

Seth, R.K. and Sehgal, S.S. (1985a) Gamma irradiation-induced changes in the nutritional profile of the larvae of *Spodoptera litura* Fabr. *Journal of Nuclear Agriculture and Biology* 14, 1– 3.

Seth, R.K. and Sehgal, S.S. (1985b) Utilization of dietary nitrogen by the larvae of *Spodoptera litura* (Fabricius) under the impact of gamma irradiation. *Journal of Nuclear Agriculture and Biology* 14, 92–95.

Shepard, M., Noblet, R. and Hunter, P.E. (1968) The effects of gamma radiation upon flight of the fall armyworm, *Spodoptera frugiperda* (J.E. Smith) (Lepidoptera, Noctuidae). *Journal of the Georgia Entomology Society* 3, 57–61.

Shipp, E. and Osborn, A.W. (1968) Irradiation of Queensland fruit fly pupae to meet quarantine requirements. *Journal of Economic Entomology* 61, 1721–1726.

Tilton, E.W. and Brower, J.H. (1983) Radiation effects on arthropods. In: Josephson, E.S. and Peterson, M.S. (eds), *Preservation of Food by Ionizing Radiation.* CRC Press, Boca Raton, Florida.

Trager, W. (1937) Cell size in relation to the growth and metamorphosis of the mosquito, *Aedes aegypti. Journal of Experimental Zoology* 76, 467–489.

Tsiropoulos, G.J. (1980) Carbohydrate utilization by normal and gamma-sterilized *Dacus oleae. Journal of Insect Physiology* 26, 633–637.

Vander Meer, R.K and Wojcik, D.P. (1982) Chemical mimicry in the myrmecophilous beetle *Myrmecaphodius excavaticollis. Science* 218, 806–808.

Vander Meer, R.K., Jouvenaz, D.P. and Wojcik, D.P. (1989) Chemical mimicry in a parasitoid (Hymenoptera, Eucharitidae) of fire ants (Hymenoptera, Formicidae), *Journal of Chemical Ecology* 15, 2247–2261.

Walder, J.M. and Calkins, C.O. (1992) Gamma radiation effects on ovarian development of the Caribbean fruit fly, *Anastrepha suspensa* (Loew) (Diptera, Tephritidae), and its relationship to sterile fly identification. *Florida Entomologist* 75, 267–271.

Heat and Cold Treatments **6**

JOHN W. ARMSTRONG
United States Department of Agriculture, Agricultural Research Service, Tropical Fruit and Vegetable Research Laboratory, PO Box 4459, Hilo, HI 96720, USA.

Introduction

Cold and heat treatments are used to disinfest fresh horticultural products of quarantine pests, primarily tephritid fruit flies from fruit, before entering export marketing channels. Both of these treatment methods were used successfully before the advent of the two most important postharvest fumigants, ethylene dibromide and methyl bromide (Armstrong, 1992). Cold treatments were first recommended as disinfestation treatments against fruit flies following the initial independent works of Hooper (1907) and Lounsbury (1907). The first heat treatment used to disinfest fruit flies from fresh fruit was developed about 22 years later during a Mediterranean fruit fly, *Ceratitis capitata* (Wiedemann), outbreak in Florida (Baker, 1952). Cold and heat disinfestation treatments are relatively easy to apply, leave no chemical residues and, in the case of heat treatments, may offer some fungicidal activity (Armstrong and Couey, 1989).

Cold or heat disinfestation treatments are conceptually simple, and consist of reducing or increasing the fruit temperature below or above the thermal tolerance limits of the target insect life stages. Unfortunately, accomplishing disinfestation using cold or heat without causing deleterious effects to fruit quality can range from difficult to impossible. Temperature and time combinations required to kill the target insects may meet or exceed those that injure the host fruit or reduce shelf-life (Couey 1982, 1989). The use of cold or heat disinfestation treatments may require the acceptance of some unavoidable damage. Generally, the research and development of cold or heat treatments is on a fruit-by-fruit basis,

From: Paull, R.E. and Armstrong, J.W. (eds) (1994) *Insect Pests and Fresh Horticultural Products: Treatments and Responses.* CAB INTERNATIONAL, Wallingford, UK.

depending on the cold or heat tolerance of both the target insects and the host fruit.

Cold Treatments

When many insects and related terrestrial arthropods from temperate and tropical areas are exposed to temperatures between 0 and 10°C, they are immobilized and enter a state of chill coma that is reversible if the duration of exposure is not prolonged. Other active insects are lethally injured by cold shock even though their body water does not freeze.

(Lee, 1991).

The function of quarantine treatments using cold is to reduce the temperature of the commodity below the threshold at which the target pest species can survive without affecting commodity quality. Lee (1991) provides information indicating that cold-induced mortality in insects is caused by both direct chilling injury, or cold shock, and indirect chilling injury. Cold shock induced by rapid cooling to low temperatures may result in cell membrane failure, induce cell leakage and lead to severe loss of membrane function. The specific mechanisms of indirect chilling injury are unknown, there being few studies of the effects of long-term cold exposure in the absence of freezing (Lee, 1991). Most research on the effects of cold on insects and other arthropods has focused on tolerance to low temperatures (Lee, 1991). Little research has attempted to determine how cold treatments kill target pest species or why some quarantine pests are more susceptible to cold treatment than others. An understanding of the basic insect physiological response to cold treatments could greatly advance the use of cold as a quarantine treatment. These data combined with data for cold-induced damage to host fruits would provide a model for rapidly determining the potential efficacy of candidate cold disinfestation treatments.

Following the work of Hooper (1907) and Lounsbury (1907), Back and Pemberton (1916) developed a cold treatment to disinfest apples of Mediterranean fruit fly, *Ceratitis capitata* (Wiedemann). Similar results were observed with infested kamani (*Terminalia catappa* L.) and peaches. Cold treatments were used to disinfest a variety of fruits during the 1928–1929 Mediterranean fruit fly outbreak in Florida. Mason and McBride (1934) reported comparable results for disinfesting citrus, guava, avocados, and mangoes of Mediterranean fruit fly, and they provided the first extensive data for the effects of cold on the life stages of this fruit fly species. The work of Mason and McBride (1934) was corroborated by the more advanced studies of Sproul (1976) that proved effective for a number of different host fruits. The temperature and exposure times required to kill Mediterranean fruit fly life stages (Sproul, 1976) are consistent with the earlier observations of Mason and McBride (1934). Nel (1936) analysed cold treatment data for nectarines, peaches, plums, and grapes infested with Mediterranean fruit fly and concluded that fruit fly control is independent of the host fruit. This means that the temperature and exposure times for disinfesting fruit fly life stages from host fruits would remain constant for each fruit fly species regardless of the host fruit, making selection of quarantine treatment schedules more precise.

Cold as a potential quarantine treatment has been studied for codling moth, *Cydia pomonella* (L.), eggs and larvae (Newcomer, 1930, 1936; Moffitt and Albano, 1972; Moffitt and Burditt, 1989a,b; Yokoyama and Miller, 1989), oriental fruit moth, *Grapholitha molesta* (Busck) (Dustan, 1963; Yokoyama and Miller, 1989). Cold also has been studied as a potential treatment against the egg stage of various stored-product insects, including almond moth, *Ephestia cautella* (Walker); Angoumois grain moth, *Sitotroga cerealella* (Olivier); cigarette beetle, *Lasioderma serricorne* (F.); confused flour beetle, *Tribolium confusum* Jacquelin duVal; cowpea weevil, *Callosobruchus maculatus* (F.); Indian meal moth, *Plodia interpunctella* (Hubner); red flour beetle, *Tribolium castaneum* (Herbst); and sawtoothed grain beetle, *Oryzaephilus surinamensis* (L.)(Adler, 1960; Mullen and Arbogast, 1979). Cold treatments exist to disinfest Caribbean fruit fly, *Anastrepha ludens* (Loew), and other *Anastrepha* species, Mediterranean fruit fly, and Queensland fruit fly, *Bactrocera tryoni* (Froggatt), from a wide range of fruits (Rippon and Smith, 1979; Benschoter, 1984; Hill *et al.*, 1988; Animal and Plant Health Inspection Service, 1992c). The eggs and larvae of these fruit fly species are killed by exposure to temperatures <10°C. However, only temperatures <3°C are practical because of the long exposure times required for complete disinfestation at higher temperatures. For example, 20, 12, or 18 days at 1.1°C are the required quarantine treatment times for Caribbean, Mediterranean, or Queensland fruit flies, respectively (Animal and Plant Health Inspection Service, 1992c). This severely limits the use of cold treatments as most tropical fruits are damaged by extended storage at temperatures <10°C (Couey, 1982). Therefore, temperate fruits, such as apples (Newcomer, 1930, 1936; Moffitt and Burditt, 1989a,b), pears, and stone fruits (Yokoyama and Miller, 1989), and those tropical and subtropical fruits that can tolerate temperatures <2°C, such as carambola (Gould and Sharp, 1990a), citrus (Adsule *et al.*, 1987; Hill *et al.*, 1988; Ismail, 1989), kiwifruit (Jessup and Baheer, 1990), litchi (J.W. Armstrong, unpublished data), and mangosteen (Burikam *et al.*, 1992) are candidates for cold treatment. However, most cold quarantine treatment research and accepted quarantine treatments (Animal and Plant Health Inspection Service, 1992c) include only Caribbean, Mediterranean, and Queensland fruit flies. Future research may show that other fruit fly and insect species are more susceptible to treatment with temperatures between 3 and 10°C.

Cold treatments are most practical when used during transit from producing areas to distant markets. For regulatory purposes, however, in-transit cold treatment facilities must be certified for reliability in providing the required treatment. Acceptable temperature monitoring and other quarantine security requirements during treatment are needed. These regulatory complexities limit quarantine cold treatments of fruit to be done prior to shipment and not in transit.

Heat Treatments

Heat has been used for over a century to disinfest host commodities of plant pathogens and hot water (100°C), live steam (116°C), or dry heat (100°C) are still used as regulatory treatments against plant pathogens (Animal and Plant Health Inspection Service, 1992f). The function of disinfestation treatments using heat is to increase the temperature of the host commodity beyond the target pest thermal tolerance threshold without adversely affecting commodity quality. In insects, thermal tolerance and heat-induced mortality are dependent on time and temperature exposures that cause a complex interaction of physiological processes and biochemical changes that are largely unknown (Jang, 1986). However, the production of heat shock proteins in heat-stressed Mediterranean fruit fly cell lines suggests that some mechanisms of heat-induced mortality may be similar to cold-induced mortality (Jang, 1992), possibly cell membrane failure.

All available evidence indicates that heat treatments, like cold treatments, function independently of the host fruit. However, rates of heating or cooling depend upon the heat transfer characteristics of the fruit (Hayes, 1984), size and shape of the fruit, type of container holding the fruit, and characteristics of the heat transfer medium. The time required to attain a uniform temperature through the fruit is proportional to the size of the fruit. In recent years, mathematical models have been developed to predict the survival of insects in fruit (Hayes *et al.*, 1984; Couey and Hayes, 1986). The temperature and exposure time effects on the individual life stages of fruit fly eggs and larvae (Jang, 1986, 1991; Hallman, 1990a; Hallman and Sharp, 1990b; Heard *et al.*, 1991; Moss and Jang, 1991; Sharp and Chew, 1987), codling moth (Yokoyama *et al.*, 1991) and oriental fruit moth (Yokoyama and Miller, 1987). These models have been used to predict the temperature and exposure time effects on fruit quality and insect disinfestation (Hayes *et al.*, 1987; Couey, 1989; Hayes and Young, 1989; Hansen, 1992). These predictive models permit more rational and rapid development of new treatments, especially when insect mortality and fruit quality models are combined to predict insect disinfestation with little or no deleterious effects on fruit quality. The extensive thermal mortality data developed by Jang (1986, 1991) suggest that the thermal responses of the fruit fly eggs and larvae follow the same kinetic formulations used by microbiologists, and that they are almost identical to the first-order kinetics used to describe monomolecular chemical reactions. In other words, thermal death of fruit fly eggs and larvae follows a logarithmic function so that the final number of survivors of any time–temperature treatment is dependent on the initial number of insects. The benefit of describing the thermal death kinetics in this fashion is that the relationship between treatment temperatures and rates of thermal death can be quantitatively assessed through Arrhenius plots (Jang, 1986, 1991). The rates of thermal death can be extrapolated between experimentally tested temperatures to provide essential information towards integrating the thermal diffusivity characteristics of the host fruit with the insect thermal death kinetics.

Four methods are currently available to heat fruit to temperatures beyond the target quarantine pest thermal tolerance limit. These methods are high-temperature forced-air (forced hot-air), hot-water immersion, microwave (dielectric) heating, and vapour heat. Each method will be discussed in the order of development.

Vapour heat

Vapour heat consists of heating fruit by forced hot air saturated with water vapour. Heat is transferred from the air to the fruit by condensation of the water vapour (heat of condensation) on the relatively cooler fruit surfaces. Fruit may be gradually heated over time (approach time) to a desired temperature that may be either the treatment end temperature (all target pests have been killed), or a holding temperature maintained for a specific time (holding time) required to kill all target pests. Gradual heating of fruit is generally more desirable than rapid heating to prevent damage to the fruit (Armstrong and Couey, 1989).

Vapour heat was first used in the United States in 1929 as a citrus quarantine disinfestation treatment against Mediterranean fruit fly and, in 1932, against Mexican fruit fly, *A. ludens* Loew (Hawkins, 1932; Baker *et al.*, 1944; Baker 1952). Further research on vapour heat disinfestation treatments was done with: citrus, stone fruits and pome fruits in Australia (Weddell, 1931); citrus, plums, and mangoes in Formosa (now Taiwan) (Koidsumi, 1937); mangoes in Mexico (Balock and Star, 1946; Stone *et al.*, 1951; Baker, 1952); mangoes and guavas in Puerto Rico (Sein, 1935); apples, avocados, bell peppers, chinese peas, citrus, cucumber, dates, eggplant, figs, grapes, green beans, lima beans, litchis, mangoes, melons, nectarines, papayas, peaches, pears, persimmons, plums, pomegranates, tomatoes, and yellow wax beans in the United States (Jones, 1940; Balock and Starr 1946; Balock and Kozuma, 1954; Claypool and Vines, 1956; Mackie, 1931; Seo *et al.*, 1974; Sinclair and Lindgren, 1955). Vapour heat treatment of papayas exported from Hawaii was approved by Japan in 1969. The treatment frequently damaged the fruit because of equipment technical problems and was discontinued in 1971 (Sugimoto and Sunagawa, 1987). Although vapour heat was initially a successful quarantine treatment technology, interest waned with the advent of ethylene dibromide and methyl bromide fumigation. Since ethylene dibromide fumigation was discontinued in 1984 (Ruckelshaus, 1984), Japan has become the leading developer of modern, computer-controlled commercial vapour heat treatment facilities that operate today in Okinawa, Philippines, Thailand, and the United States (Sugimoto *et al.*, 1983, Furasawa *et al.*, 1984, Merino *et al.*, 1985).

The older treatment of >10 and up to 16 h, depending on approach and holding time (Baker, 1952; Hawkins, 1932), was shortened (Balock and Kozuma, 1954) with a 'quick run-up' vapour heat treatment for papaya. This treatment consisted of a short preheating period to a specified temperature, then a gradual warming to 47.2°C, the end fruit temperature required for quarantine disinfestation of fruit fly eggs and larvae. The complete treatment required only 6 h. Rapid hydrocooling after treatment is required to reduce fruit heat-induced injury. This quick run-up vapour heat treatment was further modified and used for fruit fly disinfestation treatments of bitter melons, *Momordica charantia* L. (Sunagawa *et al.*, 1988), carambolas (Hallman, 1990b), eggplant (Furasawa *et al.*, 1984), green peppers (Sugimoto *et al.*, 1983), and mangoes (Heard *et al.*, 1992; Merino *et al.*, 1985; Sunagawa *et al.*, 1987). Modifications of the vapour heat treatment, primarily to decrease treatment time while maintaining fruit quality, have been facilitated by modern technology, such as computer-controlled heating regimens, and engineering for high rates of air movement that provide uniform heating throughout the fruit load.

Hot-water immersion

Water is inherently more effective than humid air as a heat transfer medium and, when properly circulated through the load of fruit, quickly establishes a uniform temperature profile. Hot-water immersion has been used for over a century as a treatment against plant pathogens. The use of hot water alone to disinfest quarantine pests from fruits began with the development of a treatment to disinfest bananas of Mediterranean fruit fly, melon fly (*Bactrocera cucurbitae* Coquillett), and oriental fruit fly (*B. dorsalis* Hendel) (Armstrong, 1982). Treatments were then developed against fruit flies in carambolas (Hallman and Sharp, 1990a), guava (Gould and Sharp, 1990b), litchis (J.W. Armstrong, unpublished data), grapefruit (Sharp, 1985), mangoes (Sharp and Spalding, 1984; Sharp,

1986; Segarra-Carmona *et al.*, 1990; Sharp *et al.*, 1988, 1989a,b,c; Sharp and Picho-Martinez, 1990; Animal and Plant Health Inspection Service, 1992a; Nascimento *et al.*, 1992), and papayas (Hayes *et al.*, 1984; Couey and Hayes, 1986). For the research and development of hot-water quarantine treatments, Sharp (1989) described a hot-water treatment appliance used to test hot-water immersion treatments on mangoes.

Potential heat-induced injury can be caused by the high water temperatures (>42°C) and exposure times necessary to achieve disinfestation of the target life stages. This injury limits hot water treatments to heat-tolerant fruit. Fruit size and shape are important, especially in obtaining fruit temperatures that ensure quarantine security. For example, exposure time modifications are required for different mango cultivars due to variation in size and shape (Animal and Plant Health Inspection Service, 1992a).

Hot-water immersion is the least difficult heat-treatment method to apply. Application requires only a large water tank, a heating system for rapid heat recovery and temperature maintenance after fruit immersion, and an acceptable temperature monitoring system for both water and fruit temperatures. Also, fungicides, insecticides or coatings may be applicable additions to hot-water immersion treatments. The leading use of hot-water immersion today is for fruit fly quarantine treatment of mangoes exported from the Caribbean Basin and Latin America. Over 30 hot-water immersion treatment facilities were certified in that region during 1992 (M.R. Williamson, personal communication).

High-temperature forced-air

The high-temperature forced-air, or forced hot-air, treatment is a modification of the vapour heat treatment and was developed as a quarantine treatment against fruit flies infesting papaya (Armstrong *et al.*, 1989; Hansen *et al.*, 1990). Hansen *et al.* (1990) observed that air saturated with water vapour was unnecessary to provide complete disinfestation of Mediterranean fruit fly, melon fly, and oriental fruit fly eggs and larvae. Furthermore, the forced hot-air treatment provided better fruit quality than either the hot-water immersion or vapour heat treatments (Armstrong *et al.*, 1990). This development led to an approved forced hot-air quarantine treatment for papaya (Animal and Plant Health Inspection Service, 1992b). Similar treatments have been developed for Caribbean fruit fly in carambolas (Sharp and Hallman, 1992), mangoes (Mangan and Ingle, 1992; Sharp, 1992), and citrus (Sharp, 1993; Shellie *et al.*, 1993; Mangan and Ingle, 1994).

Research (Gaffney and Armstrong, 1990; Sharp *et al.*, 1991) and commercial (Williamson and Winkelman, 1989; Winkelman and Williamson, 1990) forced hot-air treatment facilities have been developed. The original forced hot-air treatment method maintained relative humidity at 50 ± 10%. This relative humidity was chosen to prevent water from condensing on the fruit, providing better fruit quality than the vapour heat treatment (Armstrong *et al.*, 1989). Jones (1939) found that unsaturated air is essential to maintain papaya fruit quality during thermal treatments requiring fruit centre temperatures up to 48°C. It was suggested that covering the fruit surface impeded respiration and caused damage. To prevent water condensation on fruit surfaces, recent modifications include the manipulation of dew point (Sharp *et al.*, 1991; Shellie *et al.*, 1993). Maintaining the forced air dew point below the fruit surface temperature avoided moisture condensation on fruit even when the relative humidity is >90% (Gaffney and Armstrong, 1990; Sharp *et al.*, 1991; Mangan and Ingle, 1992, 1993; Shellie *et al.*, 1993).

Microwave heating

The use of microwaves (dielectric heating) as a quarantine treatment depends on technological advances in the field of high energy physics. Microwave heating, such as that used in commercially available cooking appliances, heats non-uniformly and from centre of the fruit outward. Therefore, microwave heating can cause damage to the commodity before killing the target pest. Microwave treatments were found to damage mangoes and papayas infested with Mediterranean fruit fly, melon fly, or oriental fruit fly with little or no effect on fruit fly survival (J.W. Armstrong, unpublished data). However, Seo *et al.* (1970) had success using interrupted or pulsated microwave treatments to disinfest mango weevil, *Cryptorhynchus mangiferae* L., from mangoes. There are indications that new microwave technologies may be available and may have potential as quarantine treatments (T.A. Batchelor, personal communication). Microwave technologies and applications privy to the armaments and defence industries may find their way into peacetime use for such purposes as quarantine treatments.

Combination Treatments

Combination treatments combine heat and cold treatments, heat with fumigation, and cold with fumigation or controlled or modified atmospheres as potential quarantine disinfestation treatments.

The feasibility of a combined heat and cold treatment was demonstrated for disinfesting papayas of Mediterranean fruit fly, melon fly, and oriental fruit fly (Couey *et al.*, 1984). The treatment consisted of a 20-min, 49°C hot-water immersion followed by 10 days' refrigeration at 9°C. Combined treatments allow the use of less severe heat and cold exposures and, although neither treatment used alone is efficacious, the combination of treatments provides complete disinfestation (Couey, 1989).

Jessup (1991) combined a 50°C fungicidal immersion treatment with 1°C cold storage to disinfest Queensland fruit fly from avocados. Probit 9 quarantine security was calculated at 11 days' cold storage at 1°C following a 3-min immersion in a 0.5% benomyl solution.

Combined heat and fumigant, or high-temperature fumigation, disinfestation treatments were developed for papaya against Mediterranean fruit fly, melon fly, and oriental fruit fly (Couey *et al.*, 1985), for stone fruits against Mediterranean fruit fly and oriental fruit fly (Armstrong and Couey, 1984; Armstrong *et al.*, 1988) and for grapefruit against Mexican fruit fly (Williamson *et al.*, 1985). Higher than ambient (>21°C) fumigation temperatures generally allow for decreased fumigant concentrations and/or time because of increased insecticidal activity and penetration into the commodity. Higher fumigation temperatures increase the respiration rate of the target pest organism increasing the rate of fumigant uptake, and result in less time to kill the target insect (Sun, 1946).

Immersion of papayas in hot water containing ethylene dibromide (Burditt *et al.*, 1962) and hot water immersion of mangoes followed by ethylene dibromide fumigation followed by refrigeration (Seo *et al.*, 1972) were tested as potential quarantine treatments against Mediterranean and oriental fruit flies.

Combined fumigation with cold disinfestation treatments, either refrigeration followed by fumigation or fumigation followed by refrigeration, are available for a number of quarantine pests in various commodities (Animal and Plant Health Inspection Service, 1992d,e). Cold combined with controlled atmospheres has been researched as a disinfestation treatment against Caribbean fruit fly infesting grapefruit (Benschoter, 1987), and against oriental fruit moth (Dustan, 1963). The interactive effects of controlled atmospheres, humidity and temperature on the mortality of Indianmeal moth, *Plodia interpunctella* (Hubner), and navel orangeworm, *Amyelois transitella* (Walker), were studied by Soderstrom *et al.* (1986).

Conclusion

Cold and heat treatments should have continued application in the research and development of quarantine disinfestation treatments. The basis for accomplishing disinfestation using cold or heat is conceptually simple and the technological capabilities for precision treatment facilities are available. The methodology is relatively easy to apply, and leaves no toxic residues. The disadvantages are the potential for fruit damage and a higher cost compared to fumigation. The use of shorter cold or heat exposures in combination with other milder treatments, as well as technological advances in new methods of cooling or heating fruit, may provide for shorter treatment times and better quality of treated fruit.

References

Adler, V.E. (1960) Effects of low temperatures on the eggs of the Angoumois grain moth, the Indian meal moth, and the confused flour beetle. *Journal of Economic Entomology* 53, 973–974.

Adsule, P.G., Ismail, M.A. and Fellers, P.J. (1987) Quality of 'Marsh' grapefruit following cold treatment as a method of disinfestation against the Caribbean fruit fly. *Journal of Food Science and Technology* 24, 107–110.

Animal and Plant Health Inspection Service (1992a) T102(a) – Hotwater treatment. *Plant Protection and Quarantine Treatment Manual.* US Department of Agriculture, Hyattsville, Maryland, pp. 5.42–5.43.

Animal and Plant Health Inspection Service (1992b) VI-T103 – High temperature forced air. *Plant Protection and Quarantine Treatment Manual.* US Department of Agriculture, Hyattsville, Maryland, p. 5.47.

Animal and Plant Health Inspection Service (1992c) T107 – Cold treatment. *Plant Protection and Quarantine Treatment Manual.* US Department of Agriculture, Hyattsville, Maryland, pp. 5.55–5.61.

Animal and Plant Health Inspection Service (1992d) T108 – Fumigation plus refrigeration. *Plant Protection and Quarantine Treatment Manual.* US Department of Agriculture, Hyattsville, Maryland, pp. 5.67–5.70.

Animal and Plant Health Inspection Service (1992e) T108 – Refrigeration plus fumigation. *Plant Protection and Quarantine Treatment Manual.* US Department of Agriculture, Hyattsville, Maryland, pp. 5.62–5.65.

Animal and Plant Health Inspection Service (1992f) T503(b), (c), (d), and T504(a), (b) – Schedules for plant diseases. *Plant Protection and Quarantine Treatment Manual.* US Department of Agriculture, Hyattsville, Maryland, p. 9.1.

Armstrong, J.W. (1982) Development of a hot-water immersion quarantine treatment for Hawaiian-grown 'Brazilian' bananas. *Journal of Economic Entomology* 75, 787–790.

Armstrong, J.W. (1992) Fruit fly disinfestation strategies beyond methyl bromide. *New Zealand Journal of Crop and Horticultural Science* 20, 181–193.

Armstrong, J.W. and Couey, H.M. (1984) Methyl bromide fumigation treatments at 30°C for California stonefruits infested with Mediterranean fruit fly (Diptera: Tephritidae). *Journal of Economic Entomology* 77, 1229–1232.

Armstrong, J.W. and Couey, H.M. (1989) Fumigation, heat, and cold. In: Robinson, A.S. and Hooper, G. (eds), *World Crop Pests, Vol. 3B, Fruit Flies, Their Biology, Natural Enemies, and Control.* Elsevier, Rotterdam, pp. 411–424.

Armstrong, J.W., Harvey, J.M., Garcia, D.L., Menezes, T.D. and Brown, S.A. (1988) Methyl bromide fumigation for control of oriental fruit fly (Diptera: Tephritidae) in California stone fruits. *Journal of Economic Entomology* 81, 1120–1123.

Armstrong, J.W., Hansen, J.D., Hu, B.K.S. and Brown, S.A. (1989) High-temperature forced-air quarantine treatment for papayas infested with tephritid fruit flies (Diptera: Tephritidae). *Journal of Economic Entomology* 82, 1667–1674.

Armstrong, J.W., Hansen, J.D., Hu, B.K.S. and Brown, S.A. (1990) Hot air disinfestion of fruit and vegetables, US Statutory Invention Registration No. H828 (Serial No. 07/270,608). US Patent Office, Washington, DC.

Back, E.A. and Pemberton C.E. (1916) Effect of cold storage upon the Mediterranean fruit fly. *Journal of Agricultural Research* 5, 657–666.

Baker, A.C. (1952) The vapor heat process. In: *US Department of Agriculture Yearbook*. US Government Printing Office, Washington, DC, pp. 401–404.

Baker, A.C., Stone, W.E., Plummer, C.C. and McPhail, M. (1944) A review of studies on the Mexican fruit fly and related Mexican species. USDA Misc. Pub. No. 531.

Balock, J.W. and Kozuma, T. (1954) Sterilization of papaya by means of vapor-heat quick run-up. Special Report No. 7, Fruit Fly Investigations in Hawaii. US Department of Agriculture, Entomology Research Branch, Honolulu, Hawaii.

Balock, J.W. and Starr, D. F. (1946) Mortality of the Mexican fruit fly in mangoes treated by the vapor-heat process. *Journal of Economic Entomolology* 38, 646–651.

Benschoter, C.A. (1984) Low-temperature storage as a quarantine treatment for the Caribbean fruit fly (Diptera: Tephritidae) in Florida citrus. *Journal of Economic Entomology* 77, 1233–1235.

Benschoter, C.A. (1987) Effects of modified atmospheres and refrigeration temperatures on survival of eggs and larvae of the Caribbean fruit fly (Diptera: Tephritidae) in laboratory diet. *Journal of Economic Entomology* 80, 1223–1225.

Burditt, A.K., Jr, Balock, J.W., Hinman, F.G. and Seo, S.T. (1962) Ethylene dibromide water dips for destroying fruit fly infestations of quarantine significance in papayas. *Journal of Economic Entomology* 56, 289–292.

Burikam, I., Sarnthoy, O., Charernsom, K., Kanno, T. and Homma, H. (1992) Cold temperature treatment for mangosteens infested with the oriental fruit fly (Diptera: Tephritidae). *Journal of Economic Entomology* 85, 2298–2301.

Claypool, L.L. and Vines, H.M. (1956) Commodity tolerance studies of deciduous fruits to moist heat and fumigants. *Hilgardia* 24, 297–354.

Couey, H.M. (1982) Chilling injury of crops of tropical and subtropical origin. *HortScience* 17, 162–165.

Couey, H.M. (1989) Heat treatment for control of postharvest diseases and insect pests of fruits. *HortScience* 24, 198–202.

Couey, H.M. and Hayes, C.F. (1986) Quarantine procedure for Hawaiian papaya using fruit selection and a two-stage hot-water immersion. *Journal of Economic Entomology* 79, 1307–1314.

Couey, H.M., Linse, E.S. and Nakamura, A.N. (1984) Quarantine procedure for Hawaiian papayas using heat and cold treatments. *Journal of Economic Entomology* 77, 984–988.

Couey, H.M., Armstrong, J.M., Hylin, J.W., Thornburg, W., Nakamura, A.N., Linse, E.S., Ogata, J. and Vetro, R. (1985) Quarantine procedure for Hawaii papaya using a hot-water treatment and high-temperature, low-dose ethylene dibromide fumigation. *Journal of Economic Entomology* 78, 879–884.

Dustan, G.G. (1963) The effect of standard cold storage and controlled atmosphere storage on survival of larvae of the oriental fruit moth, *Grapholitha molesta*. *Journal of Economic Entomology* 56, 167–169.

Furasawa, K., Sugimoto, T. and Gaja, T. (1984) The effectiveness of vapor heat treatment against the melon fly, *Dacus cucurbitae* Coquillett, in eggplant and fruit tolerance to the treatment. *Research Bulletin of the Plant Protection Service Japan* 20, 17–24.

Gaffney, J.J. and Armstrong, J.W. (1990) High-temperature forced-air research facility for heating fruits for insect quarantine treatments. *Journal of Economic Entomology* 83, 1959–1964.

Gould, W.P. and Sharp, J.L. (1990a) Cold-storage quarantine treatment for carambolas infested with the Caribbean fruit fly (Diptera: Tephritidae). *Journal of Economic Entomology* 83, 458–460.

Gould, W.P. and Sharp, J.L. (1990b) Hot-water immersion treatment for guavas infested with Caribbean fruit fly (Diptera: Tephritidae). *Journal of Economic Entomology* 85, 1235–1239.

Hallman, G.J. (1990a) Survival and reproduction of Caribbean fruit fly (Diptera: Tephritidae) adults immersed in hot water as third instars. *Journal of Economic Entomology* 83, 2331–2334.

Hallman, G.J. (1990b) Vapor-heat treatment of carambolas infested with Caribbean fruit fly (Diptera: Tephritidae). *Journal of Economic Entomology* 83, 2340–2342.

Hallman, G.J. and Sharp, J.L. (1990a) Hot-water immersion quarantine treatment for carambolas infested with Caribbean fruit fly (Diptera: Tephritidae). *Journal of Economic Entomology* 83, 1471–1474.

Hallman, G.J. and Sharp, J.L. (1990b) Mortality of Caribbean fruit fly (Diptera: Tephritidae) larvae infesting mangoes subjected to hot-water treatment, then immersion cooling. *Journal of Economic Entomology* 83, 2320–2323.

Hansen, J.D. (1992) Heating curve models of quarantine treatments against insect pests. *Journal of Economic Entomology* 85, 1846–1854.

Hansen, J.D., Armstrong, J.W., Hu, B.K.S. and Brown, S.A. (1990) Thermal death of oriental fruit fly (Diptera: Tephritidae) third instars in developing quarantine treatments for papayas. *Journal of Economic Entomology* 83, 160–167.

Hawkins, L.A. (1932) Sterilization of citrus fruit by heat. *Citriculture* 9, 7–8, 21–22.

Hayes, C.F. (1984) Thermal diffusivity of papaya fruit (*Carica papaya* L. var. Solo). *Journal of Food Science* 49, 1219.

Hayes, C.F. and Young, H. (1989) Extension of model to predict survival from heat treatment of papaya infested with oriental fruit flies (Diptera: Tephritidae). *Journal of Economic Entomology* 82, 1157–1160.

Hayes, C.F., Chingon, H.T.G., Nitta, F.A. and Wang, W.J. (1984) Temperature control as an alternative to ethylene dibromide fumigation for the control of fruit flies (Diptera: Tephritidae) in papaya. *Journal of Economic Entomology* 77, 683–686.

Hayes, C.F., Chingon, H.T.G., Nitta, F.A. and Leung, A.M.T. (1987) Calculation of survival from double hot-water immersion treatment for papayas infested with oriental fruit flies (Diptera: Tephritidae). *Journal of Economic Entomology* 80, 887–890.

Heard, T.A., Heather, N.W. and Cocoran, R.J. (1991) Dose–mortality relationships for eggs and larvae of *Bactrocera tryoni* (Diptera: Tephritidae) immersed in hot water. *Journal of Economic Entomology* 84, 1768–1770.

Heard, T.A., Heather, N.W. and Peterson, P.M. (1992) Relative tolerance to vapor heat treatment of eggs and larvae of *Bactrocera tryoni* (Diptera: Tephritidae) in mangoes. *Journal of Economic Entomology* 85, 461–463.

Hill, A.R., Rigney, C.J. and Sproul, A.N. (1988) Cold storage of oranges as a disinfestation treatment against fruit flies *Dacus tryoni* (Froggatt) and *Ceratitis capitata* (Wiedemann) (Diptera: Tephritidae). *Journal of Economic Entomology* 81, 253–257.

Hooper, T. (1907) Cold storage and fruit fly. *Journal of Department of Agriculture Western Australia* 15, 252–253.

Ismail, M. (1989) Cold treatment: a successful quarantine method for Florida grapefruit shipped to Japan. *Packinghouse Newsletter* 158, 1–2.

Jang, E.B. (1986) Kinetics of thermal death of eggs and first instar larvae of three species of fruit flies (Diptera: Tephritidae). *Journal of Economic Entomology* 79, 700–705.

Jang, E.B. (1991) Thermal death kinetics and heat tolerance in early and late third instars of the oriental fruit fly (Diptera: Tephritidae). *Journal of Economic Entomology* 84, 1298–1303.

Jang, E.B. (1992) Heat shock proteins and thermotolerance in a cultured cell line from the Mediterranean fruit fly, *Ceratitis capitata*. *Archives of Insect Biochemistry and Physiology* 19, 93–103.

Jessup, A.J. (1991) High-temperature dip and low temperatures for storage and disinfestation of avocados. *HortScience* 26, 1420.

Jessup, A.J. and Baheer, A. (1990) Low-temperature storage as a quarantine treatment for kiwifruit infested with *Dacus tryoni* (Diptera: Tephritidae). *Journal of Economic Entomology* 83, 2317–2319.

Jones, W.W. (1939) The influence of relative humidity on the respiration of papaya at high temperatures. *Proceedings of the American Society for Horticultural Science* 37, 700–705.

Jones, W.W. (1940) Vapor heat treatment for fruits and vegetables grown in Hawaii. Hawaii Agricultural Experiment Station Circ. No. 16, Honolulu.

Koidsumi, K. (1937) Heat sterilization of Formosan fruits for fruit flies (III). *Japan. Society Tropical Agriculture* 9, 275–286.

Lee, R.E. (1991) Principles of insect low temperature tolerance. In: Lee, R.E. and Denlinger, D.L. (eds), *Insects at Low Temperature*. Chapman and Hall, New York, pp. 17–46.

Lounsbury, C.P. (1907) The fruit fly (*Ceratitis capitata*). *Agricultural Journal of the Cape of Good Hope* 31, 186–187.

Mackie, D.B. (1931). Heat treatments of California fruits from the standpoint of compatibility of the Florida process. *Monographs Bulletin Department Agriculture California* 20, 211–218.

Mangan, R.L. and Ingle, S.J. (1992) Forced hot-air quarantine treatment for mangoes infested with West Indian fruit fly (Diptera: Tephritidae). *Journal of Economic Entomology* 85, 1859–1864.

Mangan, R.L. and Ingle, S.J. (1994) Forced hot-air quarantine treatment for grapefruit infested with Mexican fruit fly (Diptera: Tephritidae). *Journal of Economic Entomology* (in press).

Mason, A.C. and McBride, O.C. (1934) The effect of low temperatures on Mediterranean fruit fly in infested fruit. *Journal of Economic Entomology* 27, 897–902.

Merino, S.R., Eugenio, M.M., Ramos, A.U. and Hernandes, S.T. (1985) Fruitfly disinfestation of mangoes (*Mangifera indica* L. var. 'Manila Super') by vapor heat treatment. Ministry of Agriculture and Food, Bureau of Plant Industry, Manila.

Moffitt, H.R. and Albano, D.J. (1972) Effects of commercial fruit storage on stages of the codling moth (Lepidoptera: Tortricidae). *Journal of Economic Entomology* 65, 770–773.

Moffitt, H.R. and Burditt, A.K. (1989a) Effects of low temperatures on three embryonic stages of the codling moth (Lepidoptera: Tortricidae). *Journal of Economic Entomology* 82, 1379–1381.

Moffitt, H.R. and Burditt, A.K. (1989b) Low-temperature storage as a postharvest treatment for codling moth (Lepidoptera: Tortricidae) eggs on apple. *Journal of Economic Entomology* 82, 1679–1681.

Moss, J.I. and Jang, E.B. (1991) Effects of age and metabolic stress on heat tolerance of Mediterranean fruit fly (Diptera: Tephritidae) eggs. *Journal of Economic Entomology* 84, 537–541.

Mullen, M.A. and Arbogast, R.T. (1979) Time–temperature–mortality relationships for various stored-product insect eggs and chilling times for selected commodities. *Journal of Economic Entomology* 72, 476–478.

Nascimento, A.S., Malavasi, A., Morgante, J.S. and Duarte, A.L.A. (1992) Hot-water immersion treatment for mangoes infested with *Anastrepha fraterculus*, *A. obliqua*, and *Ceratitis capitata* (Diptera: Tephritidae) in Brazil. *Journal of Economic Entomology* 85, 456–460.

Nel, R.G. (1936) The utilization of low temperatures in the sterilization of deciduous fruit infested with the immature stages of the Mediterranean fruit fly, *Ceratitis capitata* (Wiedemann). *Scientific Bulletin of the Department of Agriculture South Africa* No. 155.

Newcomer, E.J. (1930) Experiments in killing eggs of the codling moth on harvested fruit. *Journal of Economic Entomology* 23, 798–802.

Newcomer, E.J. (1936) Effect of cold storage on eggs and young larvae of the codling moth. *Journal of Economic Entomology* 29, 1123–1125.

Rippon, L.E. and Smith, R.J. (1979) Postharvest treatment of Chinese gooseberries (kiwifruit) for the control of Queensland fruit fly. *Rural News* 73, 34–35.

Ruckelshaus, W.D. 1984. Ethylene dibromide, amendment of notice of intent to cancel registration of pesticide products containing ethylene dibromide. *Federal Register* 49, 14182–14185. US Government Printing Office, Washington, DC.

Segarra-Carmona, A.E., Franqui, R.A., Ramirez-Ramos, L.V., Santiago, L.R. and Torres-Rivera, C.N. (1990) Hot water dip treatments to destroy *Anastrepha obliqua* larvae in mangoes from Puerto Rico. *Journal of Agriculture, University of Puerto Rico* 74, 441–447.

Sein, F., Jr (1935) Heat sterilization of mangoes and guavas for fruit flies. *Journal of Agriculture, University of Puerto Rico* 19, 105–115.

Seo, S.T., Chambers, D.L., Komura, M. and Lee, C.V.L. (1970) Mortality of mango weevils (Coleoptera: Curculionidae) in mangoes treated by dielectric heating. *Journal of Economic Entomology* 63, 1977–1978.

Seo, S.T., Chambers, D.L., Akamine, E.K., Komura, M. and Lee, C.Y.L. (1972) Hot water–ethylene dibromide fumigation–refrigeration treatment for mangoes infested by oriental and Mediterranean fruit flies. *Journal of Economic Entomology* 65, 1372–1374.

Seo, S.T., Hu, B.K.S., Komura, M., Lee, C.V.L. and Harris, E.J. (1974) *Dacus dorsalis*: vapor heat treatment in papayas. *Journal of Economic Entomology* 67, 240–242.

Sharp, J.L. (1985) Submersion of Florida grapefruit in heated water to kill stages of Caribbean fruit fly, *Anastrepha suspensa*. *Proceedings of Florida State Horticultural Society* 98, 78–80.

Sharp, J.L. (1986) Hot-water treatment for control of *Anastrepha suspensa* (Diptera: Tephritidae) in mangos. *Journal of Economic Entomology* 79, 706–708.

Sharp, J.L. (1989) Hot-water immersion appliance for quarantine treatment research. *Journal of Economic Entomology* 82, 189–192.

Sharp, J.L. (1992) Hot-air quarantine treatment for mango infested with Caribbean fruit fly (Diptera: Tephritidae). *Journal of Economic Entomology* 85, 2302–2304.

Sharp, J.L. (1993) Hot-air treatment for 'Marsh' White grapefruit infested with Caribbean fruit fly (Diptera: Tephritidae). *Journal of Economic Entomology* 86, 462–464.

Sharp, J.L. and Chew, V. (1987) Time/mortality relationships for *Anastrepha suspensa* (Diptera: Tephritidae) eggs and larvae submerged in hot water. *Journal of Economic Entomology* 80, 646–649.

Sharp, J.L. and Hallman, G.J. (1992) Hot-air quarantine treatment for carambolas infested with Caribbean fruit fly (Diptera: Tephritidae). *Journal of Economic Entomology* 85, 168–171.

Sharp, J.L. and Picho-Martinez, H. (1990) Hot-water quarantine treatment to control fruit flies (Diptera: Tephritidae) in mangoes imported into the United States from Peru. *Journal of Economic Entomology* 83, 1940–1943.

Sharp, J.L. and Spalding, D.H. (1984) Hot water as a quarantine treatment for Florida mangoes infested with Caribbean fruit fly. *Proceedings of Florida State Horticultural Society* 97, 355–357.

Sharp, J.L., Ouye, M.T., Thalman, R., Hart, W., Ingle, S. and Chew, V. (1988) Submersion of 'Francis' mango in hot water as a quarantine treatment for the West Indian fruit fly and the Caribbean fruit fly (Diptera: Tephritidae). *Journal of Economic Entomology* 81, 1431–1436.

Sharp, J.L., Ouye, M.T., Hart, W., Ingle, S., Hallman, G., Gould, W. and Chew, V. (1989a) Immersion of Florida mangoes in hot water as a quarantine treatment for Caribbean fruit fly (Diptera: Tephritidae). *Journal of Economic Entomology* 82, 186–188.

Sharp, J.L., Ouye, M.T., Ingle, S.J. and Hart, W.G. (1989b) Hot-water quarantine treatment for mangoes from Mexico infested with Mexican fruit fly and West Indian fruit fly (Diptera: Tephritidae). *Journal of Economic Entomology* 82, 1657–1662.

Sharp, J.L., Ouye, M.T., Ingle, S.J., Hart, W.G., Enkerlin H., W.R., Celedonio H., H., Toledo A., J., Stevens, L., Quintero, E., Reyes F., J. and Schwarz, A. (1989c) Hot water quarantine treatment for mangoes from the State of Chiapas, Mexico infested with *Ceratitis capitata* (Wiedemann) and *Anastrepha serpentina* (Wiedemann) (Diptera: Tephritidae). *Journal of Economic Entomology* 82, 1663–1666.

Sharp, J. L., Gaffney, J.J., Moss, J.I. and Gould, W.P. (1991) Hot-air treatment device for quarantine research. *Journal of Economic Entomology* 84, 520–527.

Shellie, K.C., Firko, M.J. and Mangan, R.L. (1993) Phytotoxic response of 'Dancy' tangerine to high-temperature moist forced-air treatment for fruit fly disinfestation. *Journal of the American Society for Horticultural Science* 118, 481–485.

Sinclair, W.B. and Lindgren, D.L. (1955) Vapor heat sterilization of California citrus and avocado fruits against fruit fly insects. *Journal of Economic Entomology* 48, 133–138.

Soderstrom, E.L., Mackey, B.E. and Brandl, D.G. (1986) Interactive effects of low-oxygen atmospheres, relative humidity, and temperature on mortality of two stored-product moths (Lepidoptera: Pyralidae). *Journal of Economic Entomology* 79, 1303–1306.

Sproul, A.N. (1976) Disinfestation of Western Australian 'Granny Smith' apples by cold treatment against the egg and larval stages of the Mediterranean fruit fly (*Ceratitis capitata* (Wied.)). *Australian Journal of Experimental Agriculture and Animal Husbandry* 16, 280–285.

Stone, W.E., Servin, F. and Heins, I. (1951) A new concept in the vapor-heat sterilization of fruit for fruit flies. USDA, Agricultural Research Administration, Bureau of Entomology and Plant Quarantine, Fruit Fly and Citrus Blackfly Investigations in Mexico. USDA, PQCA-363-MS.

Sugimoto, T. and Sunagawa, K. (1987) Killing of fruit flies by vapor heat treatment of fruit. *Research Bulletin of Plant Protection Japan* 41, 124–128.

Sugimoto, T., Furasawa, K. and Mizobuchi, M. (1983) The effectiveness of vapor heat treatment against the oriental fruit fly, *Dacus dorsalis* Hendel, in green pepper and fruit tolerance to the treatment. *Research Bulletin of Plant Protection Japan* 19, 81–88.

Sun, Y.P. (1946) An analysis of some important factors affecting the results of fumigation on insects. St Paul, Minnesota Agriculture Experiment Station Technical Bulletin No. 177.

Sunagawa, K., Kume, K. and Iwaizumi, R. (1987) The effectiveness of vapor heat against the melon fly, *Dacus cucurbitae* Coquillett, in mango and fruit tolerance to the treatment. *Research Bulletin of Plant Protection Japan* 23, 13–20.

Sunagawa, K., Kume, K. and Iwaizumi, R. (1988) Sterilization of melon fly and thermal injuries to bitter momordica fruit. *Research Bulletin of Plant Protection Japan* 24, 1–5.

Weddell, J.A. (1931) Experiments with the heat treatment of fly-infested fruit. *Queensland Agricultural Journal* 36, 141–147.

Williamson, D.L., Summy, K.R., Hart, W.G., Sanchez-R., M., Wolfenbarger, D.A. and Burton, B.D. (1985) Efficacy and phytotoxicity of methyl bromide as a fumigant for the Mexican fruit fly (Diptera: Tephritidae) in grapefruit. *Journal of Economic Entomology* 79, 172–175.

Williamson, M.R. and Winkelman, P.M. (1989) Commercial scale heat treatment for disinfestation of papaya. 1989 International Winter Meeting, American Society of Agricultural Engineers at Quebec, Canada. ASAE, St Joseph, Missouri.

Winkelman, P.M. and Williamson, M.R. (1990) Advances in commercial dry heat disinfestation of papaya. 1990 International Summer Meeting, American Society of Agricultural Engineers at Columbus, Ohio. ASAE, St Joseph, Missouri.

Yokoyama, V.Y. and Miller, G.T. (1987) High temperature for control of oriental fruit moth (Lepidoptera: Tortricidae) in stone fruits. *Journal of Economic Entomology* 80, 641–645.

Yokoyama, V.Y. and Miller, G.T. (1989) Response of codling moth and oriental fruit moth (Lepidoptera: Tortricidae) immatures to low-temperature storage of stone fruits. *Journal of Economic Entomology* 82, 1152–1156.

Yokoyama, V.Y., Miller, G.T. and Dowell, R.V. (1991) Response of codling moth (Lepidoptera: Tortricidae) to high temperature, a potential quarantine treatment for exported commodities. *Journal of Economic Entomology* 84, 528–531.

Controlled Atmospheres 7

GUY J. HALLMAN
United States Department of Agriculture, Agricultural Research Service, Subtropical Horticulture Research Station, 13601 Old Cutler Road, Miami, FL 33158, USA.

Modified atmosphere treatments for pests of fresh horticultural products consist of altering the normal atmospheric gas composition to one that will kill insects. Modified atmosphere is a general term which includes all changes in the normal atmospheric composition (Calderon, 1990). Controlled atmosphere is a modified atmosphere that is maintained with little variation in gas composition by monitoring gas concentrations and adding or removing gases as needed. Other methods of modifying atmospheres include film wraps and fruit coatings. Film wraps are clear plastic sheets wrapped tightly around fruit (Houck *et al.*, 1989). Some types of film wraps include stretch wrap, which is stretched around the fruit, and shrink wrap, which is subjected to heating (150–175°C) that shrinks it to conform to the fruit surface. Coatings, derived from waxes, oils, polysaccharides, and other substances, are applied as a liquid and dried on the fruit (Nisperos-Carriedo, 1994).

Although a moderate amount of research on the use of controlled atmospheres for insect control in fresh produce has been done, treatments have rarely been applied commercially. Recent reviews discuss the use of controlled atmospheres for disinfesting fresh fruits and vegetables (Klag, 1986; Ke and Kader, 1992b; Carpenter and Potter, 1994).

Modified atmospheres have been studied for preserving the quality of fresh agricultural commodities for almost 60 years resulting in their use in long-term storage of apples and several other fruits and vegetables (Kader, 1992). Temperatures of 0–15°C and relative humidities above 85% are combined with controlled atmosphere storage of fresh commodities. Modified atmospheres for produce storage generally consist of 0–5% oxygen and up to 10% carbon dioxide.

From: Paull, R.E. and Armstrong, J.W. (eds) (1994) *Insect Pests and Fresh Horticultural Products: Treatments and Responses.* CAB INTERNATIONAL, Wallingford, UK.

Considerable research has been done on the use of controlled atmospheres to control stored product pests which has shown the technique to be feasible in many instances (Ripp, 1984; Fleurat-Lessard, 1990; Soderstrom and Brandl, 1990). Carbon dioxide concentrations required to kill insects are higher than those used for fruit storage. Klag (1986) suggested the need for increased research on controlled atmosphere quarantine treatments of insects infesting fresh fruits and vegetables. Carpenter and Potter (1994) mention only one instance (one successful shipment of asparagus from New Zealand to Japan) where a controlled atmosphere treatment was used for assuring quarantine security of a fresh commodity.

Components of Modified Atmospheres

Any significant deviation from ambient air composition (78% nitrogen, 21% oxygen, 0.033% carbon dioxide) is a modified atmosphere. To kill insects, the oxygen concentration must be reduced and/or the carbon dioxide concentration increased. Argon and atmospheric nitrogen are considered to be biologically inert and have been ignored in controlled atmosphere research, except as diluents. However, AliNiazee (1972) found that 100% mortality of adult confused flour beetle, *Tribolium confusum* Jacquelin de Val, and red flour beetle, *T. castaneum* (Herbst), occurred in half the time when the insects were exposed to 100% helium rather than 100% nitrogen at 21.1 or 26.7°C and 38±6% relative humidity (RH). The differences in lethal time were less for adults tested at 15.6°C and for pupae tested at 15.6 or 26.7°C and 38±6%. No significant differences in mortality occurred for confused or red flour beetle adults, eggs, and larvae exposed to 100% nitrogen or helium at 26.7°C and 100% RH, 26.7°C and 38±6% RH, and 15.6 or 26.7°C and 38±6% RH, respectively (AliNiazee, 1972). Carbon monoxide (1–80%) has been investigated as a component of controlled atmosphere treatments for pests of lettuce stored at 2.5°C for up to 7 days with no mortality observed (Klaustermeyer *et al.*, 1973).

Factors Influencing Efficacy of Modified Atmospheres

Fleurat-Lessard (1990) discussed factors that affected the susceptibility of stored product insects to modified atmospheres. Pests of fresh commodities, even pests found on the fruit surface, are not subjected to RH as low as those to which stored product pests are subjected. Low RH increases the susceptibility of pests to modified atmospheres (Carpenter and Potter, 1994). The adult and pupal stages of many pests of fresh commodities, such as fruit flies, are not present in the marketed commodity and need not be considered when developing treatments; generally, all stages of stored product pests are present. Stored products can tolerate greater extremes of temperature and atmospheric composition than fresh commodities. Stored product mites were more difficult to kill with controlled atmospheres than other stored product pests. Mite pests of fresh commodities also appeared harder to kill using controlled atmospheres (Lidster *et al.*, 1981, 1984). However, CO_2 levels and temperatures tested were low; higher levels might provide more rapid mortality.

Host acceptability probably affects the susceptibility of some pests of fresh commodities to controlled atmospheres. Many hosts of quarantine pests of fresh commodities are poor hosts. Consequently, pests developing on poor hosts may be more susceptible to a quarantine treatment than when developing on favourable hosts. Generally, as the O_2 concentration decreases and the CO_2 concentration increases beyond certain thresholds, pest mortality increases. However, lower mortality (64–66%) occurred to exposed blueberry maggots, *Rhagoletis mendax* Curran, in 100% CO_2 for 48 h compared with 45–90% CO_2 plus 2 or 5% O_2 (77–90% mortality) (Prange and Lidster, 1992). This same phenomenon has been observed with adult granary and rice weevils, *Sitophilus granarius* (L.) and *S. oryzae* (L.), respectively (Lindgren and Vincent, 1970). Mortality of navel orangeworm, *Amyelois transitella* Walker, decreased with increasing CO_2 content from 1 to 10% when the atmosphere contained 0.3% O_2, while at 20% CO_2, mortality did not decrease (Brandl *et al.*, 1983). Also, atmospheres containing 20% CO_2 were more lethal at 5 and 15°C to 2–24-h-old codling moth eggs than atmospheres with 40% CO_2, while the opposite occurred at 0 and 25°C (Soderstrom *et al.*, 1991).

The combined effect of low O_2 with high CO_2 on insect mortality is often synergistic; the two gases apparently have different modes of action (Fleurat-Lessard, 1990). However, Soderstrom *et al.* (1991) found no synergism for mortality of codling moth, *Cydia pomonella* (L.), eggs when combining O_2 levels between 0 and 5% with CO_2 levels between 20 and 100%.

Temperature has been positively correlated to mortality in modified atmospheres (Ke and Kader, 1992b; Carpenter and Potter, 1994). For example, as temperature increased in 2.5°C steps between 0 and 7.5°C, considerable increases occurred in mortality of overwintering European red mite, *Panonychus ulmi* (Koch), eggs exposed to 5.0% CO_2 and 3.0% O_2 or 1.5% CO_2 and 1.0% O_2 (Lidster *et al.*, 1984). However, in one study, 2–24-h-old codling moth eggs in an atmosphere of 20% CO_2 and 0% O_2 were killed with lower exposure times at 15°C than at 25°C (Soderstrom *et al.*, 1991). Holding red flour beetle, *Tribolium castaneum* (Herbst), eggs at 38°C for 24–48 h prior to treatment reduced their susceptibility to low O_2 and high CO_2 atmospheres (Soderstrom *et al.*, 1992).

Soderstrom *et al.* (1986) determined that the navel orangeworm was relatively more susceptible to controlled atmosphere treatment by reductions in RH or O_2 than the Indianmeal moth, *Plodia interpunctella* (Hübner). They hypothesized that this was because the Indianmeal moth, a storage pest, was adapted to the low RH and O_2 found in storage conditions, whereas the navel orangeworm, a field pest, was adapted to higher RH and ambient O_2 levels found in the field. Diapausing stages are usually the hardest to kill with modified atmospheres (Soderstrom *et al.*, 1990; Whiting *et al.*, 1992a).

Physiological Effects of Modified Atmospheres on Insects

Almost all of the research on modified atmospheres as quarantine treatments has been on high levels of CO_2 (>20%) and/or low levels of oxygen (<5%). Other unsuccessful work involved carbon monoxide (Ke and Kader, 1992b), and low levels of ethylene (Benschoter *et al.*, 1981), which is explosive at concentrations >3%.Recent reviews have discussed the physiological and biochemical responses of arthropods to low O_2 and high CO_2 (Nicolas and Sillans, 1989; Fleurat-Lessard, 1990; Carpenter and Potter, 1994). Low O_2 and elevated CO_2 cause different physiological lesions with CO_2 having multiple sites of action. However, resistance to CO_2 has been induced in the granary weevil and the rice weevil, indicating non-multiple modes of action in those two insects (Bond and Buckland, 1979; Navarro *et al.*, 1985). These two stored product pests responded differently to modified atmospheres than other insects studied. Therefore, they should not be used as general examples of insect responses to modified atmospheres, especially for pests of fresh commodities.

Fleurat-Lessard (1990) remarked on the difficulty of separating the effects of high CO_2 from those of low O_2. Friedlander (1983) suggested that, among other effects, high CO_2 prevented the anaplerotic production of nicotinamide adenine dinucleotide and inhibited the regeneration of choline to acetylcholine. In the absence of O_2, insect development virtually stops and survival depends on the accumulation of glycolytic products and reduction in metabolism (Fleurat-Lessard, 1990); O_2 deficiency inactivates the mechanism of selective permeability of the nerve sheaths (Miller, 1966), while O_2 deficiency and excess CO_2 cause spiracles to open resulting in water loss.

Actual death of the insect in low O_2 treatments may occur when metabolic products accumulating in the insect nervous system due to halted or slowed metabolic pathways reach toxic levels (Fleurat-Lessard, 1990). Death caused by high CO_2 may be the result of dehydration due to open spiracles which then cause changes in the ionic balance in nerve ganglia. Severe dehydration could not occur to insects inside fresh commodities and maybe not to surface pests of fresh commodities stored under high humidity, either. However, chemical changes occurring in fruit with modified atmospheres, such as an increase in the products of anaerobic metabolism, including ethanol, ethyl acetate, and acetaldehyde (Nisperos-Carriedo *et al.*, 1990; Wang, 1990), may be toxic and contribute to insect mortality. Acetaldehyde (2.0% for 4 h at 21°C) killed 100% of western flower thrips on strawberries (Aharoni *et al.*, 1979).

Modified Atmosphere Quarantine Treatments

Controlled atmospheres

External pests

Some studies that tested the effect of standard controlled atmosphere storage on pests of fresh commodities determined that the low temperatures and not the atmospheric conditions used were responsible for insect mortality (Glass *et al.*, 1961; Dickler, 1975). To my knowledge, Klaustermeyer *et al.* (1973) were the first to publish a study using controlled atmospheres to attempt control of surface pests of fresh commodities. Various controlled atmosphere combinations did not cause significant mortality to one aphid and two caterpillar species on lettuce at 20°C. However, the lowest level of oxygen used was 5%, and many of the treatments lasted only 1 or 2 h; lower oxygen levels and longer treatment times should be tested. Complete mortality of San Jose scale, *Quadraspidiotus perniciosus* (Comstock), on apples was achieved in 2 days with >90% CO_2 and <1% O_2 at ≥12°C and in one day with 96% CO_2 at 22°C (Morgan and Gaunce, 1975).

Aharoni *et al.* (1981) attained 100% mortality of western flower thrips, *Frankliniella occidentalis* (Pergrande), on strawberries after holding the berries in 88.7–90.6% CO_2, 1.9–2.3% O_2 at 2.5°C for 48 h. This procedure could be a viable method to replace methyl bromide fumigation for shipment of strawberries from California to Japan. However, a difference in flavour of treated strawberries was noted by 80% of the tasters and should be further investigated.

All apple rust mite, *Aculus schlechtendali* (Nalepa), and European red mite eggs on apples were killed by storage in 1.0% O_2 plus 1.0% CO_2 at 2.8°C for 160 days without reducing fruit quality (Lidster *et al.*, 1981). Mortality of European red mite eggs was increased significantly if the atmosphere was 1.5% CO_2 and 1.0% O_2 when the temperature was increased from 0 to 7.5°C (Lidster *et al.*, 1984). Both of these studies, and another showing a minor loss of apple quality (Lidster, 1982), suggest the possibility of providing quarantine security for apples infested with mite eggs.

Mites from five different families, the greenhouse thrips, *Heliothrips haemorrhoidalis* Bouché, and the lightbrown apple moth, *Epiphyas postvittana* Walker, on Japanese persimmons were killed within 7 days' exposure to 0.5% O_2, 5.3% CO_2 (Dentener *et al.*, 1992). However, survival of longtailed mealybug, *Pseudococcus longispinus* Targioni-Tozetti, was 0.2% after 7 days under the same conditions. Carpenter and Potter (1994) discuss a controlled atmosphere treatment that resulted in one successful shipment of asparagus from New Zealand to Japan. The asparagus were held in 60% CO_2 in air at 0–1°C for 4.5 days to kill New Zealand flower thrips, *Thrips obscuratus* (Crawford), and green peach aphid, *Myzus persicae* (Sulzer).

Internal pests

To my knowledge, Benschoter *et al.* (1981) were the first to investigate the mortality of atmospheric gases to internal pests of fresh commodities with the objective of developing a quarantine treatment. Immature stages of the Caribbean fruit fly, *Anastrepha suspensa* (Loew), were exposed to different concentrations of nitrogen and CO_2 at 22–23°C. All 5-day-old larvae in 5 ml of semi-artificial diet died after 60 h exposure to 100% nitrogen. Carbon dioxide levels of 40–100% caused 95–100% larval mortality after 48 h. The authors felt that after a certain threshold further increases in CO_2 concentrations may not result in greater mortality. Caribbean fruit flies on semi-artificial diet under different CO_2 concentrations and temperatures were used with multiple linear regression analysis to calculate expected exposure times necessary to achieve 100% mortality. However, CO_2 concentration and exposure time necessary to achieve 100% mortality might have been underestimated because of lack of an estimate for natural mortality. Oxygen concentration (2, 10, or 20%) did not affect mortality, but it may not have been sufficiently low.

Controlled atmosphere storage conditions for apples (2.5% O_2, 1–1.5% CO_2 at approximately 1°C) did not provide quarantine security of diapausing codling moth larvae within 133 days (Moffitt and Albano, 1972). A lower O_2 (0.5%), higher CO_2 (10%) atmosphere at 25°C was also unsuccessful, although 60% CO_2 might provide quarantine security for walnuts stored at 60% RH (Soderstrom *et al.*, 1990). Toba and Moffitt (1991) achieved 100% mortality of an estimated 142 021 non-diapausing larvae of codling moth in apples treated with 1.5–2.0% O_2 and >1% CO_2 at 0 ± 0.28°C, ≥95% RH for 13 weeks. These conditions could provide a feasible quarantine treatment for apples infested with codling moth where diapausing larvae were not of concern. Apple fruits from commercial orchards in the northwestern US are not considered to harbour diapausing codling moth larvae (H.R. Moffitt, personal communication).

Delate *et al.* (1990) achieved 100% mortality of sweetpotato weevil, *Cylas formicarius elegantulus* (Summers), adults in 7 days using several low O_2 and high CO_2 combinations at 25°C, 75% RH. These treatments can only be applied to cured sweet potatoes, as curing does not occur at O_2 concentrations <8%. Curing (4–7 days, 30°C, 95% RH) is required to allow wound healing prior to storage.

Maximum mortality of blueberry maggots subjected to 2, 5, or 20% O_2, plus 0.033–98% CO_2 at 5 or 21°C for 48 h was 91% (Prange and Lidster, 1992). Lower O_2 levels would probably be needed to achieve mortality required for quarantine security.

Waddell *et al.* (1990) achieved 100% mortality of three species of tortricids that attack apples in New Zealand with 3% O_2 and 3% CO_2 at 0.5°C for 60 days. Four species of tortricids were killed by 0.4% O_2 and 5% CO_2 at 40°C in 14.3 h (Whiting *et al.*, 1992a,b). It remains to be seen whether apples could tolerate these treatments.

Ke and Kader (1992b) and Carpenter and Potter (1994) discussed deficiencies with our present knowledge of controlled atmospheres and suggestions for future research. Controlled atmospheres seem to offer promise as effective insect quarantine treatments. Methods of resolving detrimental effects to commodities caused by low O_2 and high CO_2 levels recommended for insect control are being developed.

Additional research is needed on combining controlled atmospheres with other treatments. For example, controlled atmospheres and heat have been shown to be synergistic. The highest temperatures used in most controlled atmosphere studies have been ambient. Soderstrom *et al.* (1992) did use higher temperatures (38–42°C) combined with controlled atmospheres for 6 h and achieved high levels of mortality of the red flour beetle, and Whiting *et al.* (1992b) reduced the LT_{99} for fifth instar codling moth from 350 to 14.3 h in 0.4% O_2 and 5% CO_2 when the temperature was raised from 20 to 40°C. In some instances, cold may be combined favourably with controlled atmospheres. High levels of CO_2 synergized fumigants (Bond and Buckland, 1978), while high CO_2 appeared to induce increased resistance to irradiation in *Habrobracon* sp. (Clark and Herr, 1955).

Film wraps

Film wraps are used to extend the storage life of fresh commodities (Anzueto and Rizvi, 1985). Film wraps modify the atmosphere inside of a fruit; oxygen is generally lowered while CO_2 levels are increased. Shetty *et al.* (1989) were the first to use plastic film wrap to kill insects inside fruits (Table 7.1). Jang (1990) found film wrap promising as a quarantine treatment though improved wrapping techniques and wraps were needed to ensure quarantine security. Gould and Sharp (1990) obtained

99.95% mortality of Caribbean fruit fly immatures in mangoes after the fruits had been film wrapped for 15 days. Unfortunately, the fruit deteriorated after being film wrapped for only 6 days. Additionally, Sharp (1990) found that third instar Caribbean fruit flies were able to penetrate film wrap on grapefruits even when two layers were applied.

Individual film wraps have not provided quarantine security (99.9968% mortality) against any of the fruit flies studied (Table 7.1). However, the fact that 100% of the pomace fly, *Drosophila melanogaster* (Meigen), was killed after 3 days of shrink wrap is significant as this insect was much harder to kill in raisins stored in $\leqslant 0.5\%$ O_2 and 12–14% CO_2 than several beetle and moth species (Soderstrom and Brandl, 1984).

Jang (1990) suggested combining film wrap with heat or cold treatments. Combining film wrap with heat treatments may be synergistic for insect kill as was the combination of heat and controlled atmospheres (Ke and Kader, 1992b). Cold was usually not synergistic with controlled atmospheres (Gaunce *et al.*, 1982; Batchelor *et al.*, 1985). Film wrap should also be studied in combination with irradiation and fumigation as plastic films vary in their permeability to fumigants (Houck *et al.*, 1989). The mode of action of films in killing insects should be examined. All available films should be tested for quarantine purposes. Films should be designed specifically for killing insects inside wrapped commodities.

Coatings

Coatings, like film wraps, were developed to preserve the quality of fresh commodities (Nisperos-Carriedo, 1994). Thompson (1990) found that coatings applied to cherimoya killed the mite, *Brevipalpus chilensis* Baker, present on the fruit surface, and provided quarantine security (Undurraga and Lopez, 1992). The result of this study was the first use of a coating as a quarantine treatment (King, 1992). The mode of action probably involves physically restricting the mites and sealing of the tracheal and alimentary systems. Therefore, the mortality is not caused entirely by the modified atmosphere treatment.

Studies on the effect of methoprene-impregnated waxes on larvae of the Mediterranean fruit fly and oriental fruit fly in papayas and peaches presented data showing that wax alone killed insects inside fruits (Saul *et al.*, 1985, 1987). Controls comprising wax-coated fruit without methoprene yielded estimated mortalities of 20–89%. However, this observation was not examined further. Coatings varied in their ability to achieve mortality (Hallman *et al.*, 1994); however, mortality did not reach 100%. Coatings showed a synergistic reaction with hot air treatment but not with cold, irradiation, or methyl bromide treatments (Hallman, unpublished data). Available coatings should be tested for their ability to kill insects inside fruits. Coatings differ significantly in permeability to

Table 7.1. Studies of film wraps used to kill insects inside different fruits.

Insect	Fruit	Film	Mortality at	Temperature	Reference
Drosophila melanogaster	Mango	Cryovac D-955	100% at 3 days[1]	24–25°C	Shetty *et al.* (1989)
Bactrocera dorsalis	Papaya	Cryovac D-955	>95% at 6 days	27°C	Shetty *et al.* (1989)
Bactrocera cucurbitae	Papaya	Cryovac D-955	≈85% at 6 days[2]	22–24°C	Jang (1990)
Ceratitis capitata	Papaya	Cryovac D-955	≈90% at 6 days[2]	22–24°C	Jang (1990)
Anastrepha suspensa	Mango	Clysar EHC-150	99.95% at 15 days	24–26°C	Gould and Sharp (1990)
Anastrepha suspensa	Grapefruit	Clysar EHC-50-F	92.2% at 35 days	24–26°C	Sharp (1990)
Anastrepha suspensa	Grapefruit	Clysar EHC-150-F[3]	97.5% at 35 days	24–26°C	Sharp (1990)

[1] Number of insects tested not given; therefore, confidence limits for 100% mortality cannot be determined.
[2] Jang (1990) suggested that improper wrapping may have reduced mortality level.
[3] Fruit double wrapped.

atmospheric gases (Hagenmaier and Shaw, 1992). Novel materials, such as chitosan (El Ghaouth *et al.*, 1992) and soyabean proteins (Kinzel, 1992), are being investigated as coatings to maintain quality of fresh commodities. Coatings should be designed with the specific purpose of killing insects as well as preserving commodity quality; information on the mode of action of coatings would aid in their development. Besides reduction in O_2 and increase in CO_2, coatings also may cause increased production of volatiles inside the fruits, which could affect insect survival (Nisperos-Carriedo *et al.*, 1990). Combination treatments using coatings with other treatment procedures, such as forced hot air, need to be studied.

Other modified atmosphere treatments

Dentener *et al.* (1992) investigated the use of polymeric film bags containing an oxygen remover to kill insect pests found on persimmons. Complete mortality of lightbrown apple moth and longtailed mealybug, *Pseudococcus longispinus* Targioni-Tozetti, was achieved after 4 days at 20°C followed by 14 days at 0°C. Although atmospheric conditions were not precisely controlled, the treatment shows potential.

Conclusions

Controlled atmosphere quarantine treatments of pests of fresh commodities have been developed and may provide quarantine security without damaging the commodity (Table 7.2). Other potential quarantine treatments cited by Ke and Kader (1992a,b) lack the information needed to indicate whether they would work. For example, in some studies few insects were tested, mortality may have been affected by other factors not controlled (such as low RH) or 100% mortality was estimated and not achieved. Other studies lack data on the tolerance of the commodity to the treatment. All of the treatments listed (Table 7.2) require further research before they can be applied; the exception may be shipment of asparagus from New Zealand to Japan, because Carpenter and Potter (1994) report that a successful pilot shipment was made. Approval of a controlled atmosphere treatment of apples from the northwestern US for codling moth (Toba and Moffitt, 1991) depends on convincing importing countries that diapausing larvae will not be present. A controlled atmosphere quarantine treatment of Japanese persimmons grown in New Zealand could probably be developed if complete mortality of the longtailed mealybug could be achieved or if persimmons were not infested with that pest (Dentener *et al.*, 1992).

Table 7.2. Possible controlled atmosphere quarantine treatments.

Commodity	Pest	Percentage O_2	Percentage CO_2	Temperature (°C)	Time (days)	Reference
Apple	San Jose scale	<1	>90	≥12	2	Morgan and Gaunce (1975)
Apple	San Jose scale	0	96	22	1	Morgan and Gaunce (1975)
Apple	Codling moth larvae	1.5–2	<1	0	91	Toba and Moffitt (1991)
Apple	Apple rust and European red mites	1.0	1.0	2.8	160	Lidster *et al.* (1981)
Apple	Four tortricid species	0.4	5.0	40	>0.6	Whiting *et al.* (1992a,b)
Asparagus	Green peach aphid and New Zealand flower thrips	8.4[1]	60	0–1	4.5	Carpenter and Potter (1994)
Strawberry	Western flower thrips	1.9–2.3	88.7–90.6	2.5	2	Aharoni *et al.* (1981)
Sweet potato[2]	Sweet potato weevil	4	60	25	7	Delate *et al.* (1990)
Sweet potato[2]	Sweet potato weevil	2	40	25	7	Delate *et al.* (1990)
Sweet potato[2]	Sweet potato weevil	2	60	25	7	Delate *et al.* (1990)
Walnut	Codling moth larvae	8.4[1]	60	25	>14	Soderstrom *et al.* (1990)

[1] O_2 level estimated from 40% air.
[2] Cured sweet potato only.

A reduction in quarantine mortality requirements would advance the development of other potential controlled atmosphere quarantine treatments. Quarantine requirements often appear too strict for commodities with extremely low pest infestation levels (Vail *et al.*, 1993).

As with many quarantine treatment techniques, the precise mode(s) of action of modified atmospheres are not known. Also, the effects of other variables, such as host, on the efficacy of the treatments are not clear. Much of the research on modified atmospheres was done for specific, practical reasons with little control over some modifying factors, making broad comparisons and conclusions precarious.

It bears repetition that entomologists and plant physiologists should coordinate their activities to perform comparable research and increased communication is needed between scientists working in controlled atmosphere grain storage and fresh commodity storage (Ke and Kader, 1992b).

Acknowledgement

I thank Edward Soderstrom and Adel Kader for reviewing the manuscript. I thank Adel Kader for also providing me with some references.

References

Aharoni, Y., Hartsell, P.L., Stewart, J.K. and Young, D.K. (1979) Control of western flower thrips on harvested strawberries with acetaldehyde in air, 50% carbon dioxide or 1% oxygen. *Journal of Economic Entomology* 72, 820–822.

Aharoni, Y., Stewart, J.K. and Guadagni, D.G. (1981) Modified atmospheres to control western flower thrips on harvested strawberries. *Journal of Economic Entomology* 74, 338–340.

AliNiazee, M.T. (1972) Susceptibility of the confused and red flour beetles to anoxia produced by helium and nitrogen at various temperatures. *Journal of Economic Entomology* 65, 60–64.

Anzueto, C.R. and Rizvi, S.S.H. (1985) Individual packaging for shelf life extension. *Journal of Food Science* 50, 897–904.

Batchelor, T.A., O'Donnell, R.L. and Roby, J.J. (1985) The efficacy of controlled atmosphere coolstorage in controlling leafroller species. *Proceedings of the 38th New Zealand Weed and Pest Control Conference*, pp. 53–55.

Benschoter, C.A., Spalding, D.H. and Reeder, W.F. (1981) Toxicity of atmospheric gases to immature stages of *Anastrepha suspensa*. *The Florida Entomologist* 64, 543–544.

Bond, E.J. and Buckland, C.T. (1978) Control of insects with fumigants at low temperatures: toxicity of fumigants in atmospheres of carbon dioxide. *Journal of Economic Entomology* 71, 307–309.

Bond, E.J. and Buckland, C.T. (1979) Development of resistance to carbon dioxide in the granary weevil. *Journal of Economic Entomology* 72, 770–771.

Brandl, D.G., Soderstrom, E.L. and Schreiber, F.E. (1983) Effects of low-oxygen atmospheres containing different concentrations of carbon dioxide on mortality of the navel orangeworm, *Amyelois transitella* Walker (Lepidoptera: Pyralidae). *Journal of Economic Entomology* 76, 828–830.

Calderon, M. (1990) Introduction. In: Calderon, M. and Barkai-Golan, R. (eds), *Food Preservation by Modified Atmospheres*. CRC Press, Boca Raton, Florida, pp. 4–8.

Carpenter, A. and Potter, M.A. (1994) Controlled atmospheres. In: Sharp, J.L. and Hallman, G.J. (eds), *Quarantine Treatments for Pests of Food Plants*. Westview Press, Boulder, Colorado (in press).

Clark, A.M. and Herr, A.B., Jr (1955) The effects of certain gases on the radio-sensitivity of *Habrobracon* during development. *Radiation Research* 2, 538–543.

Delate, K.M., Brecht, J.K. and Coffelt, J.A. (1990) Controlled atmosphere treatments for control of sweetpotato weevil (Coleoptera: Curculionidae) in stored tropical sweet potatoes. *Journal of Economic Entomology* 83, 461–465.

Dentener, P.R., Peetz, S.M. and Birtles, D.B. (1992) Modified atmospheres for the postharvest disinfestation of New Zealand persimmons. *New Zealand Journal of Crop and Horticultural Science* 20, 203–208.

Dickler, E. (1975) Influence of standard cold storage and controlled atmosphere storage of apples from Italy on the mortality and fecundity of San Jose scale (*Quadraspidiotus perniciosus* Comst.). *Redia* 56, 401–416.

El Ghaouth, A., Ponnampalam, R., Castaigne, F. and Arul, J. (1992) Chitosan coating to extend the storage life of tomatoes. *HortScience* 27, 1016–1018.

Fleurat-Lessard, F. (1990) Effect of modified atmospheres on insects and mites infesting stored products. In: Calderon, M. and Barkai-Golan, R. (eds), *Food Preservation by Modified Atmospheres*. CRC Press, Boca Raton, Florida, pp. 21–38.

Friedlander, A. (1983) Biochemical reflection on a non-chemical control method: the effect of controlled atmosphere on the biochemical processes in stored product insects. In: *Proceedings of the 3rd International Working Conference on Stored Product Entomology*. Kansas State University, Manhattan, pp. 471–486.

Gaunce, A.P., Morgan, C.V.G. and Meheriuk, M. (1982) Control of tree fruit insects with modified atmospheres. In: Richardson, D.G. and Meheriuk, M. (eds), *Controlled Atmospheres for Storage and Transport of Perishable Agricultural Commodities*. Oregon State University, School of Agriculture Symposium Series No. 1, Beaverton, Oregon, pp. 383–390.

Glass, E.H., Chapman, P.J. and Smock, R.M. (1961) Fate of apple maggot and plum curculio larvae in apple fruits held in controlled atmosphere storage. *Journal of Economic Entomology* 54, 915–918.

Gould, W.P. and Sharp, J.L. (1990) Caribbean fruit fly (Diptera: Tephritidae) mortality induced by shrink-wrapping infested mangoes. *Journal of Economic Entomology* 83, 2324–2326.

Hagenmaier, R.D. and Shaw, P.E. (1992) Gas permeability of fruit coating waxes. *Journal of the American Society for Horticultural Science* 117, 105–109.

Hallman, G.J., Nisperos-Carriedo, M.O., Baldwin, E.A. and Campbell, C.A. (1994) Mortality of Caribbean fruit fly immatures in coated fruits. *Journal of Economic Entomology* 87 (in press).

Houck, L.G., Jenner, J.F., Moreno, D.S. and Mackey, B.E. (1989) Permeability of polymer film wraps for citrus fruit fumigated with hydrogen cyanide to control California red scale. *Journal of the American Society for Horticultural Science* 114, 287–292.

Jang, E.B. (1990) Fruit fly disinfestation of tropical fruits using semipermeable shrinkwrap film. *Acta Horticulturae* 269, 453–458.

Kader, A.A. (1992) Modified atmospheres during transport and storage. In: Kader, A.A. (ed.), *Postharvest Technology of Horticultural Crops*. Special Publication 3311. University of California, Oakland, pp. 85–92.

Ke, D. and Kader, A.A. (1992a) External and internal factors influence fruit tolerance to low-oxygen atmospheres. *Journal of the American Society for Horticultural Science* 117, 913–918.

Ke, D. and Kader, A.A. (1992b) Potential of controlled atmospheres for postharvest insect disinfestation of fruits and vegetables. *Postharvest News and Information* 3, 31N–37N.

King, L.J. (1992) Importation of cherimoyas from Chile. *Federal Register* 57, 56434–56437.

Kinzel, B. (1992) Protein-rich edible coatings for food. *Agricultural Research* 40 (5), 20–21.

Klag, N.G. (1986) Use of CA for quarantine control of insects on fresh fruits and vegetables. In: Blankenship, S.M. (ed.), *Controlled Atmospheres for Storage and Transport of Perishable Agricultural Commodities*. Horticulture Report No. 126, Department of Horticultural Sciences, North Carolina State University, Raleigh, pp. 199–206.

Klaustermeyer, J.A., Kader, A.A., Morris, L.L. and France, K.R. (1973) *Preliminary Studies Related to Insect Control in Harvested Lettuce*. Special Report, Department of Vegetable Crops, University of California, Davis, 18 pp.

Lidster, P.D. (1982) Low oxygen atmospheres to maintain apple quality in storage. In: *Controlled Atmospheres for Storage and Transport of Perishable Agricultural Commodities*. Oregon State University School of Agriculture Symposium Series 1. Timber Press, Beaverton, Oregon, pp. 109–120.

Lidster, P.D., Sanford, K.H. and McRae, K.B. (1981) Effects of modified atmosphere storage on overwintering populations of the apple rust mite and European red mite eggs. *HortScience* 16, 328–329.

Lidster, P.D., Sanford, K.H. and McRae, K.B. (1984) Effects of temperature and controlled atmosphere on the survival of overwintering populations of European red mite eggs on stored 'McIntosh' apples. *HortScience* 19, 257–258.

Lindgren, D.L. and Vincent, L.E. (1970) Effect of atmospheric gases alone or in combination on the mortality of granary and rice weevils. *Journal of Economic Entomology* 63, 1926–1929.

Miller, P.L. (1966) The regulation of breathing in insects. *Advances in Insect Physiology* 3, 279–302.

Moffitt, H.R. and Albano, D.J. (1972) Effects of commercial fruit storage on stages of the codling moth. *Journal of Economic Entomology* 65, 770–773.

Morgan, C.V.G. and Gaunce, A.P. (1975) Carbon dioxide as a fumigant against the San Jose scale (Homoptera: Diaspididae) on harvested apples. *Canadian Entomologist* 107, 935–936.

Navarro, S., Dias, R. and Donahaye, E. (1985) Induced tolerance of *Sitophilus oryzae* adults to carbon dioxide. *Journal of Stored Products Research* 21, 207–213.
Nicolas, G. and Sillans, D. (1989) Immediate and latent effects of carbon dioxide on insects. *Annual Review of Entomology* 34, 97–116.
Nisperos-Carriedo, M.O. (1994) Edible coatings and films based on polysaccharides. In: Baldwin, E.A., Nisperos-Carriedo, M.O. and Krochta, J.M. (eds), *Edible Coatings and Films to Improve Food Quality*. Tecnomic, Lancaster, Pennsylvania (in press).
Nisperos-Carriedo, M.O., Shaw, P. and Baldwin, E.A. (1990) Changes in volatile flavor components of pineapple orange as influenced by the application of lipid and composite films. *Agriculture and Food Chemistry* 38, 1382–1387.
Prange, R.K. and Lidster, P.D. (1992) Controlled-atmosphere effects on blueberry maggot and lowbrush blueberry fruit. *HortScience* 27, 1094–1096.
Ripp, B.E. (ed.) (1984) *Controlled Atmosphere and Fumigation in Grain Storages*. Elsevier, Amsterdam, Netherlands.
Saul, S.H., Mau, R.F.L. and Oi, D. (1985) Laboratory trials of methoprene-impregnated waxes for disinfesting papayas and peaches of the Mediterranean fruit fly (Diptera: Tephritidae). *Journal of Economic Entomology* 78, 652–655.
Saul, S.H., Mau, R.F.L., Kobayashi, R.M., Tsuda, D.M. and Nishina, M.S. (1987) Laboratory trials of methoprene-impregnated waxes for preventing survival of adult oriental fruit flies (Diptera: Tephritidae) from infested papayas. *Journal of Economic Entomology* 80, 494–496.
Sharp, J.L. (1990) Mortality of Caribbean fruit fly immatures in shrinkwrapped grapefruit. *Florida Entomologist* 73, 660–664.
Shetty, K.K., Klowden, M.J., Jang, E.B. and Kochan, W.J. (1989) Individual shrink wrapping: a technique for fruit fly disinfestation in tropical fruits. *HortScience* 24, 317–319.
Soderstrom, E.L. and Brandl, D.G. (1984) Low-oxygen atmosphere for postharvest insect control in bulk-stored raisins. *Journal of Economic Entomology* 77, 440–445.
Soderstrom, E.L. and Brandl, D.G. (1990) Controlled atmospheres for the preservation of tree nuts and dried fruits. In: Calderon, M. and Barkai-Golan, R. (eds), *Food Preservation by Modified Atmospheres*. CRC Press, Boca Raton, Florida, pp. 83–92.
Soderstrom, E.L., Mackey, B.E. and Brandl, D.G. (1986) Interactive effects of low-oxygen atmospheres, relative humidity, and temperature on mortality of two stored-product moths (Lepidoptera: Pyralidae). *Journal of Economic Entomology* 79, 1303–1306.
Soderstrom, E.L., Brandl, D.G. and Mackey, B. (1990) Responses of codling moth (Lepidoptera: Tortricidae) life stages to high carbon dioxide or low oxygen atmospheres. *Journal of Economic Entomology* 83, 472–475.
Soderstrom, E.L., Brandl, D.G. and Mackey, B. (1991) Responses of *Cydia pomonella* (L.) (Lepidoptera: Tortricidae) adults and eggs to oxygen deficient or carbon dioxide enriched atmospheres. *Journal of Stored Products Research* 27, 95–101.
Soderstrom, E.L., Brandl, D.G. and Mackey, B. (1992) High temperature combined with carbon dioxide enriched or reduced oxygen atmospheres for control of *Tribolium castaneum* (Herbst) (Coleoptera: Tenebrionidae). *Journal of Stored Products Research* 28, 235–238.

Thompson, A. (1990) Efecto de la aplicación de cera y aceite en frutos de chirimoya (*Annona cherimola* Mill.) cv. Concha Lisa, sobre el control de *Brevipalpus chilensis* Baker y evaluación del comportamiento de la fruta en post-cosecha. Thesis in Agronomy, Universidad Católica de Valparaíso, Valparaíso, Chile, 92pp.

Toba, H.H. and Moffitt, H.R. (1991) Controlled-atmosphere cold storage as a quarantine treatment for nondiapausing codling moth (Lepidoptera: Tortricidae) larvae in apples. *Journal of Economic Entomology* 84, 1316–1319.

Undurraga M.P. and Lopez, L.E. (1992) Use of wax and soapy water, alone or combined, as a quarantine method of control of *Brevipalpus chilensis* Baker on cherimoyas (*Annona cherimola* Mill.). (Unspecified report) Universidad Católica de Valparaíso, Valparaíso, Chile, 22pp.

Vail, P.V., Tebbets, J.S., Mackey, B.E. and Curtis, C.E. (1993) Quarantine treatments: a biological approach to decision-making for selected hosts of codling moth (Lepidoptera: Tortricidae). *Journal of Economic Entomology* 86, 70–75.

Waddell, B.C., Dentener, P.R. and Batchelor, T.A. (1990) Time–mortality response of leafrollers exposed to commercial controlled atmosphere coolstorage. *Proceedings of 43rd New Zealand Weed and Pest Control Conference*, pp. 328–333.

Wang, C. Y. (1990) Physiological and biochemical effects of controlled atmosphere on fruits and vegetables. In: Calderon, M. and Barkai-Golan, R. (eds), *Food Preservation by Modified Atmospheres*. CRC Press, Boca Raton, Florida, pp. 197–223.

Whiting, D.C., Foster, S.P., van den Huevel, J. and Maindonald, J.H. (1992a) Comparative mortality responses of four tortricid (Lepidoptera) species to a low oxygen-controlled atmosphere. *Journal of Economic Entomology* 85, 2305–2309.

Whiting, D.C., van den Huevel, J. and Foster, S.P. (1992b) Potential of low oxygen/moderate carbon dioxide atmospheres for postharvest disinfestation of New Zealand apples. *New Zealand Journal of Crop and Horticultural Science* 20, 217–222.

III

Product Physiological and Biochemical Response to Possible Disinfestation Procedures

Chemical Treatments 8

CHARLES F. FORNEY[1] and LAURIE G. HOUCK[2]
[1]*Agriculture and Agri-food Canada, Research Station, Kentville, Nova Scotia, Canada, B4N 1J5;*
[2]*United States Department of Agriculture, Agricultural Research Service, Horticultural Crops Research Laboratory, 2021 S Peach Avenue, Fresno, CA 93727, USA.*

Introduction

Chemical treatments in the form of fumigants are a practical and convenient method to eradicate insect pests from horticultural commodities. Among available commodity treatments (temperature, controlled atmospheres, and irradiation) many aspects of fumigation favour its use. Fumigants kill a broad range of pests. Methyl bromide, for example, will kill nematodes (Kohn, 1962), insects (Thompson, 1966; Munro, 1969) and fungi (Gandy and Chanter, 1976). Fumigation treatments are economical and easy to apply and can be performed under gas-proof sheets, in trucks, railcars, barges, truck–rail–ship van containers or in ships' holds (Munro, 1969; Leesch *et al.*, 1984); or simple fumigation chambers can be inexpensively built (Munro, 1969; United States Department of Agriculture, 1985). In addition, treatments can be applied in a relatively short time, usually in just a few hours.

There are, however, several concerns when using chemical treatments. The high human toxicity of some fumigants dictates that stringent safety precautions must be used to protect fumigant applicators and handlers of fumigated commodities (Thompson, 1966; Alexeeff and Kilgore, 1983). Another concern is the perception by consumers that residues of fumigants on fresh commodities pose a health risk. Although this concern is prevalent, the existence of long-term risk from pesticides is uncertain (Ames and Gold, 1990; Gold *et al.*, 1992) and generally poorly supported when quantified (Archibald and Winter, 1990; Scheuplein, 1992). However, this concern has resulted in various countries lowering acceptable residue levels or banning the use of some fumigants. This may

From: Paull, R.E. and Armstrong, J.W. (eds) (1994) *Insect Pests and Fresh Horticultural Products: Treatments and Responses.* CAB INTERNATIONAL, Wallingford, UK.

limit or prevent the use of specific fumigants on commodities destined for some domestic or export markets.

Currently, methyl bromide (MB) is the most commonly used fumigant of fresh horticultural commodities. In addition hydrogen cyanide (HCN), phosphine (PH_3), and ethylene dibromide (EDB) are used in a limited manner. Other fumigants have been shown to be effective on specific crops and/or insects but are not widely used. All of these fumigants, however, can cause injury to fresh commodities if not applied properly. How and when chemical treatments injure fresh horticultural commodities is the subject of this chapter.

This chapter explores the responses of fresh horticultural commodities, primarily fruits and vegetables, to chemical treatments, including physiological responses of fresh fruits and vegetables to chemical treatments, factors that affect the resistance of plant tissues to chemical treatments, and the physiological mechanisms that prevent the expression of injury symptoms.

Responses of Plants to Chemical Treatments

Physiological responses

Chemicals that injure plant tissues can be characterized in two groups: those that cause immediate cell death on contact; and those that alter normal metabolism leading to possible detrimental effects with time.

Contact toxicity

Many chemicals are highly toxic to plants. These chemicals, which include strong acids, strong bases, and many organic solvents, will kill plant tissues upon contact. As a result of the high toxicity, few chemicals of this type have been considered for use as a disinfestation treatment on living plant tissue.

Some useful chemical treatments, however, may become toxic to the plant cells under certain conditions. Cyanide, for example, can be used as a fumigant for scale insects and other surface pests on citrus without injuring the fruit. However, if the fruit is wet, HCN vapours will dissolve in the water and form cyanic acid that can cause fruit injury (Armitage and Steinweden, 1946).

Altered metabolism

Normal metabolism of plant cells can be disrupted by a chemical treatment in several ways. The chemical can alter enzyme activity, induce the production of specific enzymes, inhibit the action of genetic material involved with the synthesis of enzymes, or reduce membrane-associated enzyme activity due to damage to the membrane. Additionally, interaction of the chemical with the membrane may affect transportation of compounds into and out of the cell and disrupt compartmentalization of enzymes and substrates within the cell. These alterations of the normal cell by the chemical treatment can cause cellular death or a variety of disorders.

Enzyme inhibition

Some chemicals will modify proteins, including enzymes, that are directly involved in normal cellular metabolism. These modifications may inhibit the normal function of enzymes leading to abnormal metabolism and injury. Methyl bromide reacts with nucleophilic sites (–OH, SH, or NH_2 groups) in enzymes and other proteins (McKenry, 1981; Starratt and Bond, 1990a). Alterations of these molecules can inhibit their normal function in the cell. Hydrogen cyanide inhibits many oxidative enzymes, especially enzymes containing metal prosthetic groups (Price, 1985). Cytochrome oxidase, a key enzyme in aerobic respiration, is inhibited by HCN or PH_3 in insect systems (Price, 1985). Hydrogen cyanide and PH_3 also inhibit catalase (Bond, 1963), a key enzyme used to detoxify peroxides and harmful free radicals that arise during utilization of oxygen in the tissues. Phosphine also inhibits the action of catalase and peroxidase in granary weevils (Bolter and Chefurka, 1990). Ethylene dichloride (EDC) and EDB may act as respiratory inhibitors as reported in insect systems. Both EDC and EDB inhibit succinate dehydrogenase (Price, 1985) and EDB inhibits triosephosphate dehydrogenase (Morikawa, 1964). The altered metabolism caused by these effects on enzymes may lead to discoloration, changes in ripening, and other disorders of fresh produce in the weeks following fumigation.

Enzyme induction

Fumigants have been shown to induce enzyme activity. Peroxidase and cellulase activity in *Phaseolus vulgaris* plants exposed to ozone increased 2 days after fumigation (Dass and Weaver, 1972). Glutathione-*S*-transferase activity increased in barley leaves following ozone fumigation (Price *et al.*, 1990). Induction of this enzyme may provide protection from fumigant-induced injury.

Genetic alterations

Fumigants may alter or damage nucleic acids involved with the control and synthesis of vital enzymes in the plant tissue. Methyl bromide reacts with deoxyribonucleic acid (DNA) to a lesser extent than its reaction with proteins. Methyl bromide methylated DNA in wheat and maize forming 7-methylguanine, 1-methyladenine, 3-methylcytosine and 3-methyladenine and the degree of methylation increases with increasing exposure to the fumigant (Starratt and Bond, 1988). Chemical residues continue to react with cell constituents after fumigation. In apples fumigated with MB, 60% of MB residues reacted with components of the apples within 7 days following fumigation at 1.5°C (Rippon *et al.*, 1982). Although no studies have explored the effect of EDB on plant DNA, EDB induces strand breakage in DNA in rat liver tissue (Nachtomi and Sharma, 1977). The reaction of fumigants with nucleic acids in plant cells could disrupt normal metabolic processes to cause the development of various disorders, but no evidence that this occurs in plants has been found.

Membrane damage

The inhibition of catalase and peroxidase can result in the accumulation of free radicals in the cell. These free radicals react with cellular membranes causing membrane rigidification similar to that observed with natural senescence (Thompson, 1984). The induced membrane changes disrupt normal membrane function, inhibit membrane-bound enzymes and cause the membrane to leak. Ozone fumigation increases lipid peroxidation in cellular membranes of barley leaves leading to the loss of soluble proteins and chlorosis (Price *et al.*, 1990). Rigney and co-workers (1978) suggest that EDB may cause changes in the conformation and activity of intracellular membranes and their associated enzymes. They observed that EDB altered carotene synthesis, respiration, and degradation of pectic substances that are linked to membrane-associated enzymes.

Loss of cellular compartmentation can occur when a chemical, or other stress-inducing factor, impairs the integrity of cell membranes and permits uncontrolled mixing of cellular components. Methyl bromide fumigation of grapefruit callus tissue increased the rate of K^+ leakage indicating that normal plasma membrane function is impaired (Forney *et al.*, 1990). The rate of K^+ leakage increases with time after fumigation, implying that increasing leakage is not solely a result of MB reacting with the membrane. The induced membrane damage due to abnormal metabolism may contribute to the increased leakage. Ozone also induces membrane leakiness in plants (Howell, 1974). This leakiness causes oxidative enzymes to mix with and oxidize phenols to their respective quinones that polymerize with amino acids, amines, and sulfhydryl protein groups to form reddish-brown pigments. These pigments may be responsible for the discoloration observed in injured tissue.

Injury symptoms

Fumigant-induced alteration of the normal plant or plant organ metabolism can cause the development of a wide variety of injury symptoms. Most symptoms do not develop immediately after fumigation, but rather appear 1 to 3 weeks later. The rate of development is dependent on the temperature and other storage conditions after treatment. These injury symptoms include discolorations and necrotic spots, off-odours or flavours, alterations in senescence, increased susceptibility to decay, or altered sensitivity to other environmental stresses.

Discoloration and necrotic spots

Tissue discoloration is the most common detrimental response of plants to chemical fumigants. The discoloration usually occurs on the surface of fumigated fruits or vegetables. In citrus, MB can cause scald of the peel tissue (Hatton and Cubbedge, 1979), while EDB (Houck *et al.*, 1985) and PH_3 (Hatton, *et al.*, 1982) can cause brown discoloration as well as pitting of the peel. Surface browning of peaches and nectarines caused by fumigation occurs first in the suture area (Harvey and Harris, 1982; Harvey *et al.*, 1989). Apples fumigated with MB may develop brown spots or irregular brown areas on the skin that may extend into the outer flesh (Isaac, 1944). Abrasions and bruises render the surface of apples more susceptible to injury. The skin of cantaloupe and 'Honeydew' melons can become brown, reddish-tan or green after MB fumigation, although no discoloration is observed in the flesh (Lipton and Tebbets, 1984). Tomatoes fumigated with EDB developed yellow specks on the skin because either the pigment was destroyed or its synthesis was prevented (Spalding *et al.*, 1978). Methyl bromide may cause pitting and blotchiness of tomatoes by inhibiting normal ripening (Spalding *et al.*, 1978; Lipton *et al.*, 1982). Cucumbers developed yellow streaks on the skin following fumigation with MB (Armstrong and Garcia, 1985). Chrysanthemum cuttings developed leaf scorch (Helyer and Ledieu, 1989) and orchid leaves are burned and blackened by MB (Gaur *et al.*, 1984).

Internal tissues also can be affected by fumigants. Methyl bromide causes discoloration of the core area of apples, especially near the calyx (Isaac, 1944; Auda and Canessa, 1981), and internal blackening of potatoes (Ahmed and El Zorgani, 1984). Green bell peppers fumigated with MB developed a brown discoloration of the placenta near the stem-end, and a tan to brown darkening of the seeds 1 week after fumigation (Lipton *et al.*, 1982). Pink grapefruit can lose some of the pink rind blush and pink flesh colour 21 days after MB fumigation (Williamson *et al.*, 1986). Garlic fumigated with high levels of MB developed yellowing similar to 'waxy yellow breakdown', a condition that may normally develop in storage (Roth and Richardson, 1963).

Off-odours or flavours

Fumigants may affect the flavour components of fresh fruits and vegetables. Most fumigants have some effect on aroma, although these effects may by subtle and not noticed in many cases by the untrained individual. In an extensive study of 16 potential fumigants, Claypool and Vines (1956) found that all of the chemicals (with the possible exception of methyl iodide) had some effect on aroma. Some effects include the production of 1,2-dibromobutane and 1,3-dichloropropane, which cause a skunk-like odour in treated fruit, to loss of aroma in treated strawberries caused by MB fumigation. Fumigation of grapefruit with MB or EDB altered the flavour of fresh or pasteurized juice, respectively, while PH_3 affected peel aroma (Moshonas and Shaw, 1982). Apples fumigated with EDB had a mildly pungent flavour that persisted for 3–4 weeks after fumigation (Sanford *et al.*, 1976). Fruits injured by fumigation with PH_3 developed abnormal aromas while uninjured fruit did not (Seo *et al.*, 1979). Strong off-odours developed in MB-fumigated garlic whenever yellowing or decay occurred (Roth and Richardson, 1963).

Altered senescence

The effect of fumigants on the processes involving senescence, including ripening, is dependent on the chemical used, its concentration, the commodity being treated, and the storage conditions after fumigation. Pears, apples, and plums fumigated with 32 g MB m^{-3} for 2 h ripened more rapidly than untreated fruit (Claypool and Vines, 1956). When the dosage was doubled, ripening was delayed. Seo and co-workers (1979) found a similar effect with PH_3 concentration on tomato ripening. Knott and Claypool (1940) observed that ripening of MB-fumigated tomatoes is stimulated slightly at 27°C. However, MB-fumigated tomatoes held at 20°C ripened more slowly than non-fumigated fruit. Beever and Yearsley (1987) found that fumigated kiwifruit held at 12°C softened more rapidly than non-fumigated fruit at 20°C, but that softening was slower than in non-fumigated fruit at 0°C.

Several different fumigants can cause delay of fruit ripening. Mature green tomato fruit fumigated with MB ripen more slowly than non-fumigated fruit (Spalding *et al.*, 1978; Lipton *et al.*, 1982). EDB fumigation inhibits colour development of the outer-pericarp tissue of tomatoes, although the inner-pericarp turns red normally (Rigney *et al.*, 1978). MB fumigation slows softening of stone fruit in cold storage (Harvey *et al.*, 1989) and delays the colour development of plums and cherries (Harvey and Harris, 1982). However, the differences in firmness and colour are not as apparent after the fruit is warmed to 21°C for 4 days. Fumigation with either MB or EDB slows the yellowing of limes (Brown and Nickerson, 1980). Delays in ripening also are observed in fruit treated with 1,2-dibromopropane, methyl iodide, acrylonitrile, and chloroacetonitrile (Claypool and Vines, 1956).

Fumigation can also stimulate ripening. Meheriuk *et al.* (1990) found that apple softening is increased by MB fumigation. MB-fumigated avocados developed external damage and ripened 2–4 days sooner than non-fumigated controls (Ito and Hamilton, 1980). PH_3 treatments that caused no visible injury accelerated the ripening of tomatoes, bananas, and avocados (Seo *et al.*, 1979). EDB stimulated the rate of 'Honeydew' melon ripening (Lipton and Tebbets, 1984).

Alterations in ripening caused by fumigation have been correlated with changes in ethylene synthesis and responsiveness of fruit to ethylene. Higher acidity levels and lower ethylene production rates are observed in apples fumigated with MB compared with non-fumigated fruit (Drake *et al.*, 1988). Tomato fruit that demonstrated a delay in ripening due to MB fumigation produced less ethylene than non-fumigated control fruit (Brecht *et al.*, 1986). Methyl bromide also reduces the ethylene sensitivity of tomato fruit. Mature green fruit produced less 1-aminocyclopropane-1-carboxylic acid (ACC), and 'turning' fruit lacked the ability to convert ACC to ethylene. In contrast, Alumot and Chalutz (1972) found ethylene production rates were high in oranges the day after EDB fumigation. Ethylene production rates decreased over several days following fumigation, but remained four times greater after 2 weeks in fruit without peel injury than in non-fumigated fruit. Fumigated fruit that developed peel injury produced more ethylene than fruit with no injury.

Fumigants are reported to affect other aspects of senescence. MB stimulates the formation of red pigmentation of peaches and nectarines (Claypool and Vines, 1956). Peduncles on MB-fumigated cherries retain their green colour better than non-fumigated fruit after 30 days of storage at 1°C, although peduncle abscission increased 20–30% (Jessup, 1988). Also, cherries fumigated with MB are more shrivelled and pitted than controls after 30 days of storage. CS_2 and MB fumigation of potatoes stimulates softening (Pradhan *et al.*, 1960). MB stimulates leaf abscision in orchid plants (Gaur *et al.*, 1984).

Decay susceptibility

Frequently, even when no visible injury is apparent from a fumigation treatment, the commodity is made more susceptible to invasion by fungal or bacterial pathogens which results in increased rates of decay. Citrus fumigated with MB and stored for 4 weeks showed no visual indication of peel injury but decayed rapidly after 1 week at 21°C (Hatton and Cubbedge, 1986). Forney and co-workers (1990) suggested that increased membrane leakiness induced by MB fumigation may make the tissues more susceptible to invasion by decay organisms. Increased rates of decay resulting from fumigation also were reported in grapefruit (Benschoter, 1979), oranges (Hatton and Cubbedge, 1979), lemons (Houck *et al.*, 1985), avocados (Spalding *et al.*, 1977), mangoes (Spalding *et al.*, 1977), cantaloupe (Lipton and Tebbets, 1984), nectarines (Harvey and Harris, 1982), tomatoes (Spalding *et al.*, 1978; Knott and Claypool, 1940), and garlic (Roth and Richardson, 1963).

Stress sensitivity

Fumigants may alter plant tissues to make the fruit less resistant to other environmental stresses. Resistance to chilling injury may be reduced following fumigation. Early season Florida grapefruit, fumigated with MB and held at 10°C for 4 weeks, had more peel injury due to chilling than non-fumigated fruit (Benschoter, 1979). Green bell peppers fumigated with MB developed surface pitting that resembled chilling-induced sheet pitting when they were held at 10°C for 9 days after fumigation (Lipton *et al.*, 1982). Non-fumigated peppers did not develop the pitting, indicating that the fumigation may have increased chilling sensitivity.

Factors Affecting Plant Response

The varied responses of fresh horticultural commodities to fumigation indicate different levels of resistance to injury (Lindgren and Sinclair, 1951; Claypool and Vines, 1956; Benschoter, 1979; Houck *et al.*, 1985). The factors involved in the varied responses include variables in fumigant application and in the physiological status of the plant part being treated.

Treatment factors

The treatment factors include the fumigant and its concentration used, duration of exposure, load factor, environmental conditions during treatment, and environmental conditions under which the commodity is held after treatment. Each of these factors can significantly affect whether the treatment will injure the commodity.

Fumigant

Volatility of the fumigant influences how it is applied and how readily it dissipates following treatment. For example, the boiling point of EDB is 131–132°C and that of MB is 4.5°C. EDB must be heated to volatilize it, and it desorbs very slowly from the commodity and packing material. Methyl bromide, on the other hand, is very volatile and residues on treated commodities are very low within hours after treatment.

Fumigant concentration

Increasing the fumigant concentration increases the probability that the plant will be injured. When nectarines are fumigated, the percentage of fruit injured increases with increasing MB concentration above 48 g m^{-3} at 21°C (Harvey *et al.*, 1989). Citrus fruits fumigated with EDB, ethylene oxide, acrylonitrile, or acrylon developed more severe injury as the fumigant concentration increased from 16 to 48 g m^{-3} (Lindgren and Sinclair, 1951). Similar results were reported with lemons, oranges, and grapefruit fumigated with EDB at several concentrations (Houck *et al.*, 1985). The threshold concentrations of EDB required to induce injury in citrus were 12 g m^{-3} (Houck *et al.*, 1985) while under 48 g m^{-3} MB was required to induce injury under similar conditions (Lindgren and Sinclair, 1951). Treatment temperature also can alter the threshold concentration.

Plastic films reduce the concentration of fumigant reaching the commodity. There was considerable variation in permeability of 13 different plastic films to EDB, MB, and HCN (Houck and Mackey, 1989; Houck *et al.*, 1989). Some films partially restricted transmission of the fumigants to the fruit, while permeability of other films differed little from unrestricted exposure of non-wrapped fruit. For films with low fumigant permeability, increasing fumigant dosage or duration to provide desired fumigant concentration inside the film wrap may be necessary to obtain quarantine security. Some films also appear to restrict desorption of fumigant residues from the fruit.

Duration of exposure

The fumigant concentration multiplied by the exposure time is referred to as the CT product (Bond, 1984). The CT product that is efficacious for an insect pest can be obtained with numerous combinations of concentration and exposure times as long as the critical CT product is reached. However, some evidence shows plants respond differently. Citrus fruit injury increased in severity and in numbers injured when the CT product was held constant at 80 and the MB dosage increased and exposure time decreased from 2 h with 40 g MB m^{-3} to 0.17 h with 480 g MB m^{-3} (Houck *et al.*, 1987). Pine seedlings exposed to 0.18–0.20 ppm SO_2 for 24 h (CT product of 4.8) recovered faster than those exposed to 0.34–0.37 ppm for only 1 h, giving a CT product of 0.37 (Malhotra and Khan, 1978). Longer exposure times with lower concentrations of chemical in the atmosphere usually cause less injury to the commodity than short exposures with high concentrations, and are preferred whenever possible.

Load factor

The load factor is the percentage of the fumigation chamber filled by the commodity. Sorption by the commodity in the chamber will cause a decrease in the fumigant concentration over time. Additionally, the cartons containing the commodity absorb varying amounts of fumigant. Styrofoam and fibreboard absorb more EDB and MB than hardboard, waxed cardboard, or wood. Injury caused by EDB on citrus fruit decreased with increasing load factor (15>30>58%) (Houck *et al.*, 1985). The larger load factor decreased the EDB concentration in the chamber over time due to sorption of the EDB by the fruit and box material. Similarly, MB-fumigated grapefruit decayed more when fumigated in chambers containing only 3 cartons vs. chambers with 12 cartons (Benschoter, 1979). The greater load factor with 12 cartons presumably reduced the fumigant concentration and in effect reduced the concentration of MB the fruit were exposed to.

Environmental conditions during and after fumigation

Environmental factors that have been reported to affect the phytotoxic response of plants to fumigants include temperature, humidity, atmospheric pressure, and the presence of dips or coatings on the commodity.

1. *Fumigation temperature*

Cool temperatures reduce the amount of injury the fumigant produces. PH_3 residues on papayas are tenfold lower when fruit is fumigated at 13°C compared to ambient conditions of 18–26°C (Seo *et al.*, 1979). Injury on EDB-fumigated citrus is less when the fruit are fumigated at 10°C compared to 30°C (Houck *et al.*, 1985). Similarly MB-induced injury on fumigated cantaloupes (Lipton and Tebbets, 1984) and chrysanthemum cuttings (Helyer and Ledieu, 1989) is less when the fumigation is done at 20°C or less compared to above 30°C. Potatoes fumigated with MB are not injured at 7–10°C even with 72 g m^{-3} for 7 days, but potatoes are severely damaged at 30–35°C by only 18 g m^{-3} for 24 h (Ahmed and El Zorgani, 1984). However, while fumigating commodities at low temperatures is less toxic to the plant, it is also less effective in killing the target pest (Bond, 1984), hence the need to use higher temperatures.

2. Relative humidity

Bond (1984) reported more injury developed on growing plants fumigated with MB below 75% relative humidity than with plants fumigated under more humid conditions. The foliage of *Fuchsia arborescens* plants is severely burned when fumigated with MB at 40–50% relative humidity at 21°C, while no injury occurs at near saturation (Latta, 1941).

3. Atmospheric pressure

Fumigating under reduced atmospheric pressure (vacuum fumigation) may increase injury development. Fumigation of garlic cloves with MB under increasing vacuum increased injury (Roth and Richardson, 1963). The reduced atmospheric pressure may increase the penetration of the fumigant into the commodity resulting in increased injury.

4. Chemical dips and coatings

Apples dipped in the anti-scald agent, diphenylamine, prior to fumigation with MB had less injury than non-dipped control fruit (Sproul *et al.*, 1976; Meheriuk *et al.*, 1990). Waxing reduces injury to citrus fumigated with EDB (Eaks and Ludi, 1958). The fungicide, thiabendazole, incorporated into wax applied before fumigation also was found by Chalutz *et al.* (1973) to reduce EDB fumigation injury in grapefruit, lemons, and oranges.

5. Aeration

Chalutz *et al.* (1971) found that EDB residues and injury to citrus could be reduced by aerating (introducing fresh air) at higher temperatures following fumigation. Longer aeration times of 18 h compared to 2 h also help to remove MB (Rippon *et al.*, 1982). High relative humidity during aeration was reported to increase the desorption rate of EDB from wheat (Bond, 1984).

Chemical adsorbents to augment aeration may help prevent injury. Apples fumigated with MB and immediately packed into styrofoam boxes developed brownish blotchy spots on the skin and pulp with severe discoloration within 21 days after treatment (Kawakami and Soma, 1991). However, no injury developed when 40 g of activated charcoal coated with amine compounds to absorb MB was placed in the boxes. This adsorbent apparently removed the fumigant from the air as it desorbed from the fruit to reduce the need for aeration and prevent injury.

6. Holding temperature

Aeration at temperatures above 15–20°C is more effective in dissipating residual fumigant and reducing any phytotoxic response than aeration in refrigerated conditions. EDB residues on fumigated 'Shamouti' oranges decrease exponentially when fruit is held at 17 or 22°C, reaching less than 1 ppm after 4 days (Alumot and Chalutz, 1972). Residues on fruit held at 5°C decrease at a much lower rate and have residues of about 40 ppm 4 days after fumigation. Seo and co-workers (1970) also found lower concentrations of EDB residues on 'Navel' oranges held at 19°C compared with fruit held at about 6°C after fumigation. Grapefruit held at 21–27°C after MB fumigation had no injury while fruit held at 10°C had more rot and peel injury (Benschoter, 1979). Hatton and Cubbedge (1986) found grapefruit peel had less injury when held at 15°C compared to 10°C after MB fumigation. However, differences in maturity may have been responsible for these differences. Furthermore, if MB increased fruit chilling sensitivity, the increased injury observed could be chilling injury.

Storing fumigated, well-aerated produce under refrigerated conditions may retard or prevent the development of injury symptoms. 'Granny Smith' apples fumigated with MB are held at 0°C for 8 days following fumigation to avoid MB-induced internal browning (Auda and Canessa, 1981). Fuchsia plants fumigated with MB are not injured when held at cool temperatures following fumigation while plants held at ambient temperatures develop severe burning of the foliage (Latta, 1941). Citrus held at 12°C following fumigation with any one of six different fumigants developed fewer injury symptoms than fruit held at room temperature (Lindgren and Sinclair, 1951).

7. Film wrapping or bagging

Sealing fumigated produce in plastic bags or wraps after fumigation may increase injury by preventing the dissipation of fumigant residues. Avocados placed in plastic bags with less than 24 h of aeration after MB fumigation are injured with no injury on non-bagged fruit (Chaplin *et al.*, 1981). Bags apparently prevented the desorption of MB from the fruit, resulting in the bagged fruit receiving a longer exposure to MB. Similarly, fumigating film-wrapped citrus with EDB prevented dissipation of EDB residues causing more injury than observed in non-wrapped fruit (Chalutz *et al.*, 1971; Houck and Mackey, 1989).

8. Solar exposure

Exposure of plant materials to sunlight after fumigation with HCN may induce injury. Red ginger cut flowers that were placed in the sun for 1 h after treatment with 2500 ppm of HCN for 30 min developed damage on both flowers and leaves 4 days after treatment, while control flowers not exposed to the sun were undamaged (Hansen *et al.*, 1991).

9. *Atmosphere composition*

Storing fumigated produce in controlled or modified atmospheres may alter the expression of injury symptoms. CO_2 atmospheres (8–12%) inhibited the development of rind discoloration of EDB-fumigated citrus (Chalutz *et al.*, 1971). However, discoloration developed when fruit were removed from high CO_2 atmosphere and exposed to air. Reduction of oxidation reactions by lowering oxygen or elevating CO_2 concentrations may slow the development of injury symptoms. A similar effect was observed in the inhibition of chilling injury in fresh fruits and vegetables held in controlled atmospheres (Forney and Lipton, 1990). However, as with fumigated produce, injury symptoms usually develop after the commodity is placed in air.

Plant factors

The sensitivity of different plants and plant parts to chemical treatments is highly variable. Variability in tolerance has been associated with cultivar, maturity, pretreatment conditioning, and physiological state of the plant.

Cultivar

Resistance to injury from fumigation is cultivar dependent. Harvey *et al.* (1989) found that nectarine cultivars varied in their resistance to a 2-h MB fumigation. 'Red Diamond' tolerated 48 g MB m^{-3}; 'Summer Grand', 'May Grand', 'Fantasia', and 'Firebrite' tolerated 64 g MB m^{-3}; and 'Spring Red' tolerated 80 g MB m^{-3}. Additionally, 'Red Diamond' fruit exposed to $\geqslant$80 g MB m^{-3} did not soften after storage as did other cultivars treated in the same manner. The apple cultivars, 'Golden Delicious', 'Delicious', and 'Spartan' were progressively more sensitive to MB fumigation when evaluated for MB-induced surface and internal browning (Meheriuk *et al.*, 1990). In a separate report, 'Gravenstein' and 'McIntosh' apples tolerated MB fumigation while 'Delicious' and 'Golden Russet' were injured (Sanford, 1962). Citrus fruit had variable tolerance to several different fumigants with lemons and grapefruit found more resistant to injury than oranges (Lindgren and Sinclair, 1951). Of the oranges tested, 'Navel' oranges were injured more by five different fumigants than 'Valencia' oranges. Chaplin *et al.* (1981) reported that thick-skinned avocados were more tolerant to MB fumigation than thin-skinned cultivars. Orchid plants also vary in their sensitivity to MB fumigation (Gaur *et al.*, 1984).

Maturity

Both fruit maturity and stage of ripeness influence tolerance to chemical treatments. Mid-season Florida grapefruit are more tolerant of MB fumigation than immature early-season fruit or over-mature late-season fruit (Benschoter, 1979). Yellow lemons develop more injury after EDB fumigation than silver lemons, although silver lemons have also been cured several weeks prior to treatment (Houck *et al.*, 1985). De-greened grapefruit had only slight peel injury (0.5%) after EDB fumigation and 1 month of cold storage, while green fruit was not injured (Bussel and Kamburov, 1976). Immature and over-mature apples developed more injury following MB fumigation than fruit of optimum storage maturity (Phillips *et al.*, 1938). Ripe 'Honey Dew' melons were more tolerant than under-ripe melons to EDB fumigation (Lipton and Tebbets, 1984). However, mature 'Santa Rosa' plums were more sensitive to MB injury than immature fruit (Harvey and Harris, 1982).

Curing and conditioning

Conditioning citrus fruit for 1 week at 15°C imparted tolerance to chilling temperatures (Hatton and Cubbedge, 1973; Houck *et al.*, 1990) and hot water dips (Houck, 1967). Conditioning treatments also reduced injury from MB fumigation in lemons, oranges, and tangelos (Houck and Jenner, 1988) and in callus tissue grown from grapefruit peel (Forney *et al.*, 1990). Apples held 1 week at ambient temperatures prior to MB fumigation developed less injury than fruit fumigated immediately or held for 1, 3, or 6 weeks prior to fumigation (Sanford, 1962). Holding narcissus bulbs at 30°C for 1 week prior to MB fumigation reduced the adverse effect fumigation has on bulb growth (Purnell and Hague, 1965). Physiological changes that occur as plant tissues adapt to the conditioning environment appear to alter plant sensitivity to fumigants.

Physiological state of the plant

Water-stressed and senescent plant material is more readily injured by fumigants (Turner, 1984). Drake and Moffitt (1992) found that 'Bosc' pears held 5 months in controlled atmosphere storage, or that were in poor physiological condition before fumigation with MB, develop more internal discoloration than freshly harvested fruit in good condition. Two-week-old chrysanthemum cuttings are more readily injured by MB than fresh cuttings (Helyer and Ledieu, 1989). Stress caused by insect infestation can affect the response of plants to chemical fumigants. Macadamia seedlings (*Macadamia integrifolia*) heavily infested with macadamia felted coccid develop considerable leaf burn after MB fumigation (Ironside *et al.*, 1978). Moderately infested seedlings subjected to the same treatment developed only minor leaf burn.

Physiology of Chemical Resistance

Little work has been done to determine the physiological mechanisms by which plants are able to prevent injury from exposure to various chemicals. A greater understanding of how a plant can avoid or tolerate chemical exposure is needed to utilize fumigants more efficiently. Plants or plant tissues have two ways of preventing injury from chemical fumigants: avoidance and tolerance (Levitt, 1980).

The mechanisms by which plants avoid contact with a chemical are generally physical or morphological in nature. Properties of the cuticle may prevent the chemical from contacting the interior of the fruit or vegetable. The application of a wax or other barrier to the surface of the commodity may also form a barrier to the fumigant. However, in these cases where the fumigant does not contact the interior of the commodity, it would only be effective in killing surface insects. Stomatal density may influence the resistance of the plant to chemical fumigants. Tobacco varieties that are more resistant to ozone-induced injury have one-third fewer stomata per square millimetre on their leaves than susceptible varieties (Dean, 1972). Factors influencing the opening or closing of stomata, such as light, temperature, water stress, and CO_2 concentration, may also influence fumigant exposure.

At the cellular level, the cell wall may act as a barrier to prevent the fumigant from contacting the cell protoplast. The cell wall has a varying affinity for chemicals (Cumming and Taylor, 1990). This affinity could reduce the concentration of the fumigant contacting the cell protoplast depending on its chemical nature, and give the plant tissue an increased resistance to the chemical. The chemical nature of the cell wall is variable among different plants and could be one factor contributing to the different levels of resistance seen among plants.

Changes in cell membrane composition have been associated with acclimation of plants to stresses such as cold (Lynch and Steponkus, 1987) and salinity (Hurkman *et al.*, 1988; Brown and DuPont, 1989), as well as changes in ripeness (Forney, 1990). These induced changes in membrane composition may affect membrane thickness and decrease transport of chemicals across the membrane by increasing the phospholipid packing density and decreasing membrane permeability (Benz and Cros, 1978; Thompson, 1985). Membrane changes occurring during conditioning treatments, ripening, or environmental conditions in the field prior to harvest and fumigation could be responsible for altering the resistance of various horticultural commodities to injury.

The simplest mechanism for chemical tolerance is when the plant tissue has no detrimental response to the fumigant. However, this is not usually the case. Normally the plant tissue will have some type of reaction with or response to the fumigant. One mechanism for tolerance is to minimize interaction with the fumigant by slowing the metabolic rate of the plant. This slowing would reduce the rate at which the plant interacts with the fumigant, reducing the development of injury. Additionally, more time is available for the chemical to dissipate following fumigation. Chapin (1991) indicates that general characteristics of plants that are tolerant to adverse environments are slow growth rates, slow rates of photosynthesis, and a low capacity for nutrient uptake. Plants exposed to environmental stresses slow their metabolism which is often associated with increased levels of abscisic acid and decreased levels of cytokinins. The curing and conditioning procedures described earlier may impart tolerance to the commodity by slowing metabolic rates, inducing increased levels of abscisic acid, or by inducing other metabolic changes in the tissue. The observation that fumigation at low temperatures produces less damage to the plant tissue than at higher temperatures also indicates that a slow metabolism may be related to chemical tolerance.

A second mechanism for tolerance is the presence of compounds in the plant cell that will preferentially react with the fumigant and therefore prevent it from reacting with components of the cell that normally result in injury. An example is the antioxidant glutathione, which may protect the plant tissues from injury as a result of oxidative stress (Meister, 1988; Smith *et al.*, 1990). Glutathione may be methylated by MB and act to prevent other molecules from being altered. Starratt and Bond (1990b) found that glutathione content of oranges and potatoes was reduced with increasing concentrations of MB found mostly in *S*-methylglutathione, and *S*-methylcysteine. A similar response was not found with maize. Conjugation of MB with glutathione may be a major detoxification pathway. In granary weevil, MB-resistant strains contain twice as much glutathione as susceptible strains (Starratt and Bond, 1981). Glutathione synthesis in plant cells can be induced by a variety of environmental conditions such as solar radiation, heat, cold and water stress (Smith *et al.*, 1990).

The plant cell can contain enzymes that detoxify the fumigant or harmful products that are produced as a result of chemical stress. Glutathione-*S*-transferase may be a protective mechanism of the plant to reduce oxidative stress products formed in membranes and to metabolize toxins by conjugation with glutathione (Price *et al.*, 1990). Glutathione-*S*-transferase is involved in herbicide detoxifiication in plants (Ketterer, 1986) and has been implicated in the detoxification of MB by granary weevils through the formation of *S*-methylglutathione (Starratt and Bond, 1981). Glutathione reductase also can help to prevent oxidative damage to the cell by maintaining glutathione in a reduced state, which acts to scavenge hydroxyl radicals (Smith *et al.*, 1990). Catalase, peroxidase, and superoxide dismutase remove free radicals that are produced in the cell as a result of chemical stress (Thompson, 1984). The enzyme β-cyanoalanine synthase is believed to play a role in the detoxification of HCN and its production is induced in citrus by ethylene (Mizutani *et al.*, 1992).

Conclusions

Fresh horticultural commodities may be injured by fumigation treatments needed to meet quarantine requirements. Interaction of fumigants with enzymes, nucleic acids, and cellular membranes can alter normal cellular metabolism leading to various types of injury. The most common form of injury is surface discoloration, although fumigants may also cause off-odours, altered ripening and senescence, increased susceptibility to decay organisms, and lowered resistance to other stresses.

The development of deleterious responses by plant tissues is influenced by many factors.

1. During fumigation treatment, the type of fumigant, its concentration, the duration of the treatment, the load factor in the fumigation chamber, the temperature during treatment, and any chemical dips or coating on the commodity all influence the potential phytotoxic response of the commodity.
2. Following the chemical treatment, the method of aeration, holding temperature, and method of handling the commodity also can affect the commodity's response.
3. The resistance of the commodity to injury by the treatment also is influenced by the cultivar, maturity and environmental conditions the commodity experienced prior to treatment.

Plants have various mechanisms to prevent injury from a fumigation treatment. The plant can avoid contact with the chemical by the presence of a physical or chemical barrier, such as the cuticle, closed stomata, or the chemical nature of the cell wall or membrane. Mechanisms for a plant to tolerate exposure to a fumigant include a slow metabolic rate, and protective compounds or enzymes that react with the chemical to prevent it from causing injury.

Injury of fresh commodities by fumigation can be minimized if treatment conditions and parameters are carefully controlled and monitored. Identifying mechanisms that increase the resistance of commodities to fumigants may aid in making chemical treatments more useful, economical, and less damaging.

References

Ahmed, M.K. and El Zorgani, G.A. (1984) The effect of methyl bromide fumigation on stored potatoes. *Acta Horticulturae* 143, 539–544.

Alexeeff, G.V. and Kilgore, W.W. (1983) Methyl bromide. *Residue Reviews* 88, 101–153.

Alumot, E. and Chalutz, E. (1972) Fumigation of citrus fruit with ethylene dibromide: desorption of residues and ethylene evolution. *Pesticide Science* 3, 539–544.

Ames, B.N. and Gold, L.S. (1990) Too many rodent carcinogens: mitogenesis increases mutagenesis. *Science* 249, 970.

Archibald, S.O. and Winter, C.K. (1990) Pesticides in our food. Assessing the risks. In: Winter, C.K., Seiber, J.N. and Nuckton, C.F. (eds) *Chemicals in the Human Food Chain*. Van Nostrand Reinhold, New York, pp. 1–50.

Armitage, H.M. and Steinweden, J.B. (1946) Tolerance of citrus fruits to methyl bromide as a fumigant. *California Department of Agriculture Bulletin* 35, 21–29.

Armstrong, J.W. and Garcia, D.L. (1985) Methyl bromide quarantine fumigations for Hawaii-grown cucumbers infested with melon fly and oriental fruit fly (Diptera: Tephritidae). *Journal of Economic Entomology* 78(6), 1308–1310.

Auda C. and Canessa, M.A. (1981) Relacion entre fitotoxicidad del bromuro de metilo y temperatura postfumigacion en manzanas Granny Smith. *Simiente* 51, 142–144.

Beever, D.J. and Yearsley, C.W. (1987) Kiwifruit: effect of post-harvest fumigation on fruit quality. *New Zealand Journal of Experimental Agriculture* 15, 185–189.

Benschoter, C.A. (1979) Seasonal variation in tolerance of Florida 'Marsh' grapefruit to a combination of methyl bromide fumigation and cold storage. *Proceedings of the Florida State Horticulture Society* 92, 166–167.

Benz, R. and Cros, D. (1978) Influence of sterols on ion transport through lipid bilayer membranes. *Biochimica Biophysica Acta* 506, 265–280.

Bolter, C.J. and Chefurka, W. (1990) The effect of phosphine treatment on superoxide dismutase, catalase, and peroxidase in the granary weevil, *Sitophilus granarius*. *Pesticide Biochemistry and Physiology* 36, 52–60.

Bond, E.J. (1963) The action of fumigants on insects, IV. The effects of oxygen on the toxicity of fumigants to insects. *Canadian Journal of Biochemistry and Physiology* 41, 993–1004.

Bond, E.J. (1984) *Manual of Fumigation for Insect Control*. FAO Plant Production and Protection Paper No. 54. Food and Agriculture Organization of the United Nations, Rome.

Brecht, J.K., Huber, D.J., Sherman, M. and Lee, J. (1986) Methyl bromide inhibits ripening and ethylene production in tomato (*Lycopersicon esculentum* Mill.) fruit. *Journal of Plant Growth Regulation* 5, 29–35.

Brown, D.J. and DuPont, F.M. (1989) Lipid composition of plasma membranes and endo-membranes prepared from roots of barley (*Hordeum vulgare* L.). Effect of salt. *Plant Physiology* 90, 955–961.

Brown, R.E. and Nickerson, J.C. (1980) Ethylene dibromide and methyl bromide fumigation of Persian limes and lemon-limes. *Proceedings of the Florida State Horticulture Society* 93, 348–349.

Bussel, J. and Kamburov, S.S. (1976) Ethylene dibromide fumigation of citrus fruit to control the Mediterranean fruit fly, *Ceratitis capitata* (Wied.). *Journal of the American Society for Horticultural Science* 101(1), 11–14.

Chalutz, E., Schiffmann-Nadel, M., Waks, J., Alumot, E., Carmi, Y. and Bussel, J. (1971) Peel injury to citrus fruit fumigated with ethylene dibromide. *Journal of the American Society for Horticultural Science* 96(6), 782–785.

Chalutz, E., Biron, S. and Alumot, E. (1973) Reduction of ethylene dibromide peel injury in citrus fruits by thiabendazole. In: Refrigération et maturation des fruits: relation avec la qualité. *Bulletin de l'Institut International du Froid* Annexe 3, pp. 205–209.

Chapin, F.S., III (1991) Integrated responses of plants to stress. A centralized system of physiological responses. *BioScience* 41, 29–36.

Chaplin, G.R., Rippon, L.E., Singh, G. and McBride, R.L. (1981) Fruit injury and residues resulting from methyl bromide fumigation of avocados. *Australian Journal of Experimental Agriculture and Animal Husbandry* 21, 610–613.

Claypool, L.L. and Vines, H.M. (1956) Commodity tolerance studies of deciduous fruits to moist heat and fumigants. *Hilgardia* 24(12), 297–355.

Cumming, J.R. and Taylor, G.J. (1990) Mechanisms of metal tolerance in plants: physiological adaptations for exclusion of metal ions from the cytoplasm. In: Alscher, R.G., Cumming, J.R., and Allen, N.S. (eds), *Stress Responses in Plants: Adaptation and Acclimation Mechanisms*. Wiley-Liss, pp. 201–215.

Dass, H.C. and Weaver, G.M. (1972) Enzymatic changes in intact leaves of *Phaseolus vulgaris* following ozone fumigation. *Atmospheric Environment* 6, 759–763.

Dean, C.E. (1972) Stomate density and size as related to ozone-induced weather fleck in tobacco. *Crop Science* 12, 547–548.

Drake, S.R. and Moffitt, H.R. (1992) Winter pear ('Anjou' and 'Bosc') response to methyl bromide fumigation. *HortScience* 27(7), 813–816.

Drake, S.R., Moffitt, H.R., Fellman, J.K. and Sell, C.R. (1988) Apple quality as influenced by fumigation with methyl bromide. *Journal of Food Science* 53(6), 1710–1712.

Eaks, I.L. and Ludi, W.A. (1958) Response of orange and lemon fruits to fumigation with ethylene dibromide effective against eggs and larvae of the oriental and Mexican fruit flies. *Proceedings of the American Society for Horticultural Science* 72, 297–303.

Forney, C.F. (1990) Ripening and solar exposure alter polar lipid fatty acid composition of 'Honey Dew' muskmelons. *HortScience* 25(10), 1262–1264.

Forney, C.F. and Lipton, W.J. (1990) Influence of controlled atmospheres and packaging on chilling sensitivity. In: Wang, C.Y. (ed.), *Chilling Injury of Horticultural Crops*. CRC Press, Boca Raton, Florida, pp. 257–267.

Forney, C.F., Peterson, S.J. and Hartsell, P. (1990) Preconditioning grapefruit callus tissue reduces methyl bromide induced K^+ leakage. *HortScience* 25(6), 669–670.

Gandy, D.G. and Chanter, D.O. (1976) Some effects of time, temperature of treatment, and fumigant concentration on the fungicidal properties of methyl bromide. *Annals of Applied Biology* 82, 279–290.

Gaur, S.N., Raha, S.K., Masood, M.D. and Chattopadhyay, D.(1984) A note on fumigation of orchid plants with methyl bromide. *Plant Protection Bulletin* 36, 133–134.

Gold, L.S., Slone, T.H., Stern, B.R., Manley, N.B. and Ames, B.N. (1992) Rodent carcinogens: setting priorities. *Science* 258, 261–265.

Hansen, J.D., Chan H.T., Hara, A.H. and Tenbrink, V.L. (1991) Phytotoxic reaction of Hawaiian cut flowers and foliage to hydrogen cyanide fumigation. *HortScience* 26(1), 53–56.

Harvey, J.M. and Harris, C.M. (1982) Phytotoxic responses of cherries, nectarines, peaches, pears and plums fumigated with methyl bromide for control of mediterranean fruit fly. *Journal of the American Society for Horticultural Science* 107(6), 993–996.

Harvey, J.M., Harris, C.M. and Hartsell, P.L. (1989) Tolerances of California nectarine cultivars to methyl bromide quarantine treatments. *Journal of the American Society for Horticultural Science* 114(4), 626–629.

Hatton, T.T. and Cubbedge, R.H. (1973) Preferred temperature for prestorage conditioning of 'Marsh' grapefruit to prevent chilling injury at low temperatures. *HortScience* 18, 721–722.

Hatton, T.T. and Cubbedge, R.H. (1979) Phytotoxicity of methyl bromide as a fumigant for Florida citrus fruit. *Proceedings of the Florida State Horticulture Society* 92, 167–169.

Hatton, T.T. and Cubbedge, R.H. (1986) Quarantine research to disinfest Florida citrus fruits of the Caribbean fruit fly. *The Citrus Industry* 67(2), 18–26.

Hatton, T.T., Cubbedge, R.H., von Windeguth, D.L. and Spalding, D.H. (1982) Control of the Caribbean fruit fly in Florida grapefruit by phosphine fumigation. *Proceedings of the Florida State Horticultural Society* 95, 221–224.

Helyer, N.L. and Ledieu, M.S. (1989) Phytotoxicity of methyl bromide on chrysanthemum cuttings. *Annals of Applied Biology* 114, 9–13.

Houck, L.G. (1967) Hot water treatments for control of *Penicillium digitatum* green mold of Eureka lemons. *Phytopathology* 57(2), 99.

Houck, L.G. and Jenner, J.F. (1988) Phytotoxicity in California citrus fruit fumigated with methyl bromide. *International Citrus Congress Middle East 1988 Book of Abstracts*, pp.144.

Houck, L.G. and Mackey, B.E. (1989) Permeability of flexible polymer films used to wrap citrus fruit to the fumigants ethylene dibromide and methyl bromide. *Journal of the American Society for Horticultural Science* 114(1), 86–90.

Houck, L.G., Jenner, J.F., Tebbets, J.S. and Hartsell, P.L. (1985) Phytotoxic responses of citrus fruit to fumigation with ethylene dibromide. *Phytopathology* 75(5), 616–621.

Houck, L.G., Hartsell, P.L., Jenner, J.F. and Tebbets, J.C. (1987) Dosage, time, and CT product vs. injury to lemons fumigated with methyl bromide. *HortScience* 22(5), 1135.

Houck, L.G., Jenner, J.F., Moreno, D.S. and Mackey, B.E. (1989) Permeability of polymer film wraps for citrus fruit fumigated with hydrogen cyanide to control California red scale. *Journal of the American Society for Horticultural Science* 114(2), 287–292.

Houck, L.G., Jenner, J.F. and Mackey, B.E. (1990) Seasonal variability of the response of desert lemons to rind injury and decay caused by quarantine cold treatments. *Journal of Horticultural Science* 65(5), 611–617.

Howell, R.K. (1974) Phenols, ozone, and their involvement in pigmentation and physiology of plant injury. In Dugger, M. (ed.), *Air Pollution Effects on Plant Growth*. ACS Symposium Series 3. American Chemical Society, Washington, pp. 94–105.

Hurkman, W.J., Tanaka, C.K. and DuPont, F.M. (1988) The effect of salt stress on polypeptides in membrane fractions from barley roots. *Plant Physiology* 88, 1263–1273.

Ironside, D.A., Swaine, G. and Corcoran, R.J. (1978) Methyl bromide fumigation for eliminating macadamia felted coccid (*Eriococcus ironsidei* Williams) from propagating material. *Queensland Journal of Agricultural and Animal Sciences* 35(1), 29–33.

Isaac, W.E. (1944) The effect of methyl bromide fumigation on apples. *Farming in South Africa* 19, 703–710.

Ito, P.J. and Hamilton, R.A. (1980) Fumigation of avocado fruit with methyl bromide. *HortScience* 15(5), 593.

Jessup, A.J. (1988) Response of 'Lambert' and 'Ron's Seedling' sweet cherries to fumigation with methyl bromide plus cold storage. *Australian Journal of Experimental Agriculture* 28, 431–434.

Kawakami, F. and Soma, Y. (1991) Some factors causing chemical injury of apples fumigated with methyl bromide and injury protection by gas-absorbents. *Research Bulletin of Plant Protection Japan* 27, 41–46.

Ketterer, B. (1986) Detoxification reactions of glutathione and glutathione transferases. *Xenobiotica* 16, 957–973.

Knott, J.E. and Claypool, L.L. (1940) Some responses of tomato fruits to methyl-bromide fumigation. *Proceedings of the American Society for Horticultural Science* 38, 501–506.

Kohn, S. (1962) Uses of methyl bromide on our mushroom plant. *Mushroom Growers' Association Bulletin* 152, 329–333.

Latta, R. (1941) Methyl bromide fumigation of greenhouse plants at the U.S. plant introduction garden, Glenn Dale, MD. *Insect Pest Survey Bulletin* E-526, 1–31.

Leesch, J.G., Davis, R., Fons, J.G., Reeves, R., Houck, L.G. and Zehner, J.M. (1984) Ethylene dibromide fumigation of citrus in reefer compartments on a refrigerated ship. *Journal of Economic Entomology* 77, 773–783.

Levitt, J. (1980) *Responses of Plants to Environmental Stresses. Volume 1 Chilling, Freezing, and High Temperature Stresses*, 2nd edn. Academic Press, New York.

Lindgren, D.L. and Sinclair, W.B. (1951) Tolerance of citrus and avocado fruits to fumigants effective against the oriental fruit fly. *Journal of Economic Entomology* 44(6), 980–990.

Lipton, W.J. and Tebbets, J.S. (1984) Methyl bromide and ethylene dibromide as potential quarantine treatments of cantaloupe and 'Honey Dew' muskmelons against tephritid fruit flies. *Acta Horticulturae* 157, 161–168.

Lipton, W.J., Tebbets, J.S., Spitler, G.H. and Hartsell, P.L. (1982) Commodity treatments: Responses of tomatoes and green bell peppers to fumigation with methyl bromide or ethylene dibromide. *Marketing Research Report No. 1125*. United States Department of Agriculture, Washington, DC.

Lynch, D.V. and Steponkus, P.L. (1987) Plasma membrane lipid alterations associated with cold acclimation of winter rye seedlings (*Secale cereale* L. cv. Puma). *Plant Physiology* 83, 761–767.

Malhotra, S.S. and Khan, A.A. (1978) Effects of sulphur dioxide fumigation on lipid biosynthesis in pine needles. *Phytochemistry* 17, 241–244.

McKenry, M.V. (1981) The nature, mode of action, and biological activity of nematicides. In: Pimental, D. (ed.), *CRC Handbook of Pest Management in Agriculture*, Vol. III. CRC Press, Boca Raton, Florida, pp. 59–79.

Meheriuk, M., Gaunce, A.P. and Dyck, V.A. (1990) Response of apple cultivars to fumigation with methyl bromide. *HortScience* 25(5), 538–540.

Meister, A. (1988) Glutathione metabolism and its selective modification. *The Journal of Biological Chemistry* 263(33), 17205–17208.

Mizutani, F., Yamanaka, Y., Amano, S., Hino, A. and Kadaya, K. (1992) Effects of ethylene and hydrogen cyanide on β-cyanoalanine synthase activity in satsuma mandarin (*Citrus unshiu* Marc.) fruit. *Scientia Horticulturae* 49, 223–231.

Morikawa, O. (1964) Effects of nematocides on the glycolysis of the American cockroach. *Japanese Journal of Applied Entomology and Zoology* 8, 277–285.

Moshonas, M.G. and Shaw, P.E., (1982) Irradiation and fumigation effects on flavor, aroma and composition of grapefruit products. *Journal of Food Science* 47, 958–960.

Munro, H.A.U. (1969) *Manual of Fumigation for Insect Control*. FAO Agricultural Studies No. 79. FAO Plant Production and Protection Series No. 20. Food and Agriculture Organization of the United Nations, Rome.

Nachtomi E. and Sharma, D.S.R. (1977) *In vitro* repair of rat liver DNA damaged by ethylene dibromide. *Biochemical Pharmacology* 26, 1941–1945.

Phillips, W.R., Monro, H.A.U. and Allen, C.E. (1938) Some observations on the fumigation of apples with methyl bromide. *Scientific Agriculture* 19, 7–20.

Pradhan, S., Bhambhani, H.J. and Wadhi, S.R. (1960) Fumigation of infested potatoes. *Indian Journal of Entomology* 22, 181–198.

Price, N.R. (1985) The mode of action of fumigants. *Journal of Stored Product Research* 21(4), 157–164.

Price, A., Lucas, P.W. and Lea, P.J. (1990) Age dependent damage and glutathione metabolism in ozone fumigated barley: A leaf section approach. *Journal of Experimental Botany* 41(231), 1309–1317.

Purnell, R.E. and Hague, N.G.M. (1965) Fumigation of agricultural products. XIX – Methyl bromide fumigation of narcissus bulbs infested with stem nematode, *Ditylenchus dipsaci* (Kühn). *Journal of the Science of Food and Agriculture* 16, 421–424.

Rigney, C.J., Graham, D. and Lee, T.H. (1978) Changes in tomato fruit ripening caused by ethylene dibromide fumigation. *Journal of the American Society for Horticultural Science* 103(3), 420–423.

Rippon, L.E., Singh, G., Sproul, A.N. and Gilbert, W.S. (1982) Methyl bromide fumigation and cold storage for disinfestation of Granny Smith apples against Queensland and Mediterranean fruit flies. *Australian Journal of Experimental Agriculture and Animal Husbandry* 22, 116–123.

Roth, H. and Richardson, H. H. (1963) Tolerance of imported garlic bulbs to methyl bromide fumigation and hot-water dips. *Journal of Economic Entomology* 56(6), 839–842.

Sanford, K.H. (1962) Fumigation of apples to control the apple maggot, *Rhagoletis pomonella*. *Journal of Economic Entomology* 55(5), 724–727.

Sanford. K.H., Stewart, D.K.R. and Poapst, P.A. (1976) Fumigation of apples under plastic tarpaulins with ethylene dibromide to control diapausing European red mite (Acarina: Tetranychidae) eggs. *The Canadian Entomologist* 108, 217–223.

Scheuplein, R.J. (1992) Perspectives on toxicological risk – an example: foodborne carcinogenic risk. *Critical Reviews in Food Science and Nutrition* 32, 105–121.

Seo, S.T., Balock, J.W., Burditt, A.K., Jr and Ohinata, K. (1970) Residues of ethylene dibromide, methyl bromide and ethylene chlorobromide resulting from fumigation of fruits and vegetables infested with fruit flies. *Journal of Economic Entomology* 63(4), 1093–1097.

Seo, S.T., Akamine, E.K., Goo, T.T.S., Harris, E.J. and Lee, C.Y.L. (1979) Oriental and Mediterranean fruit flies: fumigation of papaya, avocado, tomato, bell pepper, eggplant, and banana with phosphine. *Journal of Economic Entomology* 72(3), 354–359.

Smith, I.K., Polle, A. and Rennenberg, H. (1990) Glutathione. In: Alscher, R.G., Cumming, J.R. and Allen, N.S. (eds), *Stress Responses in Plants: Adaptation and Acclimation Mechanisms*. Wiley-Liss, pp. 201–215.

Spalding, D.H., Benschoter, D.L., von Windeguth, D.L., King, J.R., Reeder, W.F. and Burditt, A.K., Jr (1977) Methyl bromide and phosphine fumigation injury to avocados and mangoes. *Proceedings of the Florida State Horticultural Society* 90, 268–270.

Spalding, D.H., King, J.R., Benschoter, C.A., von Windeguth, D.L., Reeder, W.F. and Burditt, A.K., Jr (1978) Ethylene dibromide, methyl bromide and phosphine fumigation of tomatoes. *Proceedings of the Florida State Horticulture Society* 91, 156–158.

Sproul, A.N., O'Loughlin, J.B., Terauds, A. and Hardisty, S. (1976) The use of diphenylamine to protect apples from methyl bromide injury. *Australian Journal of Agricultural Research* 27, 541–545.

Starratt, A.N. and Bond, E.J. (1981) Metabolism of methyl bromide by susceptible and resistant strains of the granary weevil, *Sitophilus granarius* (L.) *Pesticide Biochemistry and Physiology* 15, 275–281.

Starratt, A.N. and Bond, E.J. (1988) Methylation of DNA of maize and wheat grains during fumigation with methyl bromide. *Journal of Agricultural and Food Chemistry* 36, 1035–1039.

Starratt, A.N. and Bond, E.J. (1990a) Methylation of food commodities during fumigation with methyl bromide. In: FAO/IAEA (eds), *Studies of the Magnitude and Nature of Pesticide Residues in Stored Products*. IAEA, Vienna, pp. 97–109.

Starratt, A.N. and Bond E.J. (1990b) Role of glutathione in reducing residues of methyl bromide in fumigated commodities. *Pesticide Biochemistry and Physiology* 38, 178–185.

Thompson, G.A., Jr (1985) Mechanisms of membrane response to environmental stress. In: St John, J.B., Berlin, E. and Jackson, P.C. (eds), *Frontiers of Membrane Research*. Rowman and Allanheld, Totowa, New Jersey, pp. 347–357.

Thompson, J.E. (1984) Physical changes in the membranes of senescing and environmentally stressed plant tissues. In: Shinitzky, M. (ed.), *Physiology of Membrane Fluidity, Volume II*. CRC Press, Boca Raton, Florida, pp. 85–108.

Thompson, R.H. (1966) A review of the properties and usage of methyl bromide as a fumigant. *Journal of Stored Product Research* 1, 353–376.

Turner, J.W. (1984) Disinfestation of imported flowers and plants. In: Bailey, P. and Swincer, D. (eds), *Proceedings of the Fourth Australian Applied Entomological Research Conference*. South Australian Department of Agriculture, Adelaide, South Australia, pp. 488–493.

United States Department of Agriculture (1985) *Plant Protection and Treatment Manual, Section III*. Animal and Plant Health Inspection Service, Hyattsville, Maryland.

Williamson, D.L., Summy, K.R., Hart, W.G., Sanchez-R, M., Wolfenbarger, D.A. and Bruton, B.D. (1986) Efficacy and phytotoxicity of methyl bromide as a fumigant for the Mexican fruit fly (Diptera: Tephritidae) in grapefruit. *Journal of Economic Entomology* 79(1), 172–175.

Irradiation 9

Stephen C. Morris[1] and Andrew J. Jessup[2]
[1]*CSIRO, Division of Horticulture, PO Box 52, North Ryde, NSW 2113, Australia;*
[2]*NSW Agriculture, Horticultural Research and Advisory Station, Gosford, NSW 2250, Australia.*

Introduction

Much of the irradiation literature dealing with fresh horticultural products uses doses considerably higher than required for disinfestation. Doses of irradiation sufficient to sterilize an insect are generally considerably lower than those required to kill that insect. An alternative approach to killing or sterilization is to prevent the development of an insect from one life-stage to another – this necessitates an intermediate irradiation dose. Many importing countries argue that infesting insects can only be regarded as low quarantine risk when they are dead, either before the product leaves the exporting region or at least before it arrives on-shore in the importing region. It seems that drastic changes to quarantine inspection and product treatment certification must occur before a quarantine official will pass products infested with live, though sterile, insects, or insects incapable of proceeding to the next life stage and therefore of reaching the adult stage.

In 1986 the Food and Agriculture Organization/International Atomic Energy Agency International Consultative Group on Food Irradiation adopted an irradiation dose of 300 Gy as the minimum dose required to treat any stage of any insect likely to pose a quarantine threat, except where data were available to prove efficacy at a lower dose (FAO/IAEA, 1986). For many insects of critical quarantine importance, a dose of 300 Gy will not cause immediate death but would prevent further development. Depending on insect species, doses of between 250 Gy and about 1000 Gy are required to kill eggs, larvae or infesting adults. Consequently, this review's discussion is largely confined to research in

From: Paull, R.E. and Armstrong, J.W. (eds) (1994) *Insect Pests and Fresh Horticultural Products: Treatments and Responses.* CAB INTERNATIONAL, Wallingford, UK.

which doses below 1000 Gy were used. Other reviews discuss results using dose rates above 1000 Gy (Akamine and Moy, 1983; Basson *et al.*, 1983; Matsuyama and Umeda, 1983; Moy, 1983; Sommer and Mitchell, 1986; Urbain, 1986; Morris, 1987). Older references are cited in these reviews.

Shelf-life

Irradiation at disinfestation doses of 1000 Gy or less can extend the shelf-life of some fruit and vegetables, typically for several days (Akamine and Moy, 1983). This is achieved through irradiation-induced effects on the fruit's physiology, biochemistry and its population of spoilage micro-organisms. The effectiveness of irradiation in delaying ripening in climacteric fruit is closely allied with the fruit's physiological stage of maturity, i.e. its position on the characteristic climacteric peak of respiration and ethylene production. Irradiation delays the ripening of pears and bananas treated before the climacteric. If fruit, such as bananas and pears, are irradiated after the start of the climacteric increase in respiration, there is little effect on fruit ripening. Lack of attention to the physiological stage of maturity of various fruits at the time of irradiation can result in misleading or faulty observations concerning irradiation effects on fruit quality (Clarke, 1971; Thomas, 1986a; Urbain, 1986). Urbain (1986) suggests that maturity at the time of irradiation is the reason for so much contradictory literature regarding adverse and beneficial effects of irradiation on the shelf-life and keeping quality of fresh horticultural produce. Irradiation at higher doses can accelerate the ripening of peaches and nectarines with the degree of response depending on the time of irradiation and the dose used.

Irradiation can affect shelf-life in either a positive or a negative way. For example, irradiation of onions has a range of effects and includes the control of sprouting and pathogens (Table 9.1). Irradiation can be beneficially used at disinfestation and sprout inhibiting doses (i.e. below 1000 Gy) on many fruits and vegetable that will not tolerate doses required to control pathogens, ripening and senescence (Maxie *et al.*, 1971). As with most irradiated horticultural produce, the limiting factor is the irradiation dose tolerated before phytotoxicity. The beneficial effects of delaying ripening and senescence and controlling disease require higher doses near the phytotoxic threshold (Maxie *et al.*, 1971).

Table 9.1. Effects of irradiation on onion storage life.

Response	Dose (kGy)	Reference
Sprouting		
Control of sprouting, providing irradiated onions within 1–2 months	0.05–0.15	Takano *et al.* (1974) Fu *et al.* (1979) Thomas *et al.* (1986)
Pathogenicity		
No effect on decay	0.05–0.15	Takano *et al.* (1974) Thomas *et al.* (1986)
Reduce decay control	0.05–0.07	Thomas *et al.* (1986) Nuttall *et al.* (1961) MacQueen (1965) Wu *et al.* (1979)
Phytotoxicity		
Centre of bulbs (sprouts) become dark-brown and die	0.05–0.15	Nair *et al.* (1973) Curzio and Croci (1983)
Often internal decay	0.05–0.15	Nuttall *et al.* (1961) Nair *et al.* (1973)
Maximum of 1 month life after removal from refrigeration	0.05–0.15	Takano *et al.* (1974) Kawashima *et al.* (1984)

Source: From Morris, 1987. Reprinted with permission.

Irradiation-induced Injury

External damage

Surface pitting following irradiation is one of the typical symptoms of external damage. Pitting was found in early-season 'Washington Navels' at 75 Gy (MacFarlane and Roberts, 1968), green 'Shamouti' oranges at 140 Gy (when irradiated immediately after harvest) (Monselise and Kahan, 1966) and 'Marsh's Seedless' grapefruit at 250 Gy (Burditt *et al.*, 1981). Some mango varieties develop skin bronzing or scalding at doses as low as 500 Gy (Urbain, 1986). Avocados suffer severe injury in the forms of skin blackening and spotting at doses around 100 Gy (Kahan *et al.*, 1968). Dates, asparagus, and bananas all show skin browning when irradiated at doses greater than 350 Gy (Maxie *et al.*, 1968; Urbain, 1986). Cultivar, pre- and postharvest conditions, maturity, time between harvest and irradiation, rate of irradiation and storage conditions after irradiation all influence the incidence and severity of external phytotoxic symptoms (Kahan *et al.*, 1968; Thomas, 1986b).

Internal damage

Internal damage can result in the formation of cavities or gas fissures as seen in 'Eureka' lemons irradiated at doses greater than 500 Gy (Maxie *et al.*, 1969) or even at 75 Gy (Jessup and Baheer, unpublished data). It appears that cavitation in lemons is due to the destructive effect that irradiation has on the pectic (and other) substances that hold the fruit segments together. When lemons are irradiated at 500 Gy and juiced, the segment walls completely disintegrate, whereas segments from non-irradiated lemons remain intact. This points to a degradation of cell structural components. Cavitation can also occur in mangoes at 1500 Gy (Spalding and von Windeguth, 1988) and papaya (Urbain, 1986).

Browning reactions

One of the most common symptoms of radiation injury is a browning or discolouring of either the surface or flesh. Browning can occur with doses at 250 Gy or less for very sensitive fruits, such as avocados or apricots (Guelfat-Reich *et al.*, 1969; Maxie *et al.*, 1971; Rigney, 1986); other fruits, such as preclimacteric bananas, mangoes, pineapples, oranges and grapefruit brown at doses of 500 up to 1000 Gy (Thomas *et al.*, 1971; Jessup *et al.*, 1988; Spalding and von Windeguth, 1988). Fruit maturity influences the sensitivity to browning with preclimacteric bananas suffering skin browning at doses of 500 Gy, while ripe bananas tolerate up to 2000 Gy (Thomas *et al.*, 1971). Different tissue types can also have different susceptibility to browning with oranges and grapefruit browning at 500 Gy for the calyx but at 600 Gy for the peel (Chachin and Ogata, 1971); the cherry fruit was not affected by doses up to 1000 Gy, although at these doses there was browning of the peduncle (Jessup, 1990). The delaying effect of oxygen tension on irradiation-induced discoloration in avocados suggests that this browning damage is mediated by an oxidative means (Rigney, 1986).

Browning of potatoes occurs around the vascular bundles and involves some necrosis of cells. Browning is observed after irradiation at 100–400 Gy, reaching a maximum 3–7 days after irradiation; after 1 month for doses of 100 Gy and after 2 months for doses of 300 Gy and above. The browning is greater in immature potatoes and is not observed in sprouting potatoes treated with doses up to 500 Gy. Low temperature (5–7°C) after irradiation helped suppress browning (Ogata *et al.*, 1970). After harvest the susceptibility of potatoes to irradiation induced browning declines, until treatment after 10 weeks' storage results in no browning (Tatsumi *et al.*, 1971). Browning of potatoes after cooking by boiling is increased by irradiation, due possibly to the formation of an iron–phenolic complex, with the higher levels of darkening of boiled or steamed potatoes after irradiation due to increases in polyphenolic levels (Thomas, 1981). Darkening of irradiated potatoes after frying is due to a caramelization at high temperatures of the high levels of sucrose induced (Borsa *et al.*, 1989).

Treatment of onions and garlic with doses sufficient to control sprouting (80–150 Gy) resulted in a browning discoloration and necrosis of the meristematic growing point (Morris, 1985, 1987). This browning was minimized if the bulbs were treated as soon after harvest as possible, while still in the dormant stage (Thomas *et al.*, 1975). Interestingly, browning was reduced by irradiation for mushrooms where the typical browning associated with senescence was delayed, due to inhibition of growth and senescence by irradiation (Roy and Bahl, 1984).

Texture

General softening

Flesh softening in a large number of crops is one consequence of irradiation above a threshold (Table 9.2). The threshold varies for different crops and ranges from <300 Gy for apricots and limes to 2000–2500 Gy for strawberries, muskmelons and tomatoes. There are considerable differences between similar fruits, i.e. stone fruits, and also variations for one crop. This softening can be undesirable and can cause subsequent damage during handling, packaging, and transport (Maxie *et al.*, 1971). Care therefore needs to be taken during subsequent packing and transport; commercially unacceptable levels of softening, however, may not be experienced until doses up to 100% higher than those reported in Table 9.2.

Effects on ripening and texture

Irradiation at low doses can delay the fruit ripening process, with the result that the typical softening during ripening is also delayed. Therefore, after the initial radiation-induced softening, irradiated unripe fruit can be much firmer over time compared to soft, fully ripe non-irradiated fruit. Some fruits for which irradiation delays ripening and which retain firm texture for longer are mangoes treated at 600–1000 Gy (Boag *et al.*, 1990; Jessup *et al.*, 1988), papaya 250–750 Gy (Moy *et al.*, 1971), bananas 200 Gy (Strydom *et al.*, 1991) and tomatoes 750–2500 Gy (Abdel-Kader *et al.*, 1968). However, too high irradiation doses can accelerate fruit ripening and result in more rapid softening than unirradiated fruit (Guelfat-Reich *et al.*, 1969).

Table 9.2. Irradiation dose at which the first significant softening of several crops occurs.

≤300 Gy	300–600 Gy	600–1000 Gy	≥1000 Gy
Limes (Nyambati and Langerak, 1984)	Bananas (Eric *et al.*, 1970; Strydom *et al.*, 1991)	Papaya (Moy *et al.*, 1971)	Mangoes (Boag *et al.*, 1990; Jessup *et al.*, 1988)
Apricots (Guelfat-Reich *et al.*, 1969)	Plums (Moy *et al.*, 1983)	Oranges (Fesues *et al.*, 1981)	Raspberries (Larrigaudiere *et al.*, 1987)
Apples (Flores *et al.*, 1971)	Cherries (Eaton *et al.*, 1970)	Apples and pears (Kovacs *et al.*, 1988)	Sweet potatoes (Lu *et al.*, 1989)
	Blueberries (Eaton *et al.*, 1970)	Peaches and nectarines (Moy *et al.*, 1983)	Muskmelons (Lester, 1989)
		Cherries (Jessup, 1990)	Tomatoes (Abdel-Kader *et al.*, 1968; Yasia *et al.*, 1987)
			Mushrooms (Wahid and Kovacs, 1980)

Interactions with water loss and growth

A low dose of irradiation can control sprouting in many storage organs. This control of sprouting results in a firmer organ, since the water and carbohydrate reserves required during sprouting are not used. For potatoes, weight loss decreased with increased doses, 20 to 160 Gy, with firmness being negatively related to weight loss and sprouting (Lee and Kim, 1972). This relationship of increased firmness with control of sprouting is also found for yams at 100 Gy (Rivera *et al.*, 1974) and sweet potatoes at doses of 200–1000 Gy (Lu *et al.*, 1989). Mushrooms also retain firmness longer after irradiation, by reducing postharvest growth and opening of the pileus (head), growth of the stipe (stem) and associated weight loss (Wahid and Kovacs, 1980).

Cell wall changes affecting firmness

Irradiation-induced softening may be associated with cell wall calcium and pectin changes (see Pectins and cellulose below). Cell wall polysaccharide solubilization following irradiation of carrots was observed by Echandi *et al.* (1970). Irradiation-induced changes of apple and pear cell walls include dissolution of the middle lamellae and a creasing or wrinkling of formerly smooth cell walls (Kovacs *et al.*, 1988).

Chemical Constituents

Sugars

Irradiation at doses of up to 2000 Gy seems to have no significant effect on sugar levels, both immediately after irradiation and during subsequent storage for a range of tropical fruit (bananas, litchis, mangoes, and papayas) and also pears, grapes, dates, and strawberries (Donini and Pansolli, 1970; Eric *et al.*, 1970; Sattar *et al.*, 1971; Beyers *et al.*, 1979; Thomas and Beyers, 1979; McLauchlan *et al.*, 1992). A similar pattern of no change or a slight decline is observed over a range of doses up to 2000 Gy, for oranges (MacFarlane and Roberts, 1968; Moshonas and Shaw, 1984) and grapefruit (Garcia *et al.*, 1990). Most vegetables show no effect on sugars with doses up to 2000 Gy (Abdel-Kader *et al.*, 1968; Rivera *et al.*, 1974; Ismail and Afifi, 1976). Pistachio nuts treated at 500 and 10 kGy did not show any change in sugar levels (Kashani and Valadon, 1984).

For some crops, however, there are major changes in sugar levels. A marked decline in sugar levels occurs in green gram beans. Irradiation (2500 Gy) of seeds after soaking overnight in water resulted in over 50% decline of the flatulence-causing sugars – stachyose and raffinose (the possibility of flatulence-free beans is surely a most compelling argument on its own for food irradiation) (Rao and Vaskil, 1983). Chestnut sugar levels increased substantially after irradiation at 250–1500 Gy with the major increase being reducing sugars (Hayashi *et al.*, 1983).

Potatoes show a substantial increase in sugar content (up to tenfold) after irradiation at levels of as low as 50–150 Gy (Borsa *et al.*, 1989), to a maximum at 2000–3000 Gy (Hayashi and Kawashima, 1982). Tuber sucrose levels had the greatest increase (Burton, 1975) and peaked after 3 days, reducing sugars peaked after 7 days, both then declined over the next three months to original levels (Pratella and Menniti, 1978; Chen *et al.*, 1980). This irradiation-induced sugar increase was greatest when potatoes were subsequently stored at 5°C or lower. For potatoes with different sugar levels (induced by cool storage) before irradiation, the greatest irradiation-induced increase was in potatoes with low sucrose levels (0.14–0.37%); when tuber sucrose is greater than 1.4%, only very slight changes occurred (Burton, 1975).

Irradiation restricted the breakdown of sucrose and accelerated its synthesis in potato, especially from uridine 5′-diphosphate (UDP)-glucose and fructose-6-phosphate (Hayashi and Aoki, 1985). Potato phosphorylase, sucrose synthase and sucrose phosphate synthase activities are significantly enhanced reaching a maximum several days after irradiation (150 and 2000 Gy) and remaining high for up to 14 days. Sucrose synthase subsequently remained high while activity of the others declined (Hayashi *et al.*, 1984). Phosphoglucomutase, phosphoglucose isomerase and UDP-glucose pyrophosphorylase activities did not markedly change after irradiation (Hayashi and Kawashima, 1983a). Starch phosphorylase activity is greatly enhanced by irradiation at 50–100 Gy, while starch amylase is unaffected (Chen *et al.*, 1980).

A similar increase in sucrose levels (by 2–4 times) is found in sweet potatoes after irradiation (Hayashi and Kawashima, 1982). Sucrose was at a maximum at 5–10 days after irradiation and declined to normal within 1–3 months (Ajlouni and Hamdy, 1988). The extent of sucrose increase paralleled increasing dose rate to a maximum at 2000 Gy (Lu *et al.*, 1989). Unlike potatoes, cool storage had no effect on sucrose synthesis in sweet potato (Hayashi and Kawashima, 1982). Sweet potatoes show a similar increase in sucrose synthase and sucrose phosphate synthase activities to potatoes (Hayashi *et al.*, 1984). Irradiation of preclimacteric bananas at 350 Gy resulted in an increase of the activity of fructose-1,6-diphosphatase by fivefold after 3 days (Surendranathan and Nair, 1973).

Pectins and cellulose

Irradiation of tomatoes with doses of 500 to 2500 Gy leads to a degradation of all pectic fractions into lower molecular weight components (Yasia *et al.*,

1987). The ratio of protopectin to pectin declines and the hexametaphosphate-soluble pectin increases with no change in the water-soluble pectin. Similar degradations have been also observed in water suspensions of purified pectins (Kertesz *et al.*, 1964; Dzanic and Jankovic, 1966). Other fruits and vegetables such as oranges and grapefruit (Rouse and Dennison, 1968), apples (Sjoeberg, 1987), persimmons (Ogata *et al.*, 1968), and carrots (Ismail *et al.*, 1977) show similar pectic changes. Associated with the breakdown of pectins is the release of bound calcium from the cell wall into the cytoplasm (Echandi *et al.*, 1970). No effect of 100–1000 Gy was reported on the cellulose content in ripe raspberries (Jona and Vallania, 1976).

Changes in the wall degrading enzymes following irradiation are variable. Pectinmethylesterase is variously found to decrease in oranges following irradiation (Rouse and Dennison, 1968), to increase in apples (Flores *et al.*, 1971) and show no significant change in tomatoes (Yasia et al., 1987). Tomato polygalacturonase activity is found to be greatly reduced after irradiation, while cellulase activity slightly increases (Yasia *et al.*, 1987).

Starch

For most horticultural crops, irradiation does not seem to cause any significant changes in starch levels (Eric *et al.*, 1970; Rivera *et al.*, 1974; Beyers *et al.*, 1979). However, non-sprouted potatoes suffer significant starch losses after irradiation (Said *et al.*, 1973). There is a 25% increase in starch phosphorylase activity within the first 24 h after irradiation, then a gradual decline (Kodenchery and Nair, 1972). No toxicological effects have been found for any radiolytic products produced from a wide range of purified starches (FAO/IAEA/WHO, 1981).

Phenols

Increased polyphenol levels occur following irradiation with up to 1000 Gy in potatoes, sweet potatoes, carrots (Ogawa and Uritani, 1969), and grapefruit (Lester and Wolfenbarger, 1990). Injury symptoms frequently follow this increase. However, the several-fold increase in phenolic compounds that occurs after wounding and during suberization of potatoes is substantially reduced by irradiation before wounding (Ramamurthy *et al.*, 1992).

Phenolic enzymes

Typically, the activity of enzymes involved in the synthesis of phenolic compounds increases after irradiation. Banana polyphenoloxidase (PPO) increases with doses up to 1000 Gy, the rapid response suggests activation rather than *de novo* synthesis (Thomas and Nair, 1971). The activation may be due to conformational change suggested by isoenzyme analysis (Pendharkar and Nair, 1974). Browning of preclimacteric mangoes after irradiation at 750–1750 Gy is associated with a large dose dependent increase in PPO and peroxidase activity (Fig. 9.1). Sweet potatoes irradiated with 900 Gy show an increase in polyphenols that is paralleled by an increase in phenylalanine ammonia-lyase (PAL) and peroxidase activity and no change in *o*-diphenoloxidase activity (Ogawa and Uritani, 1969). Irradiation of potatoes (50–400 Gy) results in both PPO and peroxidase activity increasing, with PPO activity increasing more in the cortex while peroxidase activity increases more in the vascular bundle area (Tatsumi *et al.*, 1972). The pattern of phenolic enzyme activities in mushrooms is different; doses up to 2000 Gy result in a significant decrease of PPO, peroxidase and catalase during the first 4 days after irradiation. Activity then remains relatively constant in irradiated mushrooms with a delayed senescence, while levels in unirradiated mushrooms continue to decline (Mercado and Alabastro, 1980; Wahid, 1980).

Vitamins and other components

Vitamin A and β-carotene (Basson, 1983; WHO, 1988) and vitamins D and K (Basson, 1983; Thomas, 1986b) in horticultural produce do not seem to change significantly after irradiation at doses less than 1000 Gy. B-vitamins can be reduced by irradiation (<1000 Gy), the reduction varies considerably with crop, from none to 30% (Basson, 1983). Concentrations of vitamins E and K can be reduced at high doses (>1000 Gy) while lower doses cause nutritionally insignificant losses (WHO, 1988). Compared with non-irradiated fruit, there is no change in ascorbic acid in tomatoes receiving 1000 Gy (Urbain, 1986), 'Walla Walla' onions at 100–3000 Gy (Lu *et al.*, 1988), potatoes at around 150 Gy (Chachin, 1980) or grapefruit at 350 Gy (Garcia *et al.*, 1990). Thomas (1986b) in his review suggests that ascorbic acid losses in citrus are negligible or at the most 15% for doses of greater than 2000 Gy. Tatsumi *et al.* (1972) reported a decrease in ascorbic acid in irradiated potatoes immediately following harvest but none if irradiated after a normal curing period.

At doses less than 1000 Gy, little change occurs in any other components of horticultural products such as proteins, amino acids, lipids, and minerals (Table 9.3). There are no significant changes in nutritional compounds found for a wide range of other crops including mangoes, papaya, strawberries, litchis, citrus, stone fruit, pome fruit, and nuts (Maxie *et al.*, 1971; Beyers and Thomas, 1979; Beyers *et al.*, 1979; Kashani and Valadon, 1984; Morris, 1987). Volatiles of many crops also do not appear to be affected at less than 1000 Gy (Blakesley *et al.*, 1979).

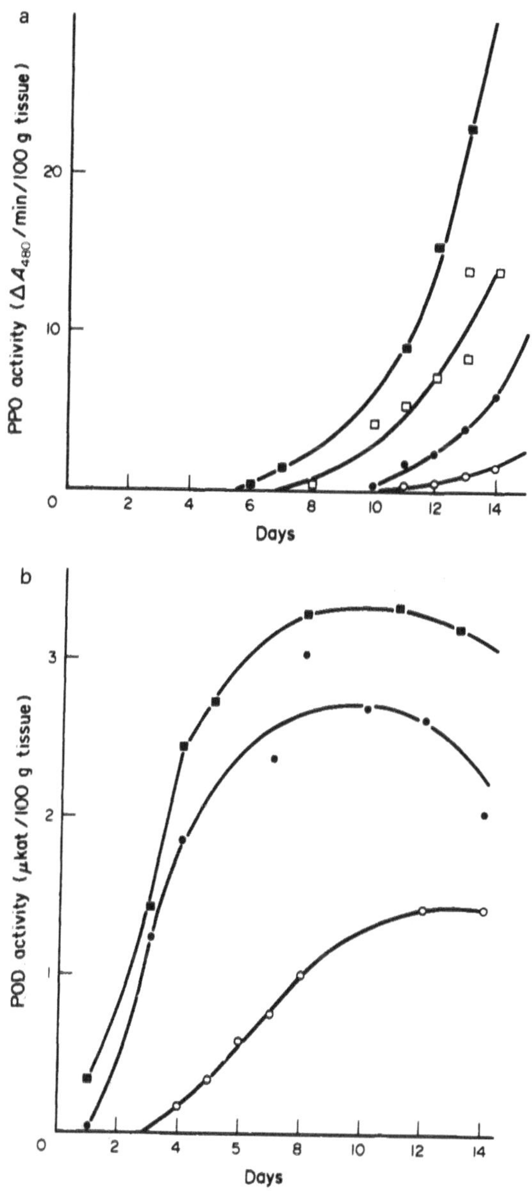

Fig. 9.1 Changes in enzyme activity during ripening of 'Kent' mangoes at 20°C, after irradiation at various doses of 1.75 kGy (■ — ■), 1.25 kGy (□ — □), 0.75 kGy (● — ●) and control (○ — ○). (a) Polyphenoloxidase activity (PPO). (b) Peroxidase activity (POD). (From Frylinck *et al.*, 1987. Reprinted with permission.)

Table 9.3. Effect of radiation (750 Gy) on composition of mangoes (4 cvs), papayas (2 cvs), and strawberries (2 cvs).

	Mango		Papaya		Strawberry	
	Control	Irrad.	Control	Irrad.	Control	Irrad.
Water (g 100 g^{-1})	80.78	80.47	86.98	86.95	90.48	90.20
Fat (g 100 g^{-1})	0.07	0.08	0.09	0.07	0.18	0.18
Ash (g 100 g^{-1})	0.36	0.36	0.45	0.46	0.45	0.46
Acidity (g 100 g^{-1})	0.30	0.34	0.10	0.12	0.68	0.70
Protein (g 100 g^{-1})	0.44	0.45	0.60	0.62	0.77	0.76
Starch (g 100 g^{-1})	1.07	1.26	0.102	0.075		
Sugar (g 100 g^{-1})	9.07	8.64	6.55	6.73	5.48	4.93
Carotene (IU 100 g^{-1})	6727	7017	3552	3970	354	492
Ascorbic acid (mg 100 g^{-1})	12.23	11.70	79.29	82.26	58.80	55.00
Riboflavin (mg 100 g^{-1})	0.07	0.06	0.03	0.03	0.03	0.03
Niacin (mg 100 g^{-1})	0.72	0.74	0.55	0.41	0.55	0.51
Thiamin (mg 100 g^{-1})	0.07	0.06	0.04	0.03	0.02	0.02
Calcium (mg 100 g^{-1})	13.11	11.67	30.90	35.59	8.76	8.99
Phosphorus (mg 100 g^{-1})	12.17	12.90	8.78	8.85	15.18	15.80

Source: From Beyers *et al.*, 1979. Reprinted with permission.

Organoleptic Attributes

Minor losses of organoleptic acceptability occur in many crops at doses of less than 1000 Gy. There is no loss of flavour in mushrooms (Urbain, 1986), oranges at 1000 Gy (Thomas, 1986b), onions at 3000 Gy (Lu *et al.*, 1988) and up to 2000 Gy for all citrus varieties reviewed by Thomas (1986b). One exception is limes treated at 500 Gy, that suffered from extensive off-flavour development and bleaching of flesh colour.

Moy and Nagai (1985) found no difference in aroma, colour, flavour or texture between irradiated (300 Gy) and control plums and nectarines. There are significant differences at higher doses (300 to 1000 Gy) in these sensory parameters with no lowering of acceptability scores. The difference may have been due to a delaying effect on ripening. Irradiated 'Eureka' lemons and 'Navel' oranges showed no change in sensory characteristics following irradiation at doses up to 1000 Gy (Moy and Nagai, 1985). Grapefruit dosed at 600 Gy produced detectable flavour and odour changes, but at 75 to 150 Gy flavour acceptability increased (Moshonas and Shaw, 1984). Ripe peaches dosed at 1000 Gy resulted in the formation of 'off', 'sour' or 'bland' flavours but irradiated green peaches were sweeter than the control fruit after ripening (Maxie *et al.*,

1965). 'Shasta' strawberries at 1000 Gy and celery at 5000 Gy also developed undesirable off flavours (Urbain, 1986). 'Daybreak' strawberries were sweeter when irradiated at 1500 Gy than unirradiated (Lovell and Flick, 1966). 'Valencia' oranges irradiated at 750 and 1000 Gy have lower flavour scores than non-irradiated fruit and first dosed at less than 500 Gy (Thomas, 1986b).

Respiration and Ethylene Production

Climacteric crops

Generally, climacteric fruits show a marked increase in respiration rate if irradiated prior to the onset of the climacteric rise (Maxie and Abdel-Kader, 1966). Irradiation after the onset of the climacteric has little effect on the degree or duration of CO_2 evolution. The response of fruit respiration to irradiation has been extensively studied with doses higher than 1000 Gy. Low-dose irradiation (<1000 Gy) causes an acceleration of the respiratory response of preclimacteric apricots and peaches (Maxie *et al.*, 1965) with an accompanying decrease in shelf-life (i.e. acceleration in ripening). Preclimacteric bananas (Thomas, 1986a) experience a delay of ripening following low-dose irradiation with no increase in respiration rate, compared with control fruit. The onset of the climacteric, however, is delayed. Irradiation (250–1000 Gy) of preclimacteric mangoes causes an immediate increase in the respiration rate, followed by a decline to about the level of non-irradiated mangoes (Akamine and Goo, 1971). The climacteric peak is delayed by about 2 days by irradiation at doses of 500–1000 Gy. However, the commencement of the climacteric is delayed by 4 days for doses greater than 500 Gy. A depression in peak carbon dioxide production has also been reported for mangoes irradiated at 75–600 Gy (Boag *et al.*, 1990).

Irradiation of preclimacteric fruit causes an increase in ethylene production (Maxie and Abdel-Kader, 1966). Generally ethylene accelerates ripening, but this acceleration does not always occur. Ethylene production after irradiation increases in pears and tomatoes while ripening is delayed (Lee *et al.*, 1968). It is postulated that the irradiation-induced increase in respiration and ethylene production is in response to tissue damage and subsequent repair. The increase in protein synthesis in irradiated preclimacteric pear tissue supports this postulate (Clarke and Fernandes, 1961). This response has been compared with ethylene production initiated by mechanical damage (Maxie and Sommer, 1968).

The irradiation-induced ethylene synthesis peaks in strawberries at 1000 Gy and then declines, while respiration increases with radiation dosage to 4000 Gy. Therefore, the membrane-bound ethylene synthesis is more sensitive to irradiation than mitochondrial respiration (Couture *et al.*, 1990). Irradiation of breaker cherry tomatoes results in a sharp burst of ethylene production during the first hour, this effect is dose dependent and reaches a maximum at 3000 Gy; this is proceeded by a sharp increase in 1-aminocyclopropane-1-carboxylic acid (ACC) synthase activity beginning 15 min after irradiation. In contrast, wounding continuously stimulated ethylene over several hours. Irradiation greatly inhibits the activity of ethylene-forming enzyme (EFE) at doses of greater than 1000 Gy (Larrigaudiere *et al.*, 1990).

Irradiated peaches and nectarines lose their sensitivity to exogenous ethylene, whereas 'Bartlett' pears do not when irradiated at the same dose (Maxie *et al.*, 1966). Bananas retain their sensitivity but require a longer exposure to ethylene to commence ripening than unirradiated fruit (Sommer and Maxie, 1966).

Non-climacteric crops

The climacteric-like rise in CO_2 production in irradiated citrus is generally proportional to the applied dose up to a certain level (Thomas, 1986b). The magnitude of the increase depends on variety and physiological age. The onset of irradiation-induced CO_2 rise in lemons and grapefruit occurs within 3 h following irradiation (Green *et al.*, 1969). This production rise then declines over the next few days but remains higher in irradiated (250–1000 Gy) than in non-irradiated fruit. Romani (1964) showed that more mature (i.e. yellower) lemons or older oranges (i.e. stored for 1 month prior to irradiation) experienced a lesser CO_2 production response than green lemons or freshly harvested oranges.

Ethylene production is also increased by irradiation in non-climacteric citrus fruits. There is also an immediate burst in ethylene within the fruit's internal atmosphere within 30 min after irradiation. The immediate ethylene burst may be mediated by the direct radiolysis of some chemical constituents of the fruit and the larger later delayed peak may be caused by the eventual production of some other metabolites. The maximal production of ethylene occurs 5 to 6 days after irradiation. The increase in ethylene production results in a more rapid breakdown of chlorophyll in irradiated lemons (Maxie et al., 1969).

Sprout and Growth Inhibition

A much quoted benefit of low-dose irradiation is the retardation of the sprouting of such crops as onions, potatoes, garlic, ginger and yams and the inhibition of growth of asparagus spears and mushrooms. Onion sprouting may be retarded at 20–150 Gy (Matsuyama and Umeda, 1983) but the practical range (60–120 Gy) is limited by commercial effectiveness and damage. Garlic sprouting is inhibited by 100–150 Gy, with dehydrated irradiated garlic being lighter in colour than non-irradiated garlic (Urbain, 1986). One problem with irradiation for sprouting control of onions and garlic is the necrosis and browning of the meristematic and sprouting tissue at the centre of the bulb. This necrotic tissue eventually decays, unless irradiated onions are cool stored (Takano *et al.*, 1974). This necrosis can be minimized if irradiation is done during dormancy (Matsuyama and Umeda, 1983; Morris 1985).

Potato sprouting is controlled at doses of 30–150 Gy. If irradiation is carried out before wound healing the potatoes are more susceptible to pathogens (Matsuyama and Umeda, 1983) and some potato products such as crisps may become blackened during frying or whole potatoes during boiling (Urbain, 1986). Mushroom cap-opening is delayed by 500–1000 Gy with the result that mushrooms retain a fresh appearance. A dose of 1000 Gy retards cap-opening by 50% depending on the period after harvest before irradiation and storage temperature (Akamine and Moy, 1983). The sooner after harvest mushrooms are irradiated the greater the delay in cap-opening. The postharvest growth and bract opening of asparagus can be slowed by irradiation of 1000 Gy (Maxie *et al.*, 1971). Sprouting of ginger rhizomes is controlled in a complex manner: irradiation with gamma doses of 25–50 Gy being effective over periods of 3 months (at 22°C) or longer (due to death of the sprouts), while doses of 500 Gy are required for control over shorter periods (Paull *et al.*, 1988).

Colour and Pigments

For many crops, irradiation at disinfestation doses does not affect the colour of fresh horticultural products. Doses of 1250 Gy accelerated the reddening of green capsicums and tomatoes and retarded the greening of white endives (Urbain, 1986). Normal colour development of mangoes is suppressed by greater than 1000 Gy (Akamine and Moy, 1983); 75–1000 Gy delay skin colour development in 'Kensington Pride' mangoes, without affecting the ability to colour, although most fruit treated at 1000 Gy never attained normal 'ripe' skin colour (Figure 9.2). Exceptions include an enhancement of anthocyanin production in peaches irradiated at 500–3000 Gy giving irradiated fruit a redder appearance. Peaches irradiated at the preclimacteric stage show an increase in anthocyanin production, but if irradiated at the firm, ripe (postclimacteric) stage there is a decrease in anthocyanins and therefore of external red coloration (Moy *et al.*, 1983). 'Tommy Atkins' mangoes irradiated at 150–1500 Gy developed less peel colour than control fruit (Spalding and von Windeguth, 1988).

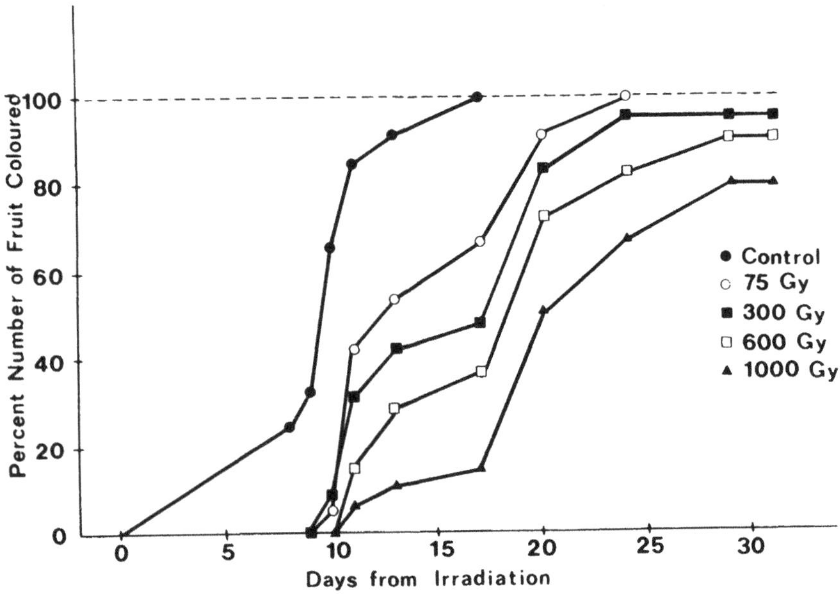

Fig. 9.2 Influence of irradiation on the skin colour of mangoes during storage at 20°C. Assessed as the percentage of fruit coloured to more than 75% or more of their skin. (From Jessup *et al.*, 1988. Reprinted with permission.)

The accelerated de-greening of 'Lisbon' and 'Eureka' lemons is possibly due to the enhancement of ethylene production by irradiation (Maxie *et al.*, 1965). Limes de-green more slowly at 300 Gy than when not irradiated (Mathur, 1963).

Effects on Pathogens

Most decay microorganisms that contribute to postharvest losses of fresh horticultural products can be inactivated only by doses of irradiation greater than 1000 Gy. The actual inactivation dose depends on the pathogen. Several researchers, however, have shown that doses at or below 1000 Gy do have some fungistatic effects. Urbain (1986) has shown that 1000 Gy will reduce rotting caused by *Penicillium italicum* and *P. digitatum* in tangerines. Rotting is also reduced in apricots by treatment at 100 Gy, in tamarillos dosed at 750–1000 Gy with unacceptable tissue softening above 1000 Gy, in guavas irradiated at 300 Gy, and in mushrooms and capsicums dosed at 800–1000 Gy (Urbain, 1986). Sommer and Maxie (1966) have reported that fungal spores can be reduced tenfold in number by irradiation doses of less than 1000 Gy, in particular the common storage pathogens of fruit such as *Rhizopus stolonifera, Alternaria citri, Geotrichum candidum,* and *Monilinia citricola.*

Generally, increased pathogen control is achieved with irradiation dose up to a level of maximum benefit. Higher doses lead to an increase in pathogen infection until infection can be far worse than for unirradiated produce. This increase is presumably due to easier access and greater extracellular nutrient availability for decay organisms resulting from tissue damage (Lu *et al.*, 1989). Sweet onions rotted more slowly than controls if treated with less than 1000 Gy (Lu *et al.*, 1988) and 50–100 Gy (Urbain, 1986) respectively, but more quickly than controls if irradiated with higher doses.

The greatest benefit seen in controlling pathogenic organisms has been when irradiation is applied in combination with heat, especially hot water dips. Decay of 'Loring' peaches is slowed by a dose of 750 Gy followed by a 7 min dip in 49°C water, whereas the same level of protection was conferred by a dose of 1500 Gy alone (Urbain, 1986). Figs irradiated at 1500 Gy and dipped in 50°C water for 5 min are slower to be infected by *Rhizopus* spp. and *Aspergillus* spp. This resulted in 3 to 4 days increased shelf-life when stored at 20 to 37°C and by 8 to 10 days when stored at 15°C (Urbain, 1986). Decay is prevented in mangoes by 750 Gy and a 5 min dip at 50 to 55°C with a 23 day extension to shelf-life at 11°C and 7 days in ambient storage. Papayas showed reduced decay if irradiated at 750 Gy prior to 20 min in 48°C water or 20 s in 60°C water (Urbain, 1986). This benefit was obtained only when fruit were at the 'colour break' (onset of climacteric rise) stage of maturity.

Many researchers have found that irradiated fruit is more likely than unirradiated fruit to be attacked by postharvest decay pathogens especially during transit. Irradiation may cause an increase in cell membrane permeability leading to localized leaking of cell contents (Hayashi and Kawashima, 1983b) providing an ideal growth medium for fungal and bacterial growth. The beneficial effects seen when irradiation is applied in combination with heat treatments may be due to the fungistatic effect of the heat treatment and a modification of the tissue chemistry due to the irradiation and/or the heat. Plums irradiated at 1500 Gy had diminished rotting compared to control plums if irradiated after a preconditioning period during which wounds caused by harvesting and handling had a chance to cure (Urbain, 1986).

A note of caution is given by Maxie *et al.* (1969), who warn that any effect irradiation has on reducing decay depends on the inoculum levels on or around the fruit. Irradiation applied before pathogen levels in the inoculum increase is a more effective control treatment than when applied during the pathogen growth phase.

Safety and Wholesomeness

Animal feeding studies designed to test for the presence of toxic or mutagenic substances in irradiated foods have been conducted since the 1950s. Many of these studies investigated responses to feeding material irradiated at very high sterilant doses. A comprehensive study organized by the FAO and the International Atomic Energy Agency in 24 countries restricted animal feeding studies to a range of foods representing typical food classes to a dose at or below 10 kGy. These studies included investigations on wheat flour, potatoes, rice, iced ocean fish, mangoes, spices, dried dates, and cocoa powder. None of the studies showed any indication that irradiation caused the formation of toxic or mutagenic substances in these foods (WHO, 1988).

Formation of radiolytic compounds

Irradiation can induce the formation of 'radiolytic' products in horticultural produce; these are most likely to be from carbohydrates. The irradiation of pure solutions of sugars commonly found in horticultural crops can result in the formation of radiolytic products with mutagenic properties such as glucosone and xylosone. However, these are not found in tropical and subtropical fruits irradiated at doses of 100–20 000 Gy and indeed the addition of fruit juice to sugar solutions markedly reduced formation of radiolytic products during irradiation (Drijver *et al.*, 1986). Cytotoxic, α,β-unsaturated carbonyl compounds produced from sugars are found in extracts from strawberries treated with doses of 2000 Gy and above, these increase with dose rate up to 30 000 Gy; mutagenic activity is only found in extracts from strawberries treated with doses greater than 5000 Gy (Schubert *et al.*, 1973).

Nine radiolytic products were isolated from irradiated starch fed to rats in daily rations at about 800 times the quantities consumed through normal daily intake – the results indicate no evidence of toxicity or mutagenicity (Truehaut and St Lebe, 1978). No toxicological effects have been found for any radiolytic products produced from sugars in various fruit products (FAO/IAEA/WHO, 1981).

Identification of Irradiated Product

A major problem with irradiation is that clear identification of irradiated product and particularly the dose received is very difficult. Various methods have been assessed including: potato sprout morphology as affected by doses of 70 Gy and above (Shimomura *et al.*, 1972); weight loss of potatoes that is proportional to dose (Magaudda, 1973); growth of extracted grapefruit seed embryos (Kawamura *et al.*, 1989); and impedance of potatoes as a ratio of two frequencies (5 and 50 kHz) that is dependent on dose over the range of 0–300 Gy and independent of cultivar, district, and length of storage (Hayashi and Kawashima, 1983b). Additional methods studied have been thermoluminescence and chemiluminescence of dried horticultural crops or seeds (Heidi and Boegl, 1986). More recent work with over 500 samples of imported spices and dried vegetables has shown chemiluminescence not to be a reliable indicator of irradiation (Meier *et al.*, 1989). Thermoluminescence, therefore, shows the greater promise (FAO/ IAEA, 1990). Another technique with promise is free radical content detected by electron spin resonance (ESR). The increase in free radical content of papaya seeds after irradiation is significant for doses of greater than 200 Gy and is related to the absorbed dose. A major problem of this technique is that over a period of a week or so levels of free radicals decline to normal (Boshard *et al.*, 1971).

Conclusion

A very large body of literature, only a small part of which is dealt with in this review, addresses the effects of irradiation on horticultural crops. However, the majority of the literature is for dose levels greater than those required for disinfestation. The literature also contains a great deal of contradiction about the effects of irradiation, particularly with regard to phytotoxic effects. The effect of maturity at harvest, stage of ripeness, or dormancy, time after harvest and even development of natural wound repair can all interact with irradiation. The lack of precision in selecting particular experimental material can therefore be a major source of contradictions. Many of the most commonly stated benefits of irradiation – disease control, delay of ripening, and senescence – occur at the same or lower doses than those causing phytotoxic symptoms (for most crops this in the range of 1000–3000 Gy).

Finally, it would seem that there are really only two uses of irradiation which can realistically achieve beneficial effects without unacceptable injury on a reasonably wide range of crops: these are sprout or growth inhibition and insect disinfestation, at doses less than 300 Gy. These dose rates are generally non-injurious to a wide range of horticultural crops. However, at this irradiation dose most insect species are not killed outright, but rather development is prevented to the next life stage or the adults are sterile. A major obstacle to practical adoption of irradiation disinfestation at these low non-injurious doses, is that dramatic changes are necessary to international quarantine inspection and product treatment certification standards. This will require quarantine officials accepting products infested with live though sterile insects, or live insects incapable of proceeding to the next life stage and therefore of reaching the adult stage.

References

Abdel-Kader, A.S., Morris, L.L. and Maxie, E.C. (1968) Physiological studies of gamma-irradiated tomato fruits. III. Effects on ascorbic acid contents, acidity and texture. *Proceedings of the American Society for Horticultural Science* 93, 843–853.

Ajlouni, S. and Hamdy, M.K. (1988) Effect of combined gamma-irradiation and storage on biochemical changes in sweet potato. *Journal of Food Science* 53, 477–481.

Akamine, E.K. and Goo, T.J. (1971) Respiration of gamma irradiated fresh fruits. *Journal of Food Science* 36, 1074–1077.

Akamine, E.K. and Moy, J.H. (1983) Delay in postharvest ripening and senescence of fruits. In: Josephson, E.S. and Peterson, M.S. (eds), *Preservation of Food by Ionizing Radiation – Volume III*. CRC Press, Boca Raton, pp. 129–158.

Basson, R.A. (1983) Recent advances in radiation chemistry of vitamins. In: Elias, P.S. and Cohen, A.J. (eds), *Recent Advances in Food Irradiation*. Elsevier Biomedical, Amsterdam, pp. 189–202.

Basson, R.A., Beyer, M., Ehlermann, D.A.E. and Van Der Linde, H.J. (1983) Chemiclearance approach to evaluation of safety of irradiated fruits. In: Elias, P.S. and Cohen, A.J. (eds), *Recent Advances in Food Irradiation*. Elsevier Biomedical, Amsterdam, pp. 59–88.

Beyers, M. and Thomas, A.C. (1979) Gamma-irradiation of subtropical fruits. IV. Changes in certain nutrients present in mangoes, papayas, and litchis during canning, freezing, and gamma-irradiation. *Journal of Agricultural and Food Chemistry* 27, 48–51.

Beyers, M., Thomas, A.C. and van Tonder, A.J. (1979) Gamma-irradiation of subtropical fruits. I. Compositional tables of mango, papaya, strawberry, and litchi fruits at the edible-ripe stage. *Journal of Agricultural and Food Chemistry* 27, 37–42.

Blakesley, C.N., Loots, J.G., du Plessis, L.M. and de Bruyn, G. (1979) Gamma irradiation of subtropical fruits. 2. Volatile components, lipids and amino acids of mango, papaya and strawberry pulp. *Journal of Agricultural and Food Chemistry* 27, 42–48.

Boag, T.S., Johnson, G.I., Izard, M., Murray, C. and Fitzsimmons, K.C. (1990) Physiological responses of mangoes cv. Kensington Pride to gamma irradiation treatment as affected by fruit maturity and ripeness. *Annals of Applied Biology* 116, 177–187.

Borsa, J., Siemens, A.J. and Mazza, G. (1989) Effects of gamma irradiation on sprout inhibition and processing quality of Norchip and Russet Burbank potatoes. *Canadian Institute of Food Science and Technology Journal* 22, 491–495.

Boshard, J.A.P., Holmes, D.E. and Piette, L.H. (1971) An inherent dosimeter for irradiated foods: papayas. *International Journal of Applied Radiation and Isotopes* 22, 316–318.

Burditt, A.K., Jr, Moshonas, M.G., Hatton, T.T., Spalding, D.H., von Windeguth, D.L. and Shaw, P.E. (1981) Low-dose irradiation treatment for grapefruit and mangoes infested with Caribbean fruit fly larvae. *Agriculture Research Results, Southern Series No. 10*. Agriculture Research Service, United States Department of Agriculture, Washington, DC, 9 pp.

Burton, W.G. (1975) The immediate effect of gamma irradiation upon the sugar content of potatoes previously at 2, 4.5, 6, 10 and 15.5 degrees. *Potato Research* 18, 109–115.

Chachin, K. (1980) Physiological and chemical aspects of irradiated potatoes. In: Chang, S.M. (ed.), *Recent Advances in Food Science and Technology, Vol. 2*. Hua Shiang Yuan Publishing, Taipei, Taiwan, pp. 434–442.

Chachin, K. and Ogata, K. (1971) Effects of delay between harvest and irradiation and of storage temperatures on the sprout-inhibition of onions by gamma irradiation. *Journal of Food Science and Technology* 18, 378–382.

Chen, C.C., Lu, T.S. and Fu, Y.K. (1980) Effect of refrigeration on the sprouting and starch–sugar conversion process in gamma-irradiated potatoes. *Journal of the Chinese Agricultural Chemical Society* 18, 88–96.

Clarke, I.D. (1971) Effects of radiation treatments. In: Hulme, A.C. (ed.), *The Biochemistry of Fruits and Their Products*. Academic Press, New York, pp. 687–705.

Clarke, I.D. and Fernandes, S.J.G. (1961) Effects of gamma-radiation on the protein content of apples and pears. *International Journal of Applied Radiation and Isotopes* 11, 186–189.

Couture, R., Makhlouf, J., Cheqour, F. and Willemont, C. (1990) Production of CO_2 and C_2H_4 after gamma irradiation of strawberry fruit. *Journal of Food Quality* 13, 385–393.

Curzio, O.A. and Croci, C.A. (1983) Extending onion storage life by gamma irradiation. *Journal of Food Processing and Preservation* 7, 19–24.

Donini, M.L.B. and Pansolli, P. (1970) Gamma irradiation of table grapes of the Hoanes variety. *Food Irradiation* 10, 15–21.

Drijver, L.D., Holzapfel, C.W. and Van der Linde, H.J. (1986) High-performance liquid chromatographic determination of D-arabino-exos-2-ulose (D-glucosone) in irradiated sugar solutions: application of the method to irradiated mango. *Journal of Agricultural and Food Chemistry* 34, 758–762.

Dzanic, M.D. and Jankovic, B.R. (1966) Radiation effects in pectins. *International Journal of Applied Radiation and Isotopes* 17, 561–566.

Eaton, G.W., Meehan, C. and Turner, N. (1970) Some physical effects of postharvest gamma radiation on the fruit of sweet cherry, blueberry and cranberry. *Canadian Institute of Food Technology Journal* 3, 152–156.

Echandi, R.J., Chase, B.R. and Massey, L.M., Jr (1970) Effect of gamma-radiation on polysaccharides and calcium distribution in carrot cell walls. *Journal of Agricultural and Food Chemistry* 18, 878–880.

Eric, B., le Compte, J., Klein, S. and Kricker, W. (1970) Study of disinfestation of bananas by gamma irradiation. *Food Technology in Australia* 22, 664–665.

FAO/IAEA (1986) Excerpts of the task force meeting on irradiation as a quarantine treatment, Suriwongse Hotel, Chiang Mai, Thailand, 17–21 February 1986. *Food Irradiation Newsletter* 10, 5–10. International Atomic Energy Agency, Vienna.

FAO/IAEA (1990) Excerpts of the research co-ordination meeting on analytical detection methods for irradiation treatments of foods, Warsaw, Poland, 25–29 June 1990. *Food Irradiation Newsletter* 10, 27–41. International Atomic Energy Agency, Vienna.

FAO/IAEA/WHO (1981) Wholesomeness of irradiated food. Report of a joint FAO/IAEA/WHO Expert Committee. *WHO Technical Report Series*, No. 659, 34pp.

Fesues, I., Kadas, L. and Kalman, B. (1981) Protection of oranges by gamma radiation against *Ceratitis capitata* Wied. *Acta Alimentaria* 10, 293–299.

Flores, M.C., Piaz, L., Ortega, M. de, Ortiz, J., Deshpande, S. and Rolz, C. (1971) Variation of soluble pectic constituents and of pectinesterase activity during ripening of some fruits. *Revista de Agroquimica y Tecnologia de Alimentos* 11, 594–603.

Frylinck, L., Dubery, I.A. and Schabort, J.C. (1987) Biochemical changes involved in stress response and ripening behaviour of gamma-irradiated mango fruit. *Phytochemistry* 26, 681–686.

Fu, Y.K., Yang, Y.S., Tsai, C.M., Chang, Y.N., Chang, M.S. and Chu, S.L. (1979) *Food Irradiation Studies on Agricultural Foodstuffs in Taiwan, ROC – Recent Advances in Food Science and Technology*, Vol. I. Hua Shiang Yuan Publishing, Taipei, Taiwan, pp. 278–295.

Garcia, Y.M., Garcia, A.A., Fernandez, M.P., Paradoa, A., Sampere, E., Castillo, E. and Serrano, G. (1990) Stability of vitamin C content in grapefruit treated with different doses of gamma irradiation. *Revista de Agroquimica y Tecnologia de Alimentos* 30, 409–415.

Green, G.F., Ahmed, F.M. and Dennison, R.A. (1969) An automated sampling system for respiratory gases and respiratory response of irradiated citrus fruits. *Journal of Food Science* 34, 627–629.

Guelfat-Reich, S., Ben-Arie, R., Kahan, R.S. and Eisenberg, E. (1969) Effect of gamma radiation on the ripening of apricots after picking. *Fruits* 24, 137–142.

Hayashi, T. and Aoki, S. (1985) Effect of irradiation on the carbohydrate metabolism responsible for sucrose accumulation in potatoes. *Journal of Agricultural and Food Chemistry* 33, 14–17.

Hayashi, T. and Kawashima, K. (1982) The effect of gamma-irradiation on the sucrose content in sweet potato roots and potato tubers. *Agricultural and Biological Chemistry* 46, 1475–1479.

Hayashi, T. and Kawashima, K. (1983a) Activities of enzymes of sugar metabolism in gamma-irradiated potato tubers. *Journal of Food Science* 48, 1242–1245.

Hayashi, T. and Kawashima, K. (1983b) Impedance measurement of irradiated potatoes. *Journal of Japanese Society of Food Science and Technology* 30, 51–54.

Hayashi, T., Ohta, H., Hayakawa, A. and Kawashima, K. (1983) Effect of gamma irradiation and cold-storage on the sucrose content of chestnuts. *Journal of Japanese Society of Food Science and Technology* 30, 557–561.

Hayashi, T., Sugimoto, T. and Kawashima, K. (1984) Effect of gamma-irradiation on the activities of sucrose synthase and sucrose phosphate synthase in potato tubers and sweet potato roots. *Journal of Japanese Society of Food Science and Technology* 31, 281–284.

Heidi, L. and Boegl, W. (1986) The identification of irradiated dried foodstuffs by luminescence measurements. *Food Laboratory Newsletter* No.5, 21–23.

Ismail, F.A. and Afifi, S.A. (1976) Control of postharvest decay in fruits and vegetables by irradiation. *Nahrung* 20, 585–592.
Ismail, F.A., Afifi, S.A. and Fahmy, A.A. (1977) Gamma-irradiation and storage on some chemical aspects of carrots. *Nahrung* 21, 303–309.
Jessup, A.J. (1990) Gamma irradiation as a quarantine treatment for sweet cherries against Queensland fruit fly. *HortScience* 25, 456–458.
Jessup, A.J., Rigney, C.J. and Wills, P.A. (1988) Effects of gamma irradiation combined with hot dipping on quality of 'Kensington Pride' mangoes. *Journal of Food Science* 53, 1486–1489.
Jona, R. and Vallania, R. (1976) Histochemical analysis of gamma-radiation effects on *Rubus* fruits. *Acta Horticulturae* 60, 81–88.
Kahan, R.S., Monselise, S.P., Riov, J., van Kooij, J.G. and Chadwick, K. (1968) Comparison of the effect of radiation of various penetration powers on the damage to citrus fruit peel. *Radiation Botany* 8, 415–423.
Kashani, G.G. and Valadon, L.R.G. (1984) Effects of gamma irradiation on the lipids, carbohydrates and proteins of Iranian pistachio kernels. *Journal of Food Technology* 19, 631–638.
Kawamura, Y., Uchiyama, S. and Saito, Y. (1989) A half-embryo test for identification of gamma-irradiated grapefruit. *Journal of Food Science* 54, 379–382.
Kawashima, K., Hayashi, T., Uozumi, J., Sugimoto, T., and Aoki, S. (1984) Irradiation of onions on a large scale. *Report of the National Food Research Institute* 40, 1–8.
Kertesz, Z.I., Glegg, R.E., Boyle, F.P., Parsons, G.F. and Massey, L.M. (1964) Effect of ionizing radiations on plant tissues. III. Softening and changes in pectins and cellulose of apples, carrots and beets. *Journal of Food Science* 29, 40–48.
Kodenchery, U.K. and Nair, M.P. (1972) Metabolic changes induced by sprout inhibition dose of gamma-irradiation in potatoes. *Journal of Agricultural and Food Chemistry* 20, 282–285.
Kovacs, E., Keresztes, A. and Kovacs, J. (1988) The effects of gamma irradiation and calcium treatment on the ultrastructure of apples and pears. *Food Microstructure* 7, 101–114.
Larrigaudiere, C., Baccaunaud, M., Raymond, J. and Pech, P.C. (1987) Effects of ionization on the physiology, fungal contamination and quality of cold-stored raspberries. *Fruits* 42, 597–602.
Larrigaudiere, C., Latche, A., Pech, J.C. and Triantaphylides, C. (1990) Short-term effects of gamma-irradiation on 1-aminocyclopropane-1-carboxylic acid metabolism in early climacteric cherry tomatoes. *Plant Physiology* 92, 577–581.
Lee, M.S. and Kim, H.L. (1972) Effects of ionizing radiation on sprout inhibition and nutritive value of potato tubers. *Korean Journal of Food Science and Technology* 4, 29–35.
Lee, T.H., McGlasson, W.B. and Edward, R.A. (1968) Physiology of disks of irradiated tomato fruit. I. Influence of cutting and infiltration on respiration, ethylene production and ripening. *Radiation Botany* 10, 521–529.
Lester, G.E. (1989) Gamma irradiation, hot water and imazalil treatments on decay organisms and physical quality of stored netted muskmelon fruit. *Journal of Food Safety* 10, 21–30.
Lester, G.E. and Wolfenbarger, D.A. (1990) Comparisons of cobalt-60 gamma irradiation dose rates on grapefruit flavedo tissue and on Mexican fruit fly mortality. *Journal of Food Protection* 53, 329–331.

Lovell, R.T. and Flick, G.J. (1966) Irradiation of Gulf Coast area strawberries. *Food Technology* July, 99–102.

Lu, J.Y., Stevens, C., Yakubu, P. and Loretan, P.A. (1988) Gamma, electron beam and ultraviolet radiation on control of storage rots and quality of Walla Walla onions. *Journal of Food Processing and Preservation* 12, 53–62.

Lu, J.Y., Miller, P. and Loretan, P.A. (1989) Gamma radiation dose rate and sweet potato quality. *Journal of Food Quality* 12, 369–376.

MacFarlane, J.J. and Roberts, E.A. (1968) Some effects of gamma irradiation on Washington Navel and Valencia oranges. *Australian Journal of Experimental Agriculture and Animal Husbandry* 8, 625–629.

MacQueen, K.F. (1965) Sprout inhibition of vegetables using gamma irradiation In: *Radiation Preservation of Foods*. National Research Council–National Academy of Sciences. Washington, DC, pp. 127–142.

Magaudda, G. (1973) The possibility of recognising irradiated and nonirradiated potatoes by their weight loss. *Journal of Food Science* 38, 1253–1254.

Mathur, P.B. (1963) Low dose gamma irradiation of fresh fruits. *Food Irradiation* 4, A26–A28.

Matsuyama, A. and Umeda, K. (1983) Sprout inhibition in tubers and bulbs. In: Josephson, E.S. and Peterson, M.S. (eds), *Preservation of Food by Ionizing Radiation – Volume III*. CRC Press, Boca Raton, pp. 159–214.

Maxie, E.C. and Abdel-Kader, A. (1966) Food irradiation – physiology of fruits as related to feasibility of the technology. In: Chichester, C.O., Mrak, E.M. and Stewart, G.F. (eds), *Advances in Food Research, Vol. 15*. Academic Press, New York, pp. 105–145.

Maxie, E.C. and Sommer, N.F. (1968) Changes in some chemical constituents of irradiated fruits and vegetables. In: *Preservation of Fruits and Vegetables by Radiation*. International Atomic Energy Agency, Vienna, pp. 39–56.

Maxie, E.C., Johnson, C.F., Boyd, C., Rae, H.L. and Stallman, R.F. (1965) Effect of gamma irradiation on peaches. Contract No. AT(11-1)-34. *Radiation Technology in Conjunction with Postharvest Procedures as a Means of Extending the Shelf Life of Fruits and Vegetables*. United States Atomic Energy Commission, Washington, DC, pp. 92–98.

Maxie, E.C., Sommer, N.F., Muller, C. and Rae, H.L. (1966) Effect of gamma radiation on the ripening of Bartlett pears. *Plant Physiology* 41, 437–442.

Maxie, E.C., Amezquita, R., Hassan, B.M. and Johnson, C.F. (1968) Effect of gamma irradiation on the ripening of banana fruits. *Journal of the American Society for Horticultural Science* 92, 235–254.

Maxie, E.C., Sommer, N.F. and Eaks, I.L. (1969) Effect of gamma irradiation on citrus fruits. *Proceedings: First International Citrus Symposium*, pp. 1375–1387.

Maxie, E.C., Sommer, N.F. and Mitchell, F.G. (1971) Infeasibility of irradiating fruits and vegetables. *HortScience* 6, 202–204.

McLauchlan, R.L., Mitchell, G.E., Johnson, G.I., Nottingham, S.M. and Hammerton, K.M. (1992) Effect of disinfestation–dose irradiation on the physiology of Tai So lychee. *Postharvest Biology and Technology* 1, 273–281.

Meier, W., Konrad-Glatt, V. and Zimmerli, B. (1989) Identification of irradiated foods, chemiluminescence measurements on spices and dried vegetables. *Beta-gamma* 2, 4–7.

Mercado, L.N. and Alabastro, E.F. (1980) Irradiation of fresh *Volvariella volvacea* mushrooms: physiological, biochemical and sensory quality aspects. *CMU Journal of Agriculture, Food and Nutrition* 2, 197–214.

Monselise, S.P. and Kahan, R.S. (1966) Changes in composition and in enzymatic activities of flavedo and juice of Shamouti oranges following gamma irradiation. *Radiation Botany* 6, 265–274.

Morris, S.C. (1985) The effect of ionizing irradiation on the postharvest quality of vegetables. *Proceedings of IAEC/RCA International Workshop on Commercialisation of Ionizing Energy Treatment of Food.* Lucas Heights, Australia, Australian Atomic Energy Commision, Publication AAEC/525.

Morris, S.C. (1987) The practical and economic benefits of ionising radiation for the postharvest treatment of fruit and vegetables: an evaluation. *Food Technology in Australia* 39, 336–341.

Moshonas, M.G. and Shaw, P.E. (1984) Effects of low-dose gamma-irradiation on grapefruit products. *Journal of Agricultural and Food Chemistry* 32, 1098–1101.

Moy, J.H. (1983) Radurization and radicidation: fruit and vegetables. In: Josephson, E.S. and Peterson, M.S. (eds), *Preservation of Food by Ionizing Radiation, Vol. III.* CRC Press, Boca Raton, pp. 83–108.

Moy, J.H. and Nagai, N.Y. (1985) Quality of fresh fruits irradiated at disinfestation doses. In: Moy, J.H. (ed.), *Radiation Disinfestation of Food and Agricultural Products.* University of Hawaii at Manoa, Honolulu, pp. 135–145.

Moy, J.H., Ross, E., Hsia, S.T. and Sawato, M. (1971) Quality evaluation of gamma irradiated papaya in shipping studies. *Food Irradiation (Saclay)* 11, 14–20.

Moy, J.H., Kaneshiro, K.Y., Ohta, A.T. and Nagai, N. (1983) Radiation disinfestation of California stone fruit infested by Medfly – effectiveness and fruit quality. *Journal of Food Science* 48, 928–931.

Nair, P.M., Thomas, P., Ussuf, K.K., Surendranrhan, K.K., Limaye, S.P., Srirangarjan, A. and Padwal-Desai, S.R. (1973) Studies of sprout inhibition of onions and potatoes and delayed ripening of bananas and mangoes. *Proceedings of Symposia on Radiation Preservation of Food.* Bombay, India, IAEA, Vienna, Austria, Public. IAEC-SM-166-11, pp. 347–366.

Nuttall, V.W., Lyall, L.H., MacQueen, K.F. (1961) Some effects of gamma irradiation on stored onions. *Canadian Journal of Plant Science* 41, 805–813.

Nyambati, M.G.O. and Langerak, D.I. (1984) Effect of mild heat, potassium metabisulfide and/or gamma irradiation on storability of lime fruits. *Acta Alimentaria* 13, 65–81.

Ogata, K., Yamanaka, H., Kato, K. and Chachin, K. (1968) Studies on maturation changes in fruits induced by ionizing radiation. III. Effects of gamma radiation on ripening and keeping quality of astringent persimmons. *Food Irradiation* 3, 131–137.

Ogata, K., Chachin, K. and Tatsumi, Y. (1970) Studies on the browning of potatoes by gamma irradiation. *Food Irradiation (Shokukin Shosha)* 5, 40–41.

Ogawa, M. and Uritani, I. (1969) Effects of gamma irradiation on metabolic changes in potato, sweet potato and carrot tissues in response to cutting. *Food Irradiation (Shokuhin Shosha)* 4, 71–76.

Paull, R.E., Chen, N.J. and Goo, T.T.C. (1988) Control of weight loss and sprouting of ginger rhizome in storage. *HortScience* 23, 734–736.

Pendharkar, M.B. and Nair, P.M. (1974) Alterations in *Solanum tuberosum* polyphenol oxidase activity induced by gamma irradiation. *Photochemistry* 13, 1373–1377.

Pratella, G.C. and Menniti, A.M. (1978) Biological effects of ionizing radiations on the metabolism of potato tubers. *Frutticoltura* 40, 61–63.

Ramamurthy, M.S., Maiti, B., Thomas, P. and Nair, P.M. (1992) High-performance liquid chromatography determination of phenolic acids in potato tubers (*Solanum tuberosum*) during wound healing. *Journal of Agricultural and Food Chemistry* 40, 569–572.

Rao, V.S. and Vaskil, U.K. (1983) Effect of gamma-irradiation on flatulence-causing oligosaccharides in green gram (*Phaseolus aureus*). *Journal of Food Science* 48, 1791–1795.

Rigney, C.J. (1986) Irradiation treatment of avocados. *Food Irradiation Newsletter* 10, 17–18.

Rivera, J.R., Gonzalez, M.A. and Cuevas-Ruiz, J. (1974) Sprout inhibition in yam by gamma irradiation. *Journal of Agriculture of the University of Puerto Rico* 58, 330–337.

Romani, R.J. (1964) Radiation physiology of fruit respiration during and immediately following kilorad doses of ionizing radiation. *Radiation Botany* 4, 299–307.

Rouse, A.H. and Dennison, R.A. (1968) Gamma radiation effects on the pectic substances in citrus fruits. *Journal of Food Science* 33, 258–261.

Roy, M.K. and Bahl, N. (1984) Studies on gamma radiation preservation of *Agaricus bisporus*. *Mushroom Journal* (144), 411–414.

Said, W.I., Khalil, M.H. and Salama, S.B. (1973) Sprout inhibition of potatoes for local consumption by treatments with different sprout inhibitors. *Agricultural Research Review* 51, 91–99.

Sattar, A., Ali, M., Khan, I. and Muhammed, A. (1971) Effect of gamma radiation on post-harvest behaviour of pears. *Science and Industry (Karachi)* 8, 330–333.

Schubert, J., Sanders, E.B., Pan, S.F. and Wald, N. (1973) Irradiated strawberries – chemical, cytogenetic, and antibacterial properties. *Journal of Agricultural and Food Chemistry* 21, 684–692.

Shimomura, H., Kuriyama, E. and Takai, Y. (1972) A method for early detection of irradiated potato tubers. I. Morphological changes by ^{60}Co gamma-rays on dormant buds. *Journal of the Food Hygienic Society of Japan* 13, 531–535.

Sjoeberg, A.M. (1987) The effects of gamma irradiation on the structure of apple pectin. *Food Hydrocolloids* 1, 271–276.

Sommer, N.F. and Maxie, E.C. (1966) Recent research on the irradiation of fruits and vegetables. In: *Food Irradiation*. International Atomic Energy Agency, Vienna, STI/PUB/127, pp. 571–587.

Sommer, N.K., and Mitchell, F.G. (1986) Gamma irradiation – a quarantine treatment for fresh fruits and vegetables? *HortScience* 21, 356–360.

Spalding, D.H. and von Windeguth, D.L. (1988) Quality and decay of mangoes. *HortScience* 23, 187–189.

Strydom, G.J., van Staden, J. and Smith, M.T. (1991) The effects of gamma radiation on the ultrastructure of the peel of banana fruits. *Environmental and Experimental Botany* 31, 43–49.

Surendranathan, K.K. and Nair, P.M. (1973) Alterations in carbohydrate metabolism of gamma-irradiated Cavendish banana. *Photochemistry* 12, 241–249.

Takano, H., Umeda, K. and Sato, T. (1974) Sprout inhibition of onions by ionizing radiation III. Effect of irradiation dose level on the sprouts of onions cv. 'Sapporoski'. *Report National Food Research Institute* 29, 55–59.

Tatsumi, Y., Chachin, K. and Ogata, K. (1971) Studies on the browning of potatoes by gamma radiation. *Food Irradiation (Shokuhin Shosha)* 6, 100–101.

Tatsumi, Y., Chachin, K. and Ogata, K. (1972) Browning of potato tubers by gamma radiation. II. Relationship between browning and changes in *o*-diphenol and ascorbic acid contents and in polyphenol oxidase and peroxidase activities in irradiated potato tubers. *Journal of Food Science and Technology* 19, 508–513.

Thomas, P. (1981) Involvement of polyphenols in the after-cooking darkening of gamma irradiated potatoes. *Journal of Food Science* 46, 1620–1621.

Thomas, P. (1986a) Radiation preservation of foods of plant origin. III. Tropical fruits: bananas, mangoes, and papayas. In: Furia, T.E. (ed.), *CRC Critical Reviews in Food Science and Nutrition, Vol. 23*. CRC Press, Boca Raton, pp. 147–205.

Thomas, P. (1986b) Radiation preservation of foods of plant origin. IV. Subtropical fruits: citrus, grapes, and avocados. In: Furia, T.E. (ed.), *CRC Critical Reviews in Food Science and Nutrition, Vol. 24*. CRC Press, Boca Raton, pp. 53–89.

Thomas, A.C. and Beyers, M. (1979) Gamma-irradiation of subtropical fruits. III. A comparison of the chemical changes occurring during normal ripening of mangoes and papayas with changes produced by gamma-irradiation. *Journal of Agricultural and Food Chemistry* 27, 157–163.

Thomas, P. and Nair, P.M. (1971) Effect of gamma irradiation on polyphenol oxidase activity and its relation to skin browning in bananas. *Phytochemistry* 10, 771–777.

Thomas, P., Dharkar, S.D. and Sreenivasan, A. (1971) Effect of gamma irradiation on the postharvest physiology of five banana varieties grown in India. *Journal of Food Science* 36, 243–247.

Thomas, P., Srirangarajan, A.N. and Limaye, S.P. (1975) Studies on sprout inhibition of onions by gamma irradiation. 1. Influence of time interval between harvest and irradiation, radiation dose and environmental conditions on sprouting. *Radiation Botany* 15, 215–222.

Thomas, P., Padwal-Desai, S.R., Srirangarjan, A.M., Joshi, M.R., Janave, M.T., Bhonde, S.R. and Qadri, S.M.H. (1986) Pilot scale tests on the efficacy of gamma irradiation for sprout inhibition of onions under commercial storage conditions. *Journal of Food Science and Technology* 23, 107–108.

Truehaut, R. and St Lebe, L. (1978) Wholesomeness of the radiolytic products of starch. In: *Food Preservation by Irradiation II. Proceedings of an International Symposium on Food Irradiation*. Wageningen, 21–25 November 1977. International Atomic Energy Agency, Vienna, pp. 31–40.

Urbain, W.M. (1986) Food irradiation. In: Schweigert, B.S. (ed.), *Food Science and Technology – a Series of Monographs*. Academic Press, Orlando, Florida, pp. 170–216.

Wahid, M. (1980) Effect of gamma irradiation and storage on the catalase and peroxidase activities of mushrooms. *Lebensmittel Wissenschaft und Technologie* 13, 291–292.

Wahid, M. and Kovacs, E. (1980) Shelf life extension of mushrooms (*Agaricus bisporus*) by gamma irradiation. *Acta Alimentaria* 9, 357–366.

WHO (1988) Effects of food irradiation. In: Kampelmacher, E.H., Rand, M.J., Young, M. and Chinsman, B. (eds), *Food Irradiation – A Technique for Preserving and Improving the Safety of Food*. World Health Organization, Geneva, pp. 24–32.

Wu, W.S., Su, C.M.H., Fu, Y.K., Tsai, C.M., Chang, M.S. and Chu, S.L. (1979) Effects of gamma radiation and chemical treatments on controlling the post-harvest diseases of potatoes and onions. In: *Food Irradiation Studies on Agricultural FoodStuffs in Taiwan, ROC – Recent Advances in Food Science and Technology*, Vol. I. Hua Shiang Yuan Publishing, Taipei, Taiwan, pp. 296–307.

Yasia, M.S., Chachin, K. and Iwata, T. (1987) Effects of gamma irradiation on tissue firmness, some cell wall degrading enzymes and pectic substances of tomato fruit. *Bulletin of the University of Osaka Prefecture B* 39, 9–20.

Heat and Cold Treatments 10

ROBERT E. PAULL[1] and ROY E. MCDONALD[2]
[1]Department of Plant Molecular Physiology, College of Tropical Agriculture and Human Resources, University of Hawaii, 3190 Maile Way, Honolulu, HI 96822, USA;
[2]United States Department of Agriculture, Agricultural Research Service, Horticultural Research Laboratory, 2120 Camden Road, Orlando, FL 32803, USA.

Introduction

Insect disinfestation of many fruits requires additional postharvest handling. These methods need to be effective, non-damaging, non-polluting, and non-carcinogenic. The withdrawal of ethylene dibromide for use as a fumigant on 1 September, 1984 has meant that alternative physical disinfestation procedures, especially cold and heat treatments (Table 10.1), have been actively studied.

The growth and development of plants that evolved in the tropics and to a lesser extent the subtropics are severely limited by temperatures between 0°C and up to 10–12°C. The injury that occurs between 0°C and 10–12°C is referred to as chilling injury. The most studied aspect of chilling injury is the postharvest storage temperature response of commercial tropical and subtropical crops (Paull, 1990a). Chilling injury limits the postharvest life of these commodities. Low temperatures, <10 to 12°C for tropical and <7°C for subtropical crops, cannot be used without caution to extend postharvest life or for insect disinfestation (Fig. 10.1). The extent of postharvest loss due to chilling injury is difficult to determine because chilling injury predisposes all commodities to pathogenic attack, especially saprophytes, as well as disrupting ripening of climacteric fruit. Cold treatments are suitable for temperate and some subtropical and a few tropical commodities. Carambola is an example of a tropical fruit that can withstand cold and a 15 day treatment at 1.1°C is used for Caribbean fruit fly disinfestation (Table 10.1).

Prestorage temperature manipulation has been used to reduce the susceptibility of many commodities to chilling injury during subsequent

From: Paull, R.E. and Armstrong, J.W. (eds) (1994) *Insect Pests and Fresh Horticultural Products: Treatments and Responses.* CAB INTERNATIONAL, Wallingford, UK.

Table 10.1. Temperature management quarantine treatments that provide quarantine security against fruit fly and do not normally reduce the condition or quality of commodities. Under commercial conditions damage can occur.

Commodity/cultivar	Treatment[a]	Approved by APHIS[b]
Caribbean fruit fly		
Carambola, 'Arkin'	VH, 43.3–43.6°C, 90 or 120 min	No
	VH, 46–46.3°C, 60 or 90 min	No
Carambola, 'Arkin'	C, 1.1°C, 15 days	Yes[c]
Grapefruit, 'Ruby Red', 'Marsh'	C, 16°C, 7 days + 1.7°C, 14 days	Yes
Grapefruit, 'Marsh'	VH, 43.3–43.7°C, 5 h	Yes
	VH, 43.5°C, 4.25 h	Yes
Grapefruit, 'Marsh'	HA, 46°C, 3 h	No
Mango, 'Tommy Atkins', 'Keitt'	HW, 46°C, 60–90 min	Yes
	HW, 46.1–46.7°C, 75 min	Yes
Mango, 'Francis'	HW, 46.1–46.7°C, 1.25–2 h	Yes
Mango, 'Tommy Atkins', 'Keitt'	HW, 46.1–46.7°C, 90 min	Yes
Mango, 'Tommy Atkins'	HA, 51.5°C, 125 min	No
Mediterranean, oriental and melon fruit flies		
Bell peppers	VH, 44.4°C, 8.75 h	Yes
Eggplant	VH, 44.4°C, 8.75 h	Yes
Papaya 'Kapoho'	HW, 42°C, 30 min + 49°C, 20 min	No[d]
	VH, 44.4°C, 8.75 h	Yes
	HA, to FCT 47.2°C	Yes
Pineapple (other than Smooth Cayenne)	VH, 44.4°C, 8.75 h	Yes
Tomato	VH, 44.4°C, 8.75 h	Yes
Zucchini squash	VH, 44.4°C, 8.75 h	Yes

[a] VH=vapour heat, C=cold, HA=hot air, HW=hot water, and FCT=fruit core temperature.
[b] Animal and Plant Health Inspection Service of the US Department of Agriculture. 1976 (revised 1985). Plant Protection and Quarantine Treatment Manual, Section VI-T100. Animal and Plant Health Inspection Service, US Department of Agriculture, Washington, DC
[c] For domestic use only.
[d] Damage and quarantine failure can occur under certain seasonal conditions.

cold storage (Hatton, 1990). One important commercial use of temperature manipulation to avoid chilling injury is the temperature conditioning of Florida grapefruit prior to an insect quarantine cold treatment (Hatton and Cubbedge, 1983). Effective conditioning of other commodities to reduce their susceptibility to chilling injury may provide for the use of a cold treatment to disinfest them of quarantine insects.

Heat treatments, like cold treatments, can place severe stress on fresh horticultural commodities. However, heat treatments have found wider applications than cold for tropical commodities (Table 10.1). For example, the initial procedure considered for papaya was a 20 min treatment of fruit

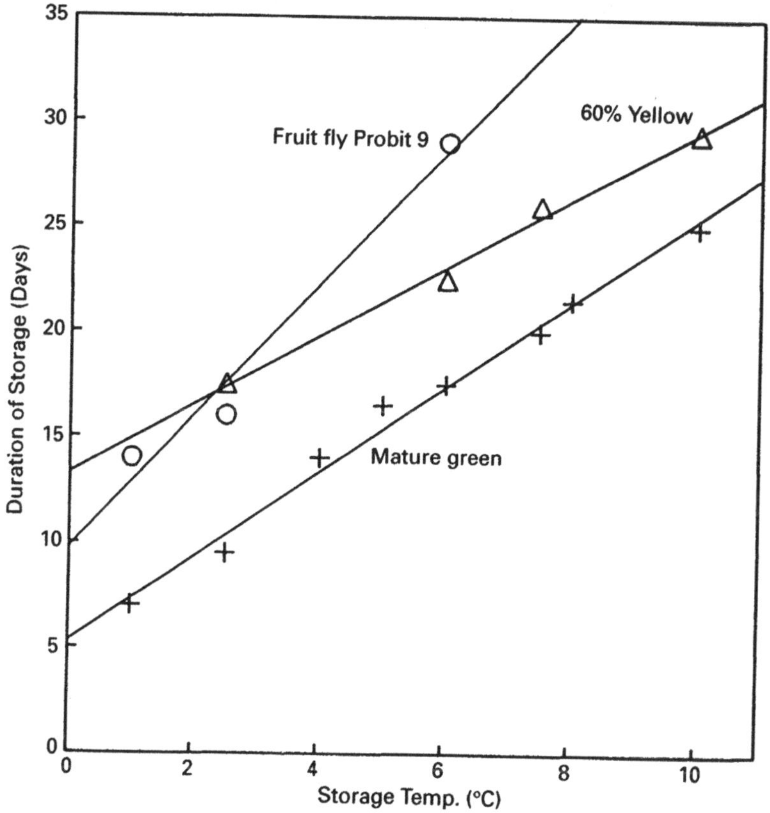

Fig. 10.1 The time and temperature relationship of incipient chilling injury symptom development in papaya fruit and relative sensitivity of fruit flies to low temperatures. Papaya chilling injury visible symptom development redrawn from Chen and Paull (1986), and Mediterrean fruit fly mortality data redrawn from Back and Pemberton (1916) and Baker (1939). Above the lines injury symptoms become more severe, while above the line fruit fly, mortality reaches probit 9.

at 49°C followed by 10 days' storage at 8.5°C (Couey *et al.*, 1984). A heat treatment in which fruit is exposed to hot air (approx. 50% RH) till the fruit core temperature reaches 47.2°C, is now being used (Armstrong *et al.*, 1989). This so-called forced air heat treatment has been developed from the currently approved vapour heat treatment (Seo *et al.*, 1974). A similar heat treatment was tested in the late 1930s (Jones *et al.*, 1939). Vapour heat disinfestation is currently the only heat treatment approved for papaya, mangoes and other fresh commodities destined for the Japanese market.

Symptoms and Development of Injury

Chilling

Symptoms

The postharvest storage life of tropical and subtropical crops does not gradually increase with decreasing storage temperature. Tropical and subtropical crops' postharvest storage life reaches a maximum at around 10–12°C and 4–7°C, respectively. Higher temperatures allow continued senescence and ripening. Storage life decreases below the 10–12°C or 4–7°C threshold due to the development of chilling injury, but ripening and senescence are severely restricted. This pattern explains why most storage recommendations for tropical commodities are just inside the range where chilling injury occurs. A number of reviews have appeared in the last ten years dealing with basic (Graham and Patterson, 1982; Wang, 1982, 1990b) and applied (Couey, 1982; Paull, 1990a) aspects of chilling injury. Symptoms of chilling injury generally develop after removal from the chilling temperature to non-chilling temperatures. The development of symptoms is very slow and the symptoms are similar to those that occur for other stresses and injury.

The chilling injury symptoms have been described by many authors and a summary has been given by Morris (1982). The symptoms most commonly seen are surface lesions, pitting, scald and large sunken areas. Internal brown discoloration of pulp, vascular strands and seeds is also common (Paull, 1990a). Water soaking of the tissue coupled with wilting and desiccation (ung choi) is a common symptom in leaves (Hirata *et al.*, 1987). The above visual symptoms frequently occur with dysfunction of a number of developmental processes. The most commonly reported is a failure of fruits to ripen and acceleration of senescence. These last symptoms often lead to a reduction of postharvest life and increased susceptibility to decay (Sommer and Mitchell, 1978). The organisms involved in this decay are usually not found in healthy tissue (Morris, 1982).

Development

Leaves of many species such as *Phaseolus vulgaris* show rapid wilting within a few hours of chilling. Initially, this effect is rapidly reversible. Sunken pits, which turn into brown necrotic areas, appear later. The extreme is the rapid leaf symptoms in *Episcia reptans* which is completely killed by as little as 4 h at 5°C (Wilson, 1984). Light (van Hasselt and van Berlo, 1980) and water stress (Wilson, 1976) are important in determining the speed of chilling injury symptom development.

The stage of fruit ripeness and age of leaf tissue at the time of low temperature storage significantly influence the plant sensitivity to visible chilling injury symptom development. The symptom and time–temperature response for tropical and subtropical crops vary widely (Paull, 1990a). For example, the *Annona* fruit is very susceptible to chilling, with injury occurring within 5 days at 4°C (Broughton and Guat, 1979). Sweetsop (*A. squamosa* L.) ripening is disrupted by storage for 13 days at 15°C, with poor taste and a dark-brown appearance (Broughton and Guat, 1979). Above 15°C the fruit ripens normally. Three types of avocado (*Persea americana* Mill.) are recognized: Mexican, Guatemalan, and West Indian (Popenoe, 1920). The Mexican is regarded as being more resistant to cold, with the Guatemalan intermediate and the West Indian fruit being most sensitive (Hatton *et al.*, 1965), with many of the commercial varieties being probably crosses.

Stage of development influences chilling sensitivity. Banana fruit of the less mature apical hands is less susceptible to chilling injury than the more mature distally borne fruit (Marriott, 1980). Avocados are most sensitive to chilling at the preclimacteric rise and climacteric peak (Zauberman *et al.*, 1973). Papaya (*Carica papaya* L.) (Chen and Paull, 1986) and tomato (*Lycopersicon esculentum* Mill) (Hobson, 1987) are most sensitive during the preclimacteric period (Fig. 10.1). By comparison, symptoms of chilling injury in the tropical fruit, carambola (*Averrhoa carambola* L.) have not been described (Oslund and Davenport, 1983). Fruit which are green and at colour break when placed into storage can be held for 3–5 weeks at 0–4.5°C (Kenney and Hull, 1986).

Heat

Symptoms

A short exposure to near-lethal temperature is a 'crisis' situation that relates to survival and the capacity to recover, while longer exposures to less extreme temperatures are more difficult to quantify. Excessive exposure leads to death and very high temperature can only be tolerated for short periods. Exposure of leafy vegetables and ornamentals to heat treatments very frequently leads to injury such as wilting and browning along with a rapid loss of postharvest life. The lethal temperature has been reported to be 45°C for tomato, 63°C for grape, and 49–52°C for apple (Huber, 1935). Even below such non-permissive temperatures, growth and metabolism can be adversely affected by heat (Table 10.2). For example, avocado and papaya fruits stored at greater than 30 and 34°C ripened abnormally, showed considerable surface pitting, and had poor flavour (Lee and Young, 1984; An and Paull, 1990). Ethylene treatment increased the threshold temperature for injury in avocado (Lee and Young, 1984).

Table 10.2. Effect of heat treatment on physiological attributes and processes occurring during tomato and papaya fruit ripening. Selected citations.

	Temperature	Duration	Response	Citation
Chlorophyll				
Tomato	33°C	48 h	Decline	Picton and Grierson (1988)
	35°	5 days	Decline	Picton and Grierson (1988)
Chloroplast activity (delayed light emission)				
Papaya	42°C	30 min	Decline	Chan and Forbus (1988)
Respiration rate at the climacteric				
Tomato	33°C	12 days	Decline	Ogura *et al.* (1976)
	35°C	3 days	Decline	Inaba and Chachin (1988)
	37°C	3 days	Decline	Cheng *et al.* (1988)
Papaya	38°C	4 days	Decline	Lurie and Klein (1990)
Ethylene production				
Tomato	35°C	2 days	Decline	Picton and Grierson (1988)
	35°C	7 days	Decline	Picton and Grierson (1988)
Ethylene forming enzyme activity				
Papaya	49°C	70 min	Decline	Chan (1986a)
	49°C	70 min	Decline	Paull and Chen (1990)
Tomato	35°C	24 h	Decline	Biggs *et al.* (1988)
Electrolyte leakage				
Tomato	45°C	166 min	Increase	Inaba and Crandall (1988)
Nucleic acid metabolism				
Tomato	35°C	48 h	Decline	Picton and Grierson (1988)
Papaya	49°C	70 min	Decline	Paull and Chen (1990)
Protein synthesis				
Tomato	35°C	1 h	Decline	Picton and Grierson (1988)
Heat shock protein synthesis				
Papaya	38°C	1 h	Increase	Paull and Chen (1990)
Tomato	35°C	4 h	Increase	Picton and Grierson (1988)
Wall degrading enzymes				
Papaya	49°C	70 min	Decline	Paull and Chen (1990); Chan *et al.* (1981)
Tomato	33°C	3 days	Decline	Ogura *et al.* (1975); Picton and Grierson (1988)

Source: After Paull, 1990b.

Problems have been experienced with heat treatments used for insect disinfestation (Paull, 1990b). The symptoms of heat injury in ripening fruit are skin scald (Fig. 10.2a) and failure of the fruit to soften fully (Fig. 10.2b) or to soften at a normal rate (Jones, 1942; Akamine and Seo, 1978; Chan *et al.*, 1981; Kerbel *et al.*, 1985; Couey, 1989; Paull and Chen, 1990). Preharvest high temperatures also can have detrimental effects on storage life (Kasmire, 1978) and quality (Liu, 1978) of fruit. Tomato fruit held for more than 6 h at temperatures greater than 41°C showed delayed ripening (Hall, 1964a,b) and increased electrolyte leakage (Inaba and Crandall, 1988). The treatment at 49°C for 20 min is near the limit for heat tolerance in papaya (Akamine, 1960).

Development

The severity of high temperature stress is primarily determined by the temperature differential and the duration of exposure. Other factors such as rapidity of the change in temperature and the previous growing conditions are also important (Levitt, 1980). The effect of exposure time, although recognized by many (Alexandrov, 1977; Levitt, 1980; Blum, 1987), has not been fully investigated and its implications are unclear.

Couey (1989) grouped fruits into heat-tolerant and heat-sensitive to hot water treatment. The heat-tolerant fruits were banana, papaya, mango, pear, and litchi, while the heat-sensitive fruits were peach, raspberry, bell pepper, and cantaloupe. The difficulty with the division is that thermotolerance is dependent upon cultivars, fruit size, maturity, and postharvest handling (Paull, 1990a).

Physiological Responses

Chilling

Factors influencing rate

The response of plant tissue to chilling injury has frequently been separated into primary and secondary responses. This separation is based on the perception of a series of events involving multiple events cascading from the primary event after a certain level of chilling injury (Lyons, 1973). Two distinct possibilities are suggested as the primary event: lipid or protein changes. Changes in either lipids or proteins as the primary event in chilling injury are thought to occur instantaneously when the tissue is subjected to chilling temperatures.

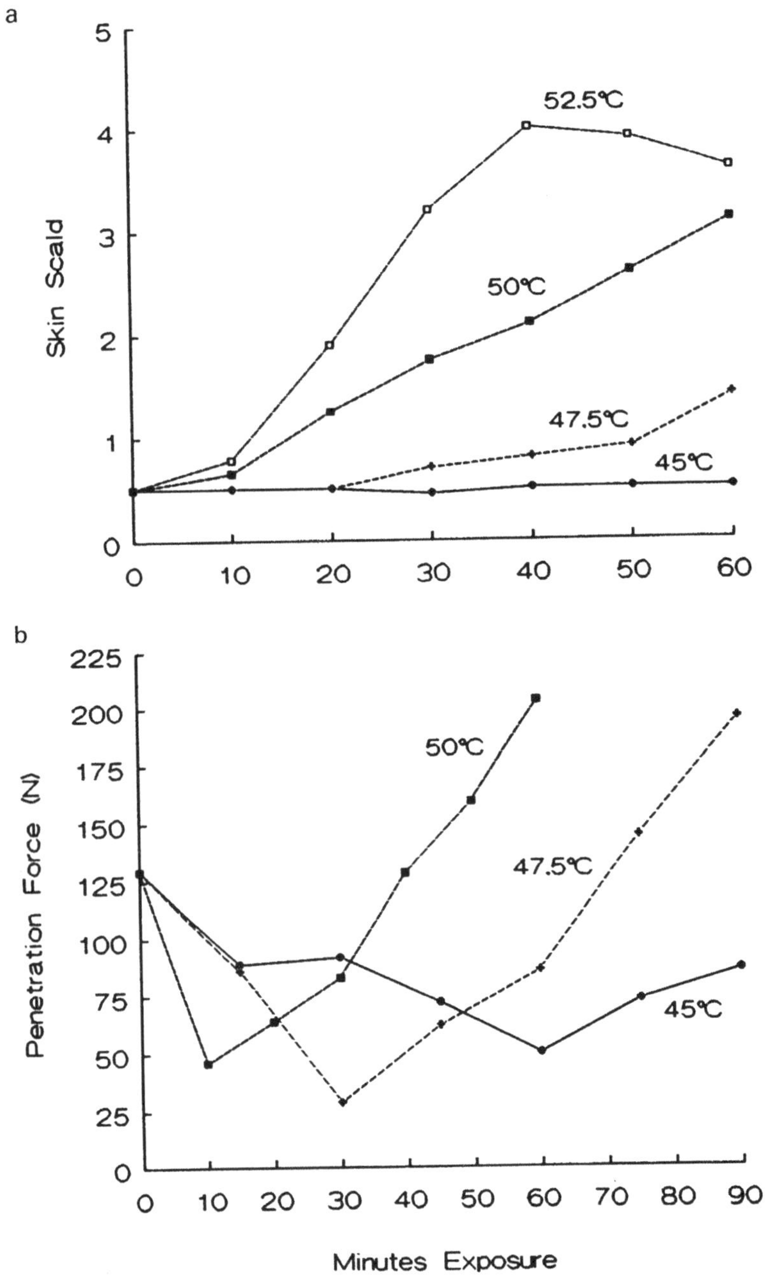

Fig. 10.2 Effect of immersion of papaya fruit for different lengths of time in hot water at different temperatures on the development of (a) skin scald (0 = no scald, 5 = 100% scald) and (b) flesh penetration force (Paull, 1990b).

The sequence of events leading to visible symptoms after primary sensing of chilling temperatures clearly varies with tissue. Physiological symptoms of chilling injury in papaya include increased ethylene production (Chan *et al.*, 1985), electrolyte leakage (Chan *et al.*, 1985), and *Alternaria* rot (Sommer and Mitchell, 1978), that occur before the appearance of visible symptoms. At 10°C, increased ethylene production occurs after 7 days, electrolyte leakage after 14 days, and visible symptoms after 24 days. This is probably not a directly connected linear sequence. Similar time sequences of physiological and visible symptom developments can be determined from the published literature for tomatoes, sweet potatoes, mung beans, and avocado (Wang, 1982). The actual sequence of events leading to visible symptoms is unknown.

The time–temperature for incipient visible symptoms of chilling injury development is the point at which irreversible dysfunction occurs; prior to this stage injury is reversible. Information on the time–temperature response to incipient chilling injury for tropical and subtropical fruit and vegetable is limited (Paull, 1990a). Papaya fruit does show a linear relationship between storage temperature and storage duration (Fig. 10.1) and chilling injury symptom development (Chen and Paull, 1986). Insect mortality plot is shown for comparison.

Electrolyte leakage

Lieberman *et al.* (1958) were the first to report an increase in electrolyte leakage from a chilling-sensitive commodity when stored at a chilling temperature. Electrical conductivity of the leachate from whole fruits of oranges, limes, and grapefruit increases in response to chilling (Pantastico *et al.*, 1968). Forney and Peterson (1990) reported that enhanced leakage of K^+ ions is an effective indicator of chilling injury in grapefruit albedo callus. Electrolyte leakage does not increase until after visible symptoms appear in papaya (Chan *et al.*, 1985) and grapefruit (McCollum and McDonald, 1991). However, electrolyte leakage is reported to be unaffected by chilling in some other fruits such as peach (Furmanski and Buescher, 1979), bell pepper and eggplant (Murata and Tatsumi, 1979), and avocado and mango (Fuchs *et al.*, 1989).

Respiration

There is an anomalous change in the rate of respiration in many plant tissues sensitive to chilling when exposed to low temperatures. Increased rates of respiration have been observed for many commodities including cucumber (Eaks and Morris, 1956), grapefruit (McCollum and McDonald, 1991), snap bean (Watada and Morris, 1966), sweet potato (Lewis and Morris, 1956), and tomato (Autio and Bramlage, 1986). Respiration rates of cucumber (Eaks and Morris, 1956) and grapefruit (McCollum and McDonald, 1991) initially increased with chilling and then tended to decrease with length of storage. These findings contrast with the pattern observed for chilled orange (Eaks, 1960, 1965) and lemon (Eaks, 1965, 1980), in which increasing the duration of chilling resulted in a greater stimulation of respiration.

While chilling generally increased respiration in whole fruits, respiration of specific tissues and organs generally decreased at chilling temperatures (Lyons and Breidenbach, 1990). Eaks (1983) reported that the respiration rate of avocado is dependent on the stage of the climacteric at the time of the chilling exposure. Although the respiratory responses of many commodities and tissues have been related to chilling temperatures, the course and rate of respiration are not uniform among these commodities and tissues.

Ethylene production

Low temperature stress enhances ethylene production of chilling-sensitive plants that normally do not produce significant amounts of ethylene (Wang, 1989b). Wang and Adams (1980) found that the pathway for ethylene biosynthesis is the same as that in ripening fruits, and the step enhanced by chilling is the synthesis of 1-aminocyclopropane-1-carboxylic acid (ACC), the immediate precursor of ethylene (Adams and Yang, 1979).

In some commodities ethylene production is stimulated during chilling; in others, ethylene production does not occur until after the chilled tissues are transferred to a warmer non-chilling temperature. Commodities in the former category include grapefruit (McCollum and McDonald, 1991) and tomato (Autio and Bramlage, 1986), with cucumber (Wang and Adams, 1982), honeydew melon (Lipton and Wang, 1987), lemon (Eaks, 1980), papaya (Chan *et al.*, 1985), pear (Stakiotakis and Dilley, 1974), and zucchini squash (Mencarelli *et al.*, 1983) in the latter category. Chilling-induced ethylene production is found to be reduced after a chilling exposure in 'Sensyu' apple (Wang *et al.*, 1990) and avocado (Eaks, 1983). The variation in tissue response to chilling temperatures leads to the conclusion that ethylene production is not a good indicator of chilling injury.

Field (1990) in his review concluded that chilling injury is not caused by ethylene and neither is ethylene consistently identified as a defence mechanism. Specific inhibitors of ethylene production do not ameliorate the development of chilling injury. The stimulation of ethylene production appears to be simply a response of tissues to chilling stress (Wang, 1989b). Wang (1989a) concluded in his review that ethylene is not directly involved in causing chilling injury nor is it an essential part of the development of chilling injury symptoms.

Heat

Factors influencing rate

The plant's response to heat has been described in detail but the literature on its repercussions on plant cell physiology is still limited (Paull, 1990b). The damage caused is very similar in most cells and tissues (Table 10.2), and leads in extreme cases to the death of cells, tissues, organs or the whole plant. Sun scald is a common problem with fruit that has limited transpirational cooling (Blum, 1987). The effect of heat on flower abscission, pollen sterility and poor fruit set is well known (Picken, 1984), and it limits the areas in which certain crops can be grown. Temperatures below the lethal level may have significant effects, such as the acceleration of plant development (Halse and Weir, 1974) and reduction in yields (Wallace, 1985).

Cellular membranes were stressed very early with respect to heat injury (Alexandrov, 1977). Much of the work on membrane lipids has been on bulk cellular extracts, and its relevance to the *in vivo* situation has been questioned (Blum, 1987). The loss of cellular membrane integrity with increasing temperatures is gradual, with repair and reversibility possible until a lethal cell temperature is reached. The role of the membrane in cell death induced by heat has recently been reviewed (Lange *et al.*, 1981; Blum, 1987).

Most of the concepts and experimental evidence for heat injury involve protein denaturation or disruption of protein synthesis. This denaturation at lethal temperatures is regarded as non-reversible, while lower temperatures can lead to a reversible type of inactivation. Heat effects on proteins and enzymes have been reviewed (Bernstam, 1978). Less than lethal temperatures can lead to disruption of transcription and translation steps in protein synthesis (Paull and Chen, 1990).

Heat shock response

Bacteria, plants, and animals exposed to higher temperatures for a brief period acquire transient thermotolerance (Lindquist and Craig, 1988). As this thermotolerance response develops, a unique group of proteins is synthesized. These proteins (heat shock proteins) are produced within 30 min after exposure to temperatures in the range 34–42°C (Kanabus *et al.*, 1984). Heat shock proteins (HSPs) and the associated thermotolerance have been reported for field-grown cotton leaves (Burke *et al.*, 1985), soyabean leaves (Kimpel and Key, 1985), sorghum (Ougham and Stoddart, 1986), and papaya fruit (Paull and Chen, 1990). The decay of HSPs occurs within 24 h, with a corresponding loss in thermotolerance. This phenomenon appears to confer a temporary, acquired heat resistance to sublethal temperatures (Altschuler and Mascarenhas, 1982). There appears to be a fundamental role for these HSPs in cellular function during high-temperature stress, but this role is still unclear (Johnson and Kucay, 1988; Yost and Lindquist, 1988). Tolerance acquired by sudden heat shock differs from tolerance acquired by sustained growth at a moderately high temperature (Wu and Wallner, 1984). This is also supported by HSPs being produced at higher temperatures when the temperature is raised slowly (3°C h^{-1}) versus a more rapid increase (Altschuler and Mascarenhas, 1982), emphasizing the importance of the heat-stress profile.

Fruit ripening

Exposure of pear (Maxie *et al.*, 1974) and tomato (Picton and Grierson, 1988) to temperatures higher than 30°C for 48 h or more results in disruption of various aspects of ripening. This disruption is characterized by a failure to develop normal pigmentation, abnormal softening, and a marked decline in ethylene production (Table 10.2) (Maxie *et al.*, 1974; Yoshida *et al.*, 1984; Picton and Grierson, 1988). Colour changes associated with ripening in tomato are inhibited with storage at temperatures above 30°C (Cheng *et al.*, 1988). Above 30°C, banana (Burg, 1962) and papaya (An and Paull, 1990) fail to ripen normally, with the pulp becoming soft and watery. Soluble solids are not affected by heat treatment while titratable acidity declined in apple (Liu, 1978; Klein and Lurie, 1990). The inhibition of ripening above 30°C has been ascribed to disruption of a protein involved in ethylene synthesis (Maxie *et al.*, 1974), of the polygalacturonase regulatory mechanism that controls the generation or activation of the enzyme (Chan *et al.*, 1981), and to suppression of ripening-related mRNA synthesis (Picton and Grierson, 1988).

Other responses: leakage, respiration, etc.

A sigmoidal curve described electrolyte leakage by tomato discs induced by increasing temperatures (Inaba and Crandall, 1988). Critical exposure times for leakage were: 34 min at 55°C, 105 min at 50°C, and 166 min at 45°C. The tomatoes did not exhibit visible heat injury. A 50% higher electrolyte leakage was found from fruit discs isolated from apple held at 38°C (Lurie and Klein, 1990). After transfer of the apple discs to 20°C leakage declined within 2 days, to control levels, suggesting the exposure did not produce irreversible injury.

Respiration rate and ethylene synthesis are affected by exposure to high temperatures (Jones, 1939; Akamine, 1966). The respiratory control ratio is higher in tomato mitochondria isolated from fruit stressed for 3 days at 37°C (Cheng *et al.*, 1988). Temperatures above 25°C reduced the extent of the respiratory climacteric in avocado (Biale and Young, 1971).

The respiratory rise between the initial preclimacteric level and the climacteric peak was 250% at 25°C, and only 30% at 30°C. Twelve days or more at approx. 33°C resulted in the suppression of respiration, ethylene synthesis, and fruit softening, and this action was not completely reversed after returning the tomato fruit to ambient temperatures (Yoshida *et al.*, 1984; Ogura *et al.*, 1976; Cheng *et al.*, 1988; Inaba and Chachin, 1988).

Heating leaves to 44°C leads to loss of chloroplast activity, especially electron transport by photosystem II (Berry *et al.*, 1975). Chlorophyll fluorescence (Smillie and Gibbons, 1981) can be used to indicate heat stress. An associated characteristic, delayed light emission, declines in papaya skin with first-order kinetics when fruit is heated to between 42 and 48°C (Chan and Forbus, 1988). At higher temperatures (49–51°C) heat inactivation is biphasic.

Biochemical Responses

Chilling

Membrane lipids

The physical phase transition of membranes from a flexible liquid-crystalline to a solid-gel structure as the primary response in chilling-sensitive plants to chilling temperatures was proposed by Lyons and Raison (1970). This concept has since received both wide support and criticism. Recently, attention has been on particular glycerolipids localized in plastid membranes. Certain molecular species of the glycerolipids phosphatidylglycerol (PG) and sulfoquinovosyldiacylglycerol (SQDG) have phase transition temperatures above 30°C and the relative amount of these molecular species correlates well with chilling sensitivity of plants (Murata and Nishida, 1990). Therefore, it is hypothesized that the PG and SQDG molecular species with high phase transition temperatures play a role in chilling sensitivity of plants. Recently, Murata *et al.* (1992) showed that genetically modifying the level of unsaturation of fatty acids in chloroplast PG changes chilling sensitivity.

Lipid peroxidation has also received attention as a determinant of chilling injury in sensitive plant tissues. A common response of plant tissues to environmental stress is lipid degradation. This has been observed for plant tissues subjected to wounding and freezing (Elstner and Konze, 1976), drought (Seneratna and McKersie, 1986), and salinity (Chrominski *et al.*, 1986) stress. When chilling-sensitive plants are chilled in the presence of light, losses in antioxidant capacities (Wise and Naylor, 1987a) and the degradation of lipids (Wise and Naylor, 1987b) are correlated with chilling injury development. Patterson *et al.* (1984) showed that antioxidant defences may be compromised when chilling-sensitive plants are exposed to low temperatures in the dark. The application of antioxidants to cucumber and pepper fruits delayed or reduced the severity of chilling injury in the dark (Wang and Baker, 1979). Omran (1980) reported that catalase activity is decreased in cucumber seedlings as a result of low temperature.

Wise and Naylor (1987b) suggested that chilling-resistant pea leaves have an endogenous mechanism(s) for the removal of toxic oxygen species before they induce lipid peroxidation. Chilling-resistant orange fruit flavedo tissue had high antioxidant activity while chilling-susceptible lime and lemon fruits had almost no active antioxidant activity (Ting and Newhall, 1965). Meir and Bramlage (1988) found that antioxidant activity of apple peel increases with later harvest dates and is highly negatively correlated with scald incidence and severity during low temperature storage.

Carbohydrates

The relationship of sucrose and soluble sugar content and chilling injury varies with the fruit. Picha (1987) found enhanced sucrose and total soluble sugar content occurs after storing sweet potato at chilling temperatures. However, coinciding with the development of chilling injury in mango, there is a decrease in sucrose and total soluble sugars (Chhatpar *et al.*, 1971). Sucrose and fructose levels in tomato decline during storage while the concentration of glucose remains fairly constant (Buescher, 1975).

High levels of reducing sugars in grapefruit peel were positively correlated with resistance to chilling injury (Purvis and Grierson, 1982). However, it was not clear whether the accumulation of reducing sugars played a direct role in reducing chilling injury or if they were a result of chilling stress. King *et al.* (1988) found that diurnal changes in chilling sensitivity of tomato seedlings were correlated with endogenous carbohydrate content. Exogenous applications of reducing sugars were shown to reduce the severity of chilling injury in excised tomato shoots (King, 1985).

Polyamines

Changes in polyamine levels in plant tissues have been reported in response to various kinds of stresses. McDonald and Kushad (1986) found putrescine levels to increase when grapefruit, lemon and pepper fruits were stored at chilling temperatures. Increased amounts of putrescine are also observed in chilled bean plants (Guye *et al.*, 1986). The level of spermidine is higher in chilled cucumber seedlings compared with non-chilled seedlings (Wang, 1987). Putrescine and spermidine levels increase in zucchini squash in response to low temperature storage (Wang and Ji, 1989). Prestorage temperature conditioning of zucchini squash leads to increases in spermidine and spermine levels during low temperature storage (Kramer and Wang, 1989a).

Since polyamines have the ability to stabilize membranes (Mager, 1959) and the apparent involvement of membrane damage in chilling injury (Wang, 1982), polyamines may play a role in reducing chilling injury. Polyamines are strong cations and may interact with the anionic components of the membrane, such as phospholipids (Roberts *et al.*, 1986). This association serves to stabilize the membrane by causing rigidification at the bilayer surface.

Other responses

Numerous other responses of plant tissue to chilling have been reported. Kozukue *et al.* (1984) found alanine increases rapidly during low temperature storage of chilling-sensitive eggplant, cucumber and okra whereas no increases are found in chilling-resistant crops. Subsequent work suggests that the accumulated alanine may be produced by glutamate-pyruvate transaminase from accumulated pyruvic acid during cold storage (Tsuchida *et al.*, 1990).

Other biochemical responses of chilling-sensitive crops to low temperature storage include the accumulation of α-keto acids (Murata, 1969), abscisic acid (Daie *et al.*, 1981), and shikimic acid (Kozukue and Ogata, 1972), increase in chlorogenic acid (Lieberman *et al.*, 1959), citric acid (Thorne and Efiuvwevwere, 1988), and dihydrophaseic acid (Daie *et al.*, 1981), changes in malic acid (Thorne and Efiuvwevwere, 1988), and decline in ascorbic acid (Lieberman *et al.*, 1959).

Heat

Cell wall hydrolases

Thermal injury (46°C, 65 to 90 min) to papaya fruit softening during ripening was correlated with a decrease to 10% of normal polygalacturonase levels (Chan *et al.*, 1981), the injury being more severe in riper fruit (one-quarter to one-half ripe) than in less ripe fruit. Heated fruit showed delayed ripening and left a 1–1.5 cm-thick area of hard tissue. They also found partial recovery to 25% of normal polygalacturonase level after 6 days at 24°C (Chan *et al.*, 1981). A similar reduction was found for tomatoes held for several days at 33°C (Ogura *et al.*, 1975). These results could be due to either denaturation or disruption of polygalacturonase synthesis. Since polygalacturonase shows first-order heat denaturation kinetics and is 60% denatured by 20 min at 70°C (Chan and Tam, 1982), exposure of fruit to 49°C would be expected to induce only minimal denaturation. The failure to soften, therefore, is possibly attributable to suppression of the mRNA for wall-softening enzymes, as found in tomatoes (Picton and Grierson, 1988). Polygalacturonase activity returned after a 6 day lag when temperature was returned to 25°C (Ogura *et al.*, 1975; Yoshida *et al.*, 1984). A similar reversibility in fruit softening has been found in apple held at 38°C for 4 days (Lurie and Klein, 1990).

The failure of papaya to recover the ability to soften (Paull and Chen, 1990) after injurious heat treatment could be related to the unique aspect of papaya where the wall-degrading enzymes polygalacturonase and xylanase have a peak activity when the fruit is 40–60% yellow (Paull and Chen, 1983). This relationship suggests that mRNA for polygalacturonase and xylanase in papaya are produced for only a short period during one stage of ripening.

In the range 35–60°C considerable (87%) loss of activity of purified glucanase from tomato fruit can occur (Hinton and Pressey, 1980), also 50% glucanase activity can be lost by exposure to 50°C for 5 min (Pressey, 1983). The role of these enzymes in fruit softening is unclear. Pectinesterase is suppressed in tomatoes held at 33°C (Ogura *et al.*, 1975). The activity several days after returning to room temperature was only one-third of that of the normal ripening fruit.

Protein synthesis

Exposure of ripening pear fruit to 40°C leads to a disruption of protein synthesis via a loss of polysomes (Romani and French, 1977). This response is readily reversible and establishes a close correlation between the protein synthesizing machinery and the progression of ripening. Apples held at 38°C for 4 days have reduced [^{35}S]methionine incorporation into proteins (Lurie and Klein, 1990). The elevated temperature treatments can lead to inhibition and/or reduced expression of softening and other ripening-related mRNAs (Picton and Grierson, 1988). Transcriptional regulation is probably modified by mRNA stability. Translated polysomal RNA confirmed that new polypeptides are synthesized by fruit following heat shock induction similar to insect disinfestation heat treatments (Paull and Chen, 1990). Continued exposure to 42°C, though still allowing HSP synthesis, causes damage. These results agree with those for tomato fruit treated for long periods at 35°C (Picton and Grierson, 1988). Apples held for 4 days at 38°C had a change in protein synthesis profile and indicate possible heat shock protein synthesis (Lurie and Klein, 1990).

Ethylene synthesis

Ethylene synthesis is reversibly inhibited at higher temperatures (Field, 1984). Temperatures greater than 35°C cause an accumulation in endogenous ACC content in apple tissue (Yu *et al.*, 1980) with a reduced ethylene production. The conversion of ACC to ethylene is apparently highly susceptible to heat damage. A rapid loss (75%) of ethylene-forming enzyme (EFE) occurs in papaya and other fruits exposed for short periods to temperatures greater than 40°C (Chan, 1986a; Paull and Chen, 1990). The heat inactivation in papaya and cucumber is biphasic with both phases following first-order kinetics (Chan, 1986b). The heat resistant EFE is about 25% of total EFE activity. Full recovery of EFE activity occurs within 3 days in papaya (Paull and Chen, 1990), apple (Klein and Lurie, 1990), and muskmelon (Dunlap *et al.*, 1990) after removal of heat. Fruit exposed to long periods at high temperature quickly recover ethylene synthesis ability (Ogura *et al.*, 1976). The loss of EFE activity due to heat treatment in cucumber fruit can be modified by conditioning for 24 h at 32.5°C (Chan and Linse, 1989). The recovery of ethylene production ability possibly requires protein synthesis (Biggs *et al.*, 1988). Wound-induced ethylene synthesis is less sensitive than climacteric-related ethylene synthesis to heat stress inactivation (Biggs *et al.*, 1988).

Amelioration of Injury

Chilling

Various treatments have been used to reduce damage to chilling-sensitive horticultural crops during exposure to low temperatures. The amelioration of chilling injury of these crops has been reviewed extensively (Lyons, 1973; Christiansen and St. John, 1981; Jackman *et al.*, 1988; Wang, 1982, 1991). Some of these approaches are described below.

Temperature manipulation

High and low temperature conditioning have been used to reduce chilling injury. Prestorage low temperature conditioning, or lowering temperatures to slightly above the chilling temperature for various periods of time, has been reported to reduce chilling injury in a number of fruits and vegetables including cucumber, 6–9 days at 18°C (Hirose, 1985), 1 day at 15°C and 10°C (Nakamura *et al.*, 1985); eggplant, 5–15 days at 10°C or 5–10 days at 20°C (Abe and Chachin, 1985), 1 day at 15°C and 10°C (Nakamura *et al.*, 1985); grapefruit, 6 days at 17°C (Chalutz *et al.*, 1985), 7 days at 16°C (Hatton and Cubbedge, 1983); lemon, 7 days at 15°C (Houck *et al.*, 1990),

3 days at 21°C (McDonald, 1986); lime, 7 days at 7–21°C (Spalding and Reeder, 1983); mango, 30 days at 10°C (Thomas and Joshi, 1988); papaya, 4 days at 12.5°C (Chen and Paull, 1986); pepper, 5 or 10 days at 10°C (McColloch, 1962), 4 or 8 days at 5 or 10°C (Thompson, 1978); and squash, 2 days at 10 or 15°C (Kramer and Wang, 1989b).

Prestorage high temperature conditioning treatments to reduce chilling injury can be divided into two categories: long term (12 h to 4 days) at 38–46°C and short term (up to 60 min) at 45–60°C (Klein and Lurie, 1992). Long-term prestorage high temperature conditioning has been reported to reduce chilling injury of some fruits and vegetables including cucumber, 24 h at 36–40°C (Hirose, 1985); grapefruit, 17–22 h at 38°C (Brooks and McColloch, 1936); mango, 24 or 48 h at 38°C (McCollum *et al.*, 1993); sweet potato, 10 days at 32°C (Picha, 1987); tomato, 3 days at 36–40°C (Lurie and Klein, 1991); and watermelon, 4 days at 26°C (Picha, 1987).

Short-term prestorage high temperature conditioning reduced chilling injury in cucumber, 42°C for 1 h (McCollum and McDonald, unpublished data); grapefruit, 53°C for 2 min (McDonald *et al.*, 1991); and orange, 53°C for 2 min (Wild and Hood, 1989). The increased tolerance to chilling injury induced by both long-term and short-term prestorage high temperature conditioning is thought to be related to the appearance of HSP which would protect the plants from low temperature injury (Lurie and Klein, 1991).

It has been demonstrated that chilling injury can be ameliorated by interrupting chilling temperatures with periodic warm periods. This process is often referred to as intermittent warming. The brief warm periods must take place before chilling damage is irreversible. Intermittent warming has been reported to reduce chilling injury of several fruits and vegetables including cucumber, 12.5°C for 18 h every 3 days (Cabrera and Saltveit, 1990), 36°C for 1 day every 3 days (Hirose, 1987), 20°C for 1 day every 3 days (Wang and Baker, 1979); grapefruit, 15.5°C for 2 days every week or 21°C for 8 h every week (Davis, 1973), 21°C for 8 h every week (Hatton *et al.*, 1981); lemon, 13°C for 1 week every 3 weeks (Cohen, 1988); nectarine, 18–20°C for 2 days every 4 weeks (Anderson, 1982), 18.3°C for 2 days every 3 weeks (Anderson and Penny, 1975); orange, 21°C for 8 h every 2 weeks (Davis and Hofmann, 1973); peach, 18–20°C for 2 days every 4 weeks (Anderson, 1982), 18.3°C for 2 days every 3 weeks (Anderson and Penny, 1975), 23–25°C for 2 days every 2 weeks (Ben-Arie *et al.*, 1970); pepper, 21°C for 1 day every 3 days (Risse and Chun, 1987), 20°C for 1 day every 3 days (Wang and Baker, 1979); potato, 15.5°C for 1 week every 3 weeks (Hruschka *et al.*, 1969); squash, 20°C for 1 day every 3 days (Kramer and Wang, 1989c); and tangor, 13°C for 7 days every 3 weeks (Cohen *et al.*, 1990).

Atmosphere modification

The effectiveness of the atmospheric composition in reducing chilling injury is dependent on the commodity, levels of O_2 and CO_2, length of storage, and temperature of storage (Wang, 1989a). Many commodities will respond favourably to low O_2 and high CO_2 concentrations. However, in some fruits modified atmospheres tend to increase the severity of chilling injury (Forney and Lipton, 1990). Beneficial effects of modified atmospheres in reducing chilling injury have been reported in avocado (Spalding and Reeder, 1975), grapefruit (Brooks and McColloch, 1936), okra (Ilker and Morris, 1975), papaya (Chen and Paull, 1986), pineapple (Paull and Rohrbach, 1985), and zucchini squash (Mencarelli *et al.*, 1983). However, modified atmosphere storage has been reported to aggravate chilling injury in asparagus (Lipton, 1965), cucumber (Eaks, 1956), pepper (Cappellini *et al.*, 1984), and tomato (Tompkins, 1963).

Chemical treatment

Several chemicals have been applied exogenously to commodities to reduce chilling injury. Vacuum infiltration with $CaCl_2$ reduced chilling injury of avocado (Chaplin and Scott, 1980), okra (Ilker and Morris, 1975), tomato (Moline, 1980), and low temperature breakdown in apple (Bangerth *et al.*, 1972). Treatment with antioxidants or with chemicals which act as free radical scavengers has been shown to reduce chilling injury. Treatment of cucumber and sweet pepper with ethoxyquin and sodium benzoate reduced the development of chilling injury symptoms (Wang and Baker, 1979). The effectiveness of diphenylamine, butylated hydroxyanisole and ethoxyquin in preventing scald of apples stored at low temperatures is thought to be due to antioxidative properties (Wills *et al.*, 1981). Squalene, a naturally occurring compound in the epicuticular wax of grapefruit, was also reported to be effective in protecting grapefruit against chilling injury (Nordby and McDonald, 1990).

Prestorage treatment with dimethylpolysiloxane, safflower oil, and mineral oil prevented chilling-induced underpeel discoloration of banana (Jones *et al.*, 1978). Aljuburi and Huff (1984) found a reduction in chilling injury symptoms when grapefruit were coated with vegetable oil.

Pre- and postharvest application of several plant growth regulators has been shown to alter the resistance of commodities to chilling injury (Wang, 1990a). Benzyladenine, gibberellic acid, and 2,4-dichlorophenoxyacetic (2,4-D) acid applied postharvest and benzyladenine and 2,4-D applied preharvest modified the susceptibility of grapefruit to chilling injury (Ismail and Grierson, 1977). Treatment of zucchini squash with ABA (Wang, 1990b) and muskmelon with ethylene reduced the incidence of chilling injury (Lipton and Aharoni, 1979).

Polyamine levels change in response to several types of stress including chilling stress (McDonald and Kushad, 1986; Kramer and Wang, 1989a). Pressure infiltration with putrescine, spermidine or spermine reduced brown core of apple, a symptom of chilling injury (Wang and Kramer, 1990). Treatment of zucchini squash with spermine prior to storage reduced chilling injury (Kramer and Wang, 1989a).

The fungicide thiabendazole has been reported to reduce the incidence of chilling injury in grapefruit (Schiffmann-Nadel *et al.*, 1972). Wild and Hood (1989) reported that dipping oranges in suspensions of thiabendazole or benomyl at 53°C reduced chilling injury even more. The fungicide imazalil was also found to be effective in reducing chilling injury of grapefruit (McDonald *et al.*, 1991).

Heat

Conditioning and rate of heating

Many procedures have been empirically developed to reduce the injury caused by heat treatment. Most seem to rely on the heat shock response. The disruption of papaya fruit softening by the injurious heat treatment was reduced or prevented by a 4 h pretreatment at 42°C or a 1 h pretreatment at temperatures higher than 35°C followed by 3 h at 22°C (Paull and Chen, 1990). The 8 h approach time to 44°C for the papaya vapour heat treatment (Seo *et al.*, 1974) is designed to reduce subsequent injury when the fruit is exposed to 44°C for 8.75 h. This approach time would be expected to provide the conditions necessary to develop HSP and associated thermotolerance. The same would be likely to occur in deciduous fruits (Claypool and Vines, 1956). The quick rise of temperature to 49°C using vapour heat always damaged Valencia oranges and grapefruit (Sinclair and Lindgren, 1955), but they were tolerant when heated for 8 h at 43°C before the 49°C treatment. Cucumber can be preconditioned by a 24 h treatment at 32.5°C to tolerate subsequent exposure for 50 min to hot water at 46°C (Chan and Linse, 1989). Though this 32.5°C is lower than the temperature normally used to induce HSP, the 24 h duration would probably induce HSP. The extent of protection against softening disruption was clearly dependent upon season, stage of fruit ripeness at exposure, and pretreatment time and temperature. The seasonal variation could be due to field-induced thermotolerance.

Conclusions

If cold and heat treatments are to gain wider use as a means to disinfest horticultural commodities of quarantine insects, then better ways must be found to reduce chilling and heat injury when commodities are subjected to stress temperatures. Various approaches have been taken to achieve this objective. These approaches include temperature manipulation, atmosphere modification, and chemical treatments. Some techniques described in this review are effective for several commodities, but there is also considerable variation among crops in their response to chilling stress, with less variation for heat stress. An understanding of the conditions necessary to induce HS proteins and thermotolerance is essential to optimize heat treatments used for insect disinfestation and disease control.

The effectiveness of any treatment is influenced by crop maturity and environmental factors. No single method has been found which is universally applicable. Although considerable strides have been made towards elucidating the physiological and biochemical responses of horticultural commodities to temperature stress, a complete understanding of the mechanism(s) is still needed. The better we understand the physiological and biochemical responses of crops to temperature stress, the closer we may be to the objective of reducing stress effects on plants. The greatest promise of reducing chilling and heat injury probably lies in genetic modification of chilling- and heat-sensitive plants by genetic engineering as compared with conventional breeding. It is, therefore, essential that there be a thorough understanding of the physiological and biochemical processes involved in the temperature stress injury phenomenon to provide a sound foundation from which to devise effective stress control strategies.

References

Abe, K. and Chachin, K. (1985) Influence of conditioning on the occurrence of chilling injury and the changes of surface structure of eggplant fruits. *Journal of the Japanese Society of Horticultural Science* 54, 247–256.

Adams, D.O. and Yang, S.F. (1979) Ethylene biosynthesis: identification of 1-aminocyclopropane-1-carboxylic acid as an intermediate in the conversion of methionine to ethylene. *Proceedings of the National Academy of Sciences (USA)* 76, 170–174.

Akamine, E.K. (1960) Temperature effects in fresh papayas processed for shipment. *Hawaii Agricultural Experimental Station Bulletin* 122.

Akamine, E.K. (1966) Respiration of fruits of papaya (*Carica papaya* L., var. solo) with reference to the effect of quarantine disinfestation treatments. *Proceedings of the American Society for Horticultural Science* 89, 231–236.

Akamine, E.K. and Seo, S.S. (1978) Progress report on search for alternative disinfestation treatments for export papaya. *Proceedings l4th Annual Hawaii Papaya Industry Association Conference.* University of Hawaii. Cooperative Extension Service. Miscellaneous Publication 167.

Alexandrov, V. Ya. (1977) *Cells, Molecules and Temperature.* Translated from the Russian by V.A. Bernstam. Springer-Verlag, Berlin, Heidelberg, New York.

Aljuburi, H.J. and Huff, A. (1984) Reduction in chilling injury to stored grapefruit (*Citrus paradisi* Macf.) by vegetable oils. *Scientia Horticulturae* 24, 53–58.

Altschuler, M. and Mascarenhas, J.P. (1982) Heat shock proteins and effects of heat shock in plants. *Plant Molecular Biology* 1, 103–115.

An, R.F. and Paull, R.E. (1990) Storage temperature and ethylene on the ripening of papaya fruit. *Journal of the American Society for Horticultural Science* 115, 949–953.

Anderson, R.E. (1982). Long-term storage of peaches and nectarines intermittently warmed during controlled-atmosphere storage. *Journal of the American Society for Horticultural Science* 107, 214–216.

Anderson, R.E. and Penny, R.W. (1975) Intermittent warming of peaches and nectarines stored in a controlled atmosphere or air. *Journal of the American Society for Horticultural Science* 100, 151–153.

Armstrong, J.W., Hansen, J.D., Hu, B.K.S. and Brown, S.A. (1989) High-temperature, forced air quarantine treatment for papaya infested with tephritid fruit flies (Diptera:Tephritidae). *Journal of Economic Entomology* 82, 1667–1674.

Autio, W.R. and Bramlage, W.J. (1986) Chilling sensitivity of tomato fruit in relation to ripening and senescence. *Journal of the American Society for Horticultural Science* 111, 201–204.

Back, E.A. and Pemberton, C.E. (1916) Effect of cold-storage temperatures upon the Mediterranean fruit fly. *Journal of of Agricultural Research* V(15), 657–666.

Baker, A.C. (1939) The basis for treatment of products where fruit flies are involved as a condition for entry into the United States. US Department of Agriculture Circular No. 551, 7pp.

Bangerth, F., Dilley, D.R. and Dewey, D.H. (1972) Effect of postharvest calcium treatments on internal breakdown and respiration of apple fruits. *Journal of the American Society for Horticultural Science* 97, 679–682.

Ben-Arie, R., Lavee, S. and Guelfat-Reich, S. (1970) Control of woolly breakdown of 'Elberta' peaches in cold storage, by intermittent exposure to room temperature. *Journal of the American Society for Horticultural Science* 95, 801–803.

Bernstam, V.A. (1978) Heat effects on protein biosynthesis. *Annual Review of Plant Physiology* 29, 25–46.

Berry, J.A., Fork, D.C. and Garrison, S. (1975) Mechanistic studies of thermal damage to leaves. *Carnegie Institute Washington Year Book* 74, 751–759.

Biale, J.B. and Young, R.E. (1971) The avocado pear. In: Hulme, A.C. (ed.), *The Biochemistry of Fruits and Their Products.* Academic Press, New York, pp. 1–63.

Biggs, M.S., Woodson, W.R. and Handa, A.K. (1988) Biochemical basis of high temperature inhibition of ethylene biosynthesis in ripening tomato fruit. *Physiologia Plantarum* 72, 572–578.

Blum, A. (1987) *Plant Breeding for Stress Environments.* CRC Press, Boca Raton, Florida.

Brooks, C. and McColloch, L.P. (1936) Some storage diseases of grapefruit. *Journal of Agricultural Research* 69, 319–351.

Broughton, W.J. and Guat, T. (1979) Storage conditions and ripening of the custard apple *Annona squamosa* L. *Scientia Horticulturae* 10, 73–82.

Buescher, R.W. (1975) Organic acid and sugar levels in tomato pericarp as influenced by storage at low temperature. *HortScience* 10, 158–159.

Burg, S.P. (1962) The physiology of ethylene formation. *Annual Review of Plant Physiology* 13, 265–302.

Burke, J.J., Hatfield, J.L., Klein, R.R. and Mullet, J.E. (1985) Accumulation of heat shock proteins in field grown cotton. *Plant Physiology* 78, 394–398.

Cabrera, R.M. and Saltveit, M.E. (1990) Physiological response to chilling temperatures of intermittently warmed cucumber fruit. *Journal of the American Society for Horticultural Science* 115, 256–261.

Cappellini, M.C., Lachance, P.A. and Hudson, D.E. (1984) Effect of temperature and carbon dioxide atmospheres on the market quality of green bell peppers. *Journal of Food Quality* 7, 17–25.

Chalutz, E., Waks, J. and Shiffmann-Nadel, M. (1985) Reducing susceptibility of grapefruit to chilling injury during cold treatment. *HortScience* 20, 226–228.

Chan, H.T. (1986a) Effects of heat treatments on the ethylene-forming enzyme system in papaya. *Journal of Food Science* 51, 581–583.

Chan, H.T. (1986b) Heat inactivation of the ethylene forming enzyme system in cucumber. *Journal of Food Science* 51, 1491–1493.

Chan, H.T. and Forbus, W.B. (1988) Delayed light emission as a biochemical indicator of papaya heat treatment. *Journal of Food Science* 53, 1490–1492.

Chan, H.T. and Linse, E. (1989) Conditioning cucumbers for quarantine heat treatments. *HortScience* 24, 985–989.

Chan, H.T., Jr and Tam, S.Y.T. (1982) Partial separation and characterization of papaya endo- and exo-polygalacturonase. *Journal of of Food Science* 47, 1478–1483.

Chan, H.T., Jr, Tam, S.Y.T. and Seo, S.T. (1981) Papaya polygalacturonase and its role in thermally injured ripening fruit. *Journal of Food Science* 46, 190–197.

Chan, H.T., Sanxter, S. and Couey, H.M. (1985) Electrolyte leakage and ethylene production induced by chilling injury of papayas. *HortScience* 20, 1070–1072.

Chaplin, G.R. and Scott, K.J. (1980) Association of calcium in chilling injury susceptibility of stored avocado. *HortScience* 15, 514–515.

Chen, N.M. and Paull, R.E. (1986) Development and prevention of chilling injury in papaya fruit (*Carica papaya* L.). *Journal of the American Society for Horticultural Science* 111, 639–643.

Cheng, T.S., Floros, J.D., Shewfelt, R.L. and Chang, C.J. (1988) The effect of high-temperature stress on ripening of tomatoes (*Lycopersicon esculentum*). *Journal of Plant Physiology* 132, 459–464.

Chhatpar, H.S., Mattoo, A.K. and Modi, V.V. (1971) Biochemical studies on chilling injury in mangoes. *Phytochemistry* 10, 1007–1009.

Christiansen, M.N. and St John, J.B. (1981) The nature of chilling injury and its resistance in plants. In: Olien, C.R. and Smith, M.N. (eds), *Analysis and Improvement of Plant Cold Hardiness.* CRC Press, Boca Raton, FL, pp. 1–16.

Chrominski, A., Khan, M.A., Weber, D.J. and Smith, B.N. (1986) Ethylene and ethane production in response to salinity stress. *Plant Cell Environment* 9, 687–691.

Claypool, L.L. and Vines, H.M. (1956) Commodity tolerance studies of deciduous fruits to moist heat and fumigants. *Hilgardia* 24, 297–355.

Cohen, E. (1988) Commercial use of long-term storage of lemon with intermittent warming. *HortScience* 23, 400.

Cohen, E., Shalom, Y. and Rosenberger, I. (1990) Post-harvest behaviour of

'Ortanizue' ('Topaz') tangor citrus fruit during long-term storage at various temperatures. *Scientia Horticulturae* 44, 235–240.

Couey, H.M. (1982) Chilling injury of crops of tropical and subtropical origin. *HortScience* 17, 162–165.

Couey, H.M. (1989) Heat treatment for control of postharvest disease and insect pests of fruit. *HortScience* 24, 198–202.

Couey, H.M., Linse, E.S. and Nakamura, A.N. (1984) Quarantine procedure for Hawaiian papayas using heat and cold treatments. *Journal of Economic Entomology* 77, 984–988.

Daie, J., Campbell, W.F. and Seeley, S.D. (1981) Temperature- and stress-induced production of abscisic acid and dihydrophaseic acid in warm- and cool-season crops. *Journal of the American Society for Horticultural Science* 106, 11–13.

Davis, P.L. (1973) Intermittent warming of grapefruit to avoid rind injury during storage. *Proceedings of the Florida State Horticultural Society* 86, 280–283.

Davis, P.L. and Hofmann, R.C. (1973) Reduction of chilling injury of citrus fruits in cold storage by intermittent warming. *Journal of Food Science* 38, 871–873.

Dunlap, J.R., Lingle, S.E. and Lester, G.E. (1990) Ethylene production in netted muskmelon subjected to postharvest heating and refrigerated storage. *HortScience* 25, 207–209.

Eaks, I.L. (1956) Effect of modified atmospheres on cucumbers at chilling and non-chilling temperatures. *Proceedings of the American Society for Horticultural Science* 67, 473–478.

Eaks, I.L. (1960) Physiological studies of chilling injury in citrus fruits. *Plant Physiology* 35, 632–636.

Eaks, I.L. (1965) Effect of chilling on the respiration of oranges and lemons. *Proceedings of the American Society for Horticultural Science* 87, 181–186.

Eaks, I.L. (1980) Effect of chilling on respiration and volatiles of California lemon fruit. *Journal of the American Society for Horticultural Science* 105, 865–869.

Eaks, I.L. (1983) Effects of chilling on respiration and ethylene production of 'Haas' avocado fruit at 20°C. *HortScience* 18, 235–237.

Eaks, I.L. and Morris, L.L. (1956) Respiration of cucumber fruits associated with physiological injury at chilling temperatures. *Plant Physiology* 31, 308–314.

Elstner, E.F. and Konze, J.R. (1976) Effect of point freezing on ethylene and ethane production by sugar beet leak disks. *Nature* 263, 351–352.

Field, R.J. (1984) The effect of temperature on ethylene production by plant tissue. In: Roberts, J.A. and Tucker, G.A. (eds.), *Ethylene and Plant Development*. Butterworths, London, pp. 47–69.

Field, R.J. (1990) Influence of chilling stress on ethylene production. In: Wang, C.Y. (ed.), *Chilling Injury of Horticultural Crops*. CRC Press, Boca Raton, pp. 235–253.

Forney, C.F. and Lipton, W.J. (1990) Influence of controlled atmospheres and packaging on chilling sensitivity. In: Wang, C.Y. (ed.), *Chilling Injury of Horticultural Crops*. CRC Press, Boca Raton, pp. 257–267.

Forney, C.F. and Peterson, J.J. (1990) Chilling-induced potassium leakage of cultured citrus cells. *Physiologia Plantarum* 78, 193–196.

Fuchs, Y., Zauberman, G., Rot, I. and Weksler, A. (1989) Chilling injury and electrolyte leakage in cold stored mango and avocado fruits. *Acta Horticulturae* 258, 303–308.

Furmanski, R.J. and Buescher, R.W. (1979) Influence of chilling on electrolyte leakage and internal conductivity of peach fruits. *HortScience* 14, 167–168.

Graham, D. and Patterson, B.D. (1982) Responses of plants to low nonfreezing temperatures: Proteins, metabolism and acclimation. *Annual Review of Plant Physiology* 33, 347–372.

Guye, M.G., Vigh, L. and Wilson, J.M. (1986) Polyamine titre in relation to chill-sensitivity in *Phaseolus* sp. *Journal of Experimental Botany* 37, 1036–1043.

Hall, C.B. (1964a) The ripening response of detached tomato fruits to daily exposures to high temperatures. *Proceedings of the Florida State Horticultural Society* 76, 252–256.

Hall, C.B. (1964b) The effect of short periods of high temperature on the ripening of detached tomato fruits. *Proceedings of the American Society for Horticultural Science* 84, 501–506.

Halse, N.J. and Weir, R.N. (1974) Effect of temperature on spikelet number in wheat. *Australian Journal of Agricultural Research* 25, 687–701.

Hatton, T.T. (1990) Reduction of chilling injury with temperature manipulation. In: Wang, C.Y. (ed.), *Chilling Injury of Horticultural Crops*. CRC Press, Boca Raton, pp. 269–280.

Hatton, T.T. and Cubbedge, R.H. (1983) Preferred temperature for prestorage conditioning of 'Marsh' grapefruit to prevent chilling injury at low temperatures. *HortScience* 18, 721–722.

Hatton, T.T., Reeder, W.F. and Campbell, C.W. (1965) Ripening and storage of Florida avocados. US Department of Agriculture Research Service Marketing Research Report No. 697, 13pp.

Hatton, T.T., Davis, P.L., Cubbedge, R.H. and Munroe, K.A. (1981) Temperature management and carbon dioxide treatments that reduce chilling injury in grapefruit stored at low temperatures. *Proceedings of the International Society for Citriculture* 2, 728–731.

Hinton, D.M. and Pressey, R. (1980) Glucanases in fruits and vegetables. *Journal of the American Society of Horticultural Science* 105, 499–502.

Hirata, K., Chachin, K. and Iwata, T. (1987) Changes of K^+ leakage, free amino acid contents and phenyl propanoid metabolism in water convolvulus (*Ipomoea aquatica* Forsk) with reference to chilling injury. *Journal of the Jappananese Society of Horticultural Science* 55, 516–523.

Hirose, T. (1985) Effects of pre- and interposed warming on chilling injury, respiratory rate and membrane permeability of cucumber fruits during cold storage. *Journal of the Japanese Society of Horticultural Science* 53, 459–466.

Hirose, T. (1987) Effects of intermittent warming on chilling injury and contents of α-keto acid of cucumber fruits during cold storage. *Scientific Report of the Faculty of Agriculture Kobe University* 17, 171–174.

Hobson, G.E. (1987) Low-temperature injury and the storage of ripening tomatoes. *Journal of Horticultural Science* 62, 55–62.

Houck, L.G., Jenner, J.F. and Mackey, B.E. (1990) Seasonal variability of the response of desert lemons to rind injury and decay caused by quarantine cold treatments. *Journal of Horticultural Science* 65, 611–617.

Hruschka, H., Smith, W.L. and Baker, J.E. (1969) Reducing chilling injury of potatoes by intermittent warming. *American Potato Journal* 46, 38–53.

Huber, H. (1935) Der Warmehaushalt der Pflanzen. *Naturwissenschaftliche Landwirtschaft* 17, 1–148.

Ilker, Y. and Morris, L.L. (1975) Alleviation of chilling injury of okra. *HortScience* 10, 324.
Inaba, M. and Chachin, K. (1988) Influence of and recovery from high-temperature stress on harvested mature green tomatoes. *HortScience* 23, 190–192.
Inaba, M. and Crandall, P.G. (1988) Electrolyte leakage as an indicator of high temperature injury to harvested mature green tomatoes. *Journal of the American Society for Horticultural Science* 113, 96–99.
Ismail, M.A. and Grierson, W. (1977) Seasonal susceptibility of grapefruit to chilling injury as modified by certain growth regulators. *HortScience* 12, 118–120.
Jackman, R.L., Yada, R.Y., Marangoni, A., Parkin, K.L. and Stanley, D.W. (1988) Chilling injury. A review of quality factors. *Journal of Food Quality* 11, 253–278.
Johnson, R.N. and Kucay, B.L. (1988) Competitive inhibitor of hsp 70 gene expression causes thermosensitivity. *Science* 242, 1552–1554.
Jones, R.L., Freebairn, H.T. and McDonnell, J.F. (1978) The prevention of chilling injury, weight loss reduction, and ripening retardation in banana. *Journal of the American Society for Horticultural Science* 103, 219–221.
Jones, W.W. (1939) The influence of relative humidity on the respiration of papaya at high temperature. *Proceedings of the American Society for Horticultural Science* 37, 109–124.
Jones, W.W. (1942) Respiration and chemical changes of the papaya fruit in relation to temperature. *Plant Physiology* 17, 481–486.
Jones, W.W., Holtzman, J.J. and Galloway, A.G. (1939) The effect of high temperature sterilization on Solo papaya. *Hawaii Agricultural Experimental Station Circular* 14, 8pp.
Kanabus, J., Pikaad, C.S. and Cherry, J.H. (1984) Heat shock proteins in tobacco cell suspension during growth cycle. *Plant Physiology* 75, 639–644.
Kasmire, R.F. (1978) Handling cantaloupes in extreme hot weather. *Western Grower and Shipper* 49, 10–11.
Kenney, P. and Hull, L. (1986) Effects of storage condition on carambola quality. *Proceedings of the Florida State Horticultural Society* 99, 222–224.
Kerbel, E.L., Mitchell, F.G. and Mayer, G. (1985) Effect of postharvest heat treatments for insect control on the quality and market life of peaches. *HortScience* 20, 725–727.
Kimpel, J.A. and Key, J.L. (1985) Presence of heat shock mRNAs in field grown soybeans. *Plant Physiology* 79, 672–678.
King, A.I. (1985) Regulation of chilling sensitivity in seedlings of tomato. PhD Thesis. University of California, Davis.
King, A.I., Joyce, D.C. and Reid, M.S. (1988) Role of carbohydrates in diurnal chilling sensitivity of tomato seedlings. *Plant Physiology* 86, 764–768.
Klein, J.D. and Lurie, S. (1990) Prestorage heat treatment as a means of improving poststorage quality of apples. *Journal of the American Society for Horticultural Science* 115, 265–269.
Klein, J.D. and Lurie, S. (1992) Heat treatments for improved postharvest quality of horticultural crops. *HortTechnology* 2, 316–320.
Kozukue, N. and Ogata, K. (1972) Physiological and chemical studies of chilling injury in pepper fruits. *Journal of Food Science* 37, 708–711.
Kozukue, N., Kozukue, E., Hirose, T. and Mizuno, S. (1984) Accumulation of alanine in chilling-sensitive crops. *HortScience* 19, 498–501.

Kramer, G.F. and Wang, C.Y. (1989a) Correlation of reduced chilling injury with increased spermine and spermidine levels in zucchini squash. *Physiologia Plantarum* 76, 479–484.

Kramer, G.F. and Wang, C.Y. (1989b) Effects of chilling and temperature preconditioning on the activity of polyamine biosynthetic enzymes in zucchini squash. *Journal of Plant Physiology* 136, 115–119.

Kramer, G.F. and Wang, C.Y. (1989c) Reduction of chilling injury in zucchini squash by temperature management. *HortScience* 24, 995–996.

Lange, O.L., Nobel, P.S., Osmond, C.B. and Ziegler, H. (eds) (1981) *Physiological Plant Ecology. I. Responses to the Physical Environment.* Springer-Verlag, Berlin, 625 pp.

Lee, S.K. and Young, R.E. (1984) Temperature sensitivity of avocado fruit in relation to C_2H_4 treatment. *Journal of the American Society for Horticultural Science* 109, 689–692.

Lewis, D.A. and Morris, L.L. (1956) Effects of chilling storage on respiration and deterioration of several sweet potato varieties. *Proceedings of the American Society for Horticultural Science* 68, 421–428.

Levitt, J. (1980) *Responses of Plants to Environmental Stresses.* Vol. I. 2nd edn. Academic Press, New York.

Lieberman, M., Craft, C.C., Audia, W.V. and Wilcox, M.S. (1958) Biochemical studies of chilling injury in sweet potatoes. *Plant Physiology* 33, 307–311.

Lieberman, M., Craft, C.C. and Wilcox, M.S. (1959) Effect of chilling on the chlorogenic acid and ascorbic acid content of Porto Rico sweet-potatoes. *Proceedings of the American Society for Horticultural Science* 74, 642–648.

Lindquist, S. and Craig, E.A. (1988) The heat shock proteins. *Annual Review of Genetics* 22, 631–677.

Lipton, W.J. (1965) Postharvest responses of asparagus spears to high carbon dioxide and low oxygen atmospheres. *Proceedings of the American Society for Horticultural Science* 86, 347–356.

Lipton, W.J. and Aharoni, Y. (1979) Chilling injury and ripening of 'Honey Dew' muskmelons stored at 2.5 or 5°C after ethylene treatment at 20°C. *Journal of the American Society for Horticultural Science* 104, 327–330.

Lipton, W.J. and Wang, C.Y. (1987) Chilling exposures and ethylene treatment change the level of ACC in 'Honey Dew' melons. *Journal of the American Society for Horticultural Science* 112, 109–112.

Liu, F.W. (1978) Modification of apple quality by high temperature. *Journal of the American Society for Horticultural Science* 103, 730–732.

Lurie, S. and Klein, J.D. (1990) Heat treatment of ripening apples: differential effects on physiology and biochemistry. *Physiologia Plantarum* 78, 181–186.

Lurie, S. and Klein, J.D. (1991) Acquistion of low-temperature tolerance in tomatoes by exposure to high-temperature stress. *Journal of the American Society for Horticultural Science* 116, 1007–1012.

Lyons, J.M. (1973) Chilling injury in plants. *Annual Review of Plant Physiology* 24, 445–466.

Lyons, J.M. and Breidenbach, R.W. (1990) Relation of chilling stress to respiration. In: Wang, C.Y. (ed.), *Chilling Injury of Horticultual Crops.* CRC Press, Boca Raton, pp. 223–233.

Lyons, J.M. and Raison, J.K. (1970) Oxidative activity of mitochondria isolated from plant tissue sensitive and resistant to chilling injury. *Plant Physiology* 45, 386–389.

Mager, J. (1959) The stabilizing effect of spermine and related polyamines on bacterial protoplasts. *Biochimica et Biophysica Acta* 36, 529–531.

Marriott, J. (1980) Bananas physiology and biochemistry of storage and ripening for optimum quality. *CRC Critical Reviews of Food Science and Nutrition* 13, 41–88.

Maxie, E.C., Mitchell, F.G., Somner, N.F., Snyder, R.G. and Rae, H.L. (1974) Effects of elevated temperatures on ripening of 'Bartlett' pears *Pyrus communis* L. *Journal of the American Society for Horticultural Science* 99, 344–349.

McColloch, L.P. (1962) Chilling injury and *Alternaria* rot of bell peppers. US Department of Agriculture Market Research Report 536.

McCollum, T.G. and McDonald, R.E. (1991) Electrolyte leakage, respiration, and ethylene production as indices of chilling injury in grapefruit. *HortScience* 26, 1191–1192.

McCollum, T.G., D'Aquino, S. and McDonald, R.E. (1993) Heat treatment inhibits chilling injury on mangos. *HortScience* 28, 197–198.

McDonald, R.E. (1986) Effects of vegetable oils, CO_2, and film wrapping on chilling injury and decay of lemons. *HortScience* 21, 476–477.

McDonald, R.E. and Kushad, M.M. (1986) Accumulation of putrescine during chilling injury of fruits. *Plant Physiology* 82, 324–326.

McDonald, R.E., Miller, W.R., McCollum, T.G. and Brown, G.E. (1991) Thiabendazole and imazalil applied at 53°C reduce chilling injury and decay of grapefruit. *HortScience* 26, 397–399.

Meir, S. and Bramlage, W.J. (1988) Antioxidant activity in 'Cortland' apple peel and susceptibility to superficial scald after storage. *Journal of the American Society for Horticultural Science* 113, 412–418.

Mencarelli, F., Lipton, W.J. and Peterson, S.J. (1983) Responses of 'Zucchini' squash to storage in low-O_2 atmospheres at chilling and nonchilling temperatures. *Journal of the American Society for Horticultural Science* 108, 884–890.

Moline, H.E. (1980) Effects of vacuum infiltration of calcium chloride on ripening rate and chilling injury of tomato fruit. *Phytopathology* 70, 691.

Morris, L.L. (1982) Chilling injury of horticultural crops: an overview. *HortScience* 17, 161–162.

Murata, N. and Nishida, I. (1990) Lipids in relation to chilling sensitivity of plants. In: Wang, C.Y. (ed.), *Chilling Injury of Horticultural Crops*. CRC Press, Boca Raton, pp. 181–199.

Murata, N., Ishizaki-Nishizama, O., Higashi, S., Hayashi, H., Tasaka, Y. and Nishida, I. (1992) Genetically engineered alteration in the chilling sensitivity of plants. *Nature* 356, 710–713.

Murata, T. (1969) Physiological and biochemical studies of chilling injury in bananas. *Physiologia Plantarum* 22, 401–411.

Murata, T. and Tatsumi, Y. (1979) Ion leakage in chilled plant tissues. In: Lyons, J.M., Graham, D. and Raison, J.K. (eds), *Low Temperature Stress in Crop Plants*. Academic Press, New York, pp. 141–151.

Nakamura, R., Inaba, A. and Ito, T. (1985) Effect of cultivating conditions and postharvest stepwise cooling on the chilling sensitivity of eggplant and cucumber fruits. *Scientific Report of the Faculty of Agriculture, Okayama University* 66, 19–29.

Nordby, H.E. and McDonald, R.E. (1990) Squalene in grapefruit wax as a possible natural protectant against chilling injury. *Lipids* 25, 807–810.
Ogura, N., Nakagawa, H. and Takehana, H. (1975) Effect of storage temperature of tomato fruits on changes of their polygalacturonase and pectinesterase activities accompanied with ripening. *Journal of the Agricultural Chemical Society of Japan* 49, 271–274.
Ogura, N., Hayashi, R., Ogishima, T., Abe, Y., Nakagawa, H. and Takehana, H. (1976) Ethylene production by tomato fruits at various temperatures and effect of ethylene treatment on fruits. *Journal of the Agricultural Chemical Society of Japan* 50, 519–523.
Omran, R.G. (1980) Peroxide levels and the activities of catalase, peroxidase, and indolacetic acid oxidase during and after chilling cucumber seedlings. *Plant Physiology* 65, 407–408.
Oslund, C.R. and Davenport, T.L. (1983) Ethylene and carbon dioxide in ripening of *Averrhoa carambola*. *HortScience* 18, 229–230.
Ougham, H.J. and Stoddart, J.L. (1986) Synthesis of heat shock proteins and acquisition of thermotolerance in high temperature tolerant and high temperature susceptible lines of sorghum. *Plant Science* 44, 163–167.
Pantastico, E.B., Soule, J. and Grierson, W. (1968) Chilling injury in tropical and subtropical fruits: II. Limes and grapefruits. *Proceedings of the Tropical Region, American Society for Horticultural Science* 12, 171–183.
Patterson, B.D., Payne, L.A., Chen, Y-Z. and Graham, D. (1984) An inhibition of catalase induced by cold in chilling-sensitive plants. *Plant Physiology* 76, 1014–1018.
Paull, R.E. (1990a) Chilling injury in tropical and subtropical crops. In: Wang, C.Y. (ed.), *Chilling Injury of Horticultural Crops*. CRC Press, Boca Raton, pp. 17–36.
Paull, R.E. (1990b) Postharvest heat treatments and fruit ripening. *Postharvest News and Information* 1, 355–363.
Paull, R.E. and Chen, N.J. (1983) Postharvest variation in cell wall-degrading enzymes of papaya (*Carica papaya* L.) during fruit ripening. *Plant Physiology* 72, 382–385.
Paull, R.E. and Chen, N.J. (1990) Heat shock response in field grown ripening papaya fruit. *Journal of the American Society for Horticultural Science* 115, 623–631.
Paull, R.E. and Rohrbach, K.G. (1985) Symptom development of chilling injury in pineapple fruit. *Journal of the American Society for Horticultural Science* 110, 100–105.
Picha, D.H. (1987) Chilling injury, respiration, and sugar changes in sweet potatoes stored at low temperature. *Journal of the American Society for Horticultural Science* 112, 497–502.
Picken, A.J.F. (1984) A review of pollination and fruit set in the tomato (*Lycopersicon esculentum* Mill). *Journal of Horticultural Science* 59, 1–13.
Picton, S. and Grierson, D. (1988) Inhibition of expression of tomato-ripening genes at high temperature. *Plant, Cell and Environment* 11, 265–272.
Popenoe, W. (1920) *Manual of Tropical and Subtropical Fruits*. Hafner Press, 474pp.
Pressey, R. (1983) Beta-Galactosidases in ripening tomatoes. *Plant Physiology* 71, 132–135.

Purvis, A.C. and Grierson, W. (1982) Accumulation of reducing sugar and resistance of grapefruit peel to chilling injury as related to winter temperatures. *Journal of the American Society for Horticultural Science* 107, 139–142.

Risse, L.A. and Chun, D. (1987) Influence of various conditioning times and injury decay of nonwrapped and film wrapped peppers. *Proceedings of the Florida Horticultural Society* 100, 29–32.

Roberts, D.R., Dumbroff, E.B. and Thompson, J.E. (1986) Exogenous polyamines alter membrane fluidity in bean leaves – a basis for potential misinterpretation of their true physiological role. *Planta* 167, 395–401.

Romani, R. and French, K. (1977) Temperature-dependent changes in the polysomal population of senescent (ripening) pear fruit. *Plant Physiology* 60, 930–932.

Schiffmann-Nadel, M., Chalutz, M.E., Waks, J. and Lattar, F.S. (1972) Reduction of pitting of grapefruit by thiabendazole during long-term cold storage. *HortScience* 7, 394–395.

Seneratna, T. and McKersie, B.D. (1986) Loss of desiccation tolerance during seed germination: a free radical mechanism of injury. In: Leopold, A.C. (ed.), *Membranes, Metabolism and Dry Organisms*. Cornell University Press, Ithaca, New York, pp. 85–101.

Seo, S.T., Hu, B.K.S., Komura, M., Lee, C.Y.L. and Harris, E.J. (1974) *Dacus dorsalis*. Vapor heat treatment in papaya. *Journal of Economic Entomology* 67, 240–242.

Sinclair, W.B. and Lindgren, D.L. (1955) Vapor heat sterilization of California citrus and avocado fruits against fruit-fly insects. *Journal of Economic Entomology* 48, 133–138.

Smillie, R.M. and Gibbons, G.C. (1981) Heat tolerance and heat hardening in crop plants measured by chlorophyll fluorescence. *Carlsberg Research Communication* 46, 395–403.

Sommer, N.F. and Mitchell, F.G. (1978) Relation of chilling temperatures to postharvest alternaria rot of papaya fruit. *Proceedings of the Tropical Region, American Society for Horticultural Science* 22, 40–47.

Spalding, D.H. and Reeder, W.F. (1975) Low-oxygen high-carbon dioxide controlled atmosphere storage for control of antracnose and chilling injury of avocados. *Phytopathology* 65, 458–460.

Spalding, D.H. and Reeder, W.F. (1983) Conditioning 'Tahiti' limes to reduce chilling injury. *Proceedings of the Florida State Horticultural Society* 96, 231–232.

Stakiotakis, E.M. and Dilley, D.R. (1974) Induction of ethylene production in 'Bosc' pears by postharvest cold stress. *HortScience* 9, 336–338.

Thomas, P. and Joshi, M.R. (1988) Reduction of chilling injury in ripe Alphonso mango fruit in cold storage by temperature conditioning. *International Journal of Food Science and Technology* 23, 447–455.

Thompson, B.D. (1978) Chilling injury and physical damage of peppers. *Proceedings of the Florida State Horticultural Society* 91, 144–146.

Thorne, S.N. and Efiuvwevwere, B.J.O. (1988) Changes in organic acids in chilled tomato fruit (*Lycopersicon esculentum* Mill). *Journal of the Science of Food and Agriculture* 44, 309–319.

Ting, S.V. and Newhall, W.F. (1965) The occurrence of a natural antioxidant in citrus fruit. *Journal of Food Science* 30, 57–63.

Tompkins, R.B. (1963) The effects of temperature, extent of evaporation and restriction of ventilation on the storage life of tomatoes. *Journal of Horticultural Science* 38, 335–347.

Tsuchida, H., Dan-Hong, C., Inoue, K., Kozukue, N. and Mizuno, S. (1990) Changes in pyruvic acid content and GPT acivity in chilling-sensitive and nonsensitive crops. *HortScience* 25, 952–953.

van Hasselt, Ph.R. and van Berlo, H.A.C. (1980) Photooxidation damage to the photosynthetic apparatus during chilling. *Physiologia Plantarum* 50, 52–56.

Wallace, D.H. (1985) Physiological genetics of plant maturity adaptation and yield. *Plant Breeding Review* 3, 21–54.

Wang, C.Y. (1982) Physiological and biochemical responses of plants to chilling stress. *HortScience* 17, 173–186.

Wang, C.Y. (1987) Changes of polyamines and ethylene in cucumber seedlings in response to chilling stress. *Physiologia Plantarum* 69, 253–257.

Wang, C.Y. (1989a) Chilling injury of fruits and vegetables. *Food Reviews International* 5, 209–236.

Wang, C.Y. (1989b) Relation of chilling stress to ethylene production. In: Li, P. (ed.), *Low Temperature Stress Physiology in Crops.* CRC Press, Boca Raton, pp. 177–189.

Wang, C.Y. (1990a) Alleviation of chilling injury of horticultural crops. In: Wang, C.Y. (ed.), *Chilling Injury of Horticultural Crops.* CRC Press, Boca Raton, pp. 281–302.

Wang, C.Y. (1990b) Involvement of abscisic acid in reducing chilling injury of zucchini squash. *HortScience* 25, 1096.

Wang, C.Y. (1991) Reduction of chilling injury in fruits and vegetables. *Postharvest News and Information* 2, 165–168.

Wang, C.Y. and Adams, D.O. (1980) Ethylene production by chilled cucumbers (*Cucumis sativus* L.). *Plant Physiology* 66, 841–843.

Wang, C.Y. and Adams, D.O. (1982) Chilling-induced ethylene production in cucumbers (*Cucumis sativus* L.). *Plant Physiology* 69, 424–427.

Wang, C.Y. and Baker, J.E. (1979) Effects of two free radical scavengers and intermittent warming on chilling injury and polar lipid composition of cucumber and sweet pepper fruits. *Plant Cell Physiology* 20, 243–251.

Wang, C.Y. and Ji, Z.O. (1989) Effect of low-oxygen storage on chilling injury and polyamines in zucchini squash. *Scientia Horticulturae* 39, 1–7.

Wang, C.Y. and Kramer, G.F. (1990) Polyamines reduce chilling injury in McIntosh apples and zucchini squash. *HortScience* 25, 1096.

Wang, H-G., Gemma, H., Oogaki, C. and Kurokawa, M. (1990) Influence of harvest maturity on chilling sensitivity, respiratory rate and ethylene production during storage in 'Sensyu' apple fruit. *Journal of the Japanese Society of Horticultural Science* 59, 163–169.

Watada, A.E. and Morris, L.L. (1966) Effect of chilling and nonchilling temperatures on snap beans fruits. *Proceedings of the American Society for Horticultural Science* 89, 368–374.

Wild, B.L. and Hood, C.W. (1989) Hot dip treatments reduce chilling injury in long-term storage of 'Valencia' oranges. *HortScience* 24, 109–110.

Wills, R.B.H., Hopkirk, G. and Scott, K.J. (1981) Reduction of soft scald in apples with antioxidants. *Journal of the American Society for Horticultural Science* 106, 569–571.

Wilson, J.M. (1976) The mechanism of chill- and drought-hardening of *Phaseolus vulgaris* leaves. *New Phytologist* 76, 257–270.

Wilson, J.M. (1984) The economic importance of chilling injury. *Outlook on Agriculture* 14, 197–203.

Wise, R.R. and Naylor, A.W. (1987a) Chilling-enhanced photooxidation: evidence for the role of singlet oxygen and superoxide in the breakdown of pigments and endogenous antioxidants. *Plant Physiology* 83, 278–282.

Wise, R.R. and Naylor, A.W. (1987b) Chilling-enhanced photooxidation: the peroxidative destruction of lipids during chilling injury to photosynthesis and ultrastructure. *Plant Physiology* 83, 272–277.

Wu, M.T. and Wallner, S.J. (1984) Heat stress responses in cultured plant cells. Heat tolerance induced by heat shock versus elevated growing temperatures. *Plant Physiology* 75, 778–780.

Yoshida, O., Nakagawa, H., Ogura, N. and Sato, T. (1984) Effect of heat treatment on the development of polygalacturonase in tomato fruit during ripening. *Plant and Cellular Physiology* 25, 505–509.

Yost, H.J. and Lindquist, S. (1988) Translation of unspliced transcripts after heat shock. *Science* 242, 1544–1548.

Yu, Y.B., Adams, D.O. and Yang, S.F. (1980) Inhibition of ethylene production by 2,4-dinitrophenol and high temperature. *Plant Physiology* 66, 286–290.

Zauberman, G., Schiffmann-Nadel, M. and Yanko, U. (1973) Susceptibility to chilling injury of three avocado cultivars at various stages of ripening. *HortScience* 8, 511–513.

Controlled Atmospheres 11

ADEL A. KADER and DANGYANG KE
Department of Pomology, University of California, Davis, CA 95616, USA.

Controlled atmosphere (CA) storage conditions are used to supplement exposure to the proper temperatures and relative humidities required to maintain the quality of fresh horticultural products. The benefits of CA include retardation of senescence and undesirable compositional changes, reduction of respiration and ethylene production rates, alleviation of some physiological disorders, and reduction of the incidence and severity of decay. Carbon monoxide (5–10%) added to CA with less than 5% O_2 is an effective fungistat that can be used on commodities that do not tolerate the fungistatic concentrations of CO_2 (above 15%). Exposure of horticultural products to O_2 and/or CO_2 levels beyond their optimum range can result in the initiation and/or aggravation of certain physiological disorders, irregular ripening, increased susceptibility to decay, and development of off-flavours.

Although much information is available on CA requirements of and recommendations for fresh horticultural crops (Blanpied, 1990; Hatton and Spalding, 1990; Leshuk and Saltveit, 1990), the extent of commercial use of CA for long-term storage, other than for apples and pears, is still limited. Refinements in CA storage that have been made in recent years include low-O_2 (1.0–1.5%) storage, low-ethylene CA storage, rapid CA (rapid establishment of the optimum levels of O_2 and CO_2), and sequential CA storage (e.g. storage in 1% O_2 for 2–6 weeks followed by storage in 2–3% O_2 for the remainder of the storage period). Pre-CA-storage treatments with 20–40% CO_2 for 5–10 days may be beneficial for some commodities. Other recent developments, which have expanded use of CA or MA (modified atmospheres; less precise than CA) during transport and short-term storage, include the use of polymeric films and edible coatings to create a desired MA within the commodity.

From: Paull, R.E. and Armstrong, J.W. (eds) (1994) *Insect Pests and Fresh Horticultural Products: Treatments and Responses.* CAB INTERNATIONAL, Wallingford, UK.

Fresh horticultural products vary greatly in their relative tolerance to low O_2 concentrations (Table 11.1) and elevated CO_2 concentrations (Table 11.2) during long-term storage (Kader *et al.*, 1989). These tolerance limits are the levels below which (O_2) or above which (CO_2) physiological

Table 11.1. Classification of fruits and vegetables according to their tolerance to low O_2 concentrations during transport and/or storage within the optimum range of temperatures and relative humidities.

Minimum O_2 concentration tolerated (%)	Commodities
0.5	Tree nuts, dried fruits, and vegetables
1.0	Some cultivars of apples and pears, broccoli, mushroom, garlic, onion, most cut or sliced (minimally processed) fruit and vegetables
2.0	Most cultivars of apples and pears, kiwifruit, apricot, cherry, nectarine, peach, plum, strawberry, papaya, pineapple, olive, cantaloupe, sweet corn, green bean, celery, lettuce, cabbage, cauliflower, Brussels sprouts
3.0	Avocado, persimmons, tomato, pepper, cucumber, artichoke
5.0	Citrus fruits, green peas, asparagus, potato, sweet potato

Table 11.2. Classification of fruits and vegetables according to their tolerance to elevated CO_2 concentrations during transport and/or storage within the optimum range of temperatures and relative humidities.

Maximum CO_2 concentration tolerated (%)	Commodities
2	Apple (Golden Delicious), Asian pear, European pear, apricot, grape, olive, tomato, pepper (sweet), lettuce, endive, Chinese cabbage, celery, artichoke, sweet potato
5	Apple (most cultivars), peach, nectarine, plum, orange, avocado, banana, mango, papaya, kiwifruit, cranberry, pea, pepper (chili), eggplant, cauliflower, cabbage, Brussels sprouts, radish, carrot
10	Grapefruit, lemon, lime, persimmon, pineapple, cucumber, summer squash, snap bean, okra, asparagus, broccoli, parsley, leek, green onion, dry onion, garlic, potato
15	Strawberry, raspberry, blackberry, blueberry, cherry, fig, cantaloupe, sweet corn, mushroom, spinach, kale, Swiss chard

damage would be expected, and can be different at temperatures above or below those recommended for each commodity. A given commodity may tolerate much higher CO_2 concentrations and/or lower O_2 concentrations than those shown in Tables 11.1 and 11.2 for a short duration. The tolerance to low O_2 concentrations decreases with the increase in temperature and/or duration since O_2 requirements for aerobic respiration of the commodity increase with rise in temperature (Boersig *et al.*, 1988). Depending on the commodity, damage associated with elevated CO_2 may either increase or decrease with an increase in temperature. CO_2 production increases but its solubility decreases with temperature. Further, the physiological effects of CO_2 could be temperature dependent (Herner, 1987). Tolerance limits to elevated CO_2 decrease with a reduction in O_2 level, and similarly the tolerance limits to reduced O_2 decrease with the increase in CO_2 level (Little and Peggie, 1987; Lougheed, 1987).

Although much research has been conducted to evaluate the beneficial and detrimental effects of CA and MA on fresh horticultural crops, the mode of action of O_2 and CO_2 on these commodities is still not fully understood (Burton, 1978; Kader, 1986; Weichmann, 1986; Zagory and Kader, 1989; Wang, 1990). Generally the effect of reduced O_2 and/or elevated CO_2 on lowering the respiration rate has been assumed to be the primary reason for the beneficial effects of CA/MA on fresh produce. Kader (1986) indicated that postharvest deterioration of fresh produce can be caused by many factors in addition to high respiration rates, including biochemical changes associated with respiratory metabolism, ethylene biosynthesis and action, compositional changes, anatomical changes associated with growth and development, physical injuries, water loss, physiological disorders, and pathological breakdown. CA containing less than 1% O_2 and/or more than 60% CO_2 have been shown to be effective in controlling insects of stored products, such as nuts and dried fruits and vegetables. Such atmospheres provide an alternative to chemical fumigants (e.g. methyl bromide).

To meet quarantine requirements for fresh horticultural products, there has been an increased interest in recent years in studying the potential of short-term exposure to O_2 concentrations below 1% and/or CO_2 levels above 20% for postharvest insect disinfestation. Ke and Kader (1992a) reviewed the insecticidal effects of very low O_2 and/or very high CO_2 atmospheres at various temperatures and compared them with the responses and tolerance of fresh fruits and vegetables to similar CA conditions. The time required for 100% mortality varies with insect species and its developmental stage, temperature, O_2 and CO_2 concentrations, and relative humidity. A comparison of the time required for 100% mortality varies with insect species and its developmental stage, temperature, O_2 and CO_2 concentrations, and relative humidity. A comparison of the time required for 100% mortality of insects and commodity tolerances to the insecticidal CA indicates that many insects can be killed before detrimental effects occur in the host fruits and vegetables (Ke and Kader, 1992a).

Responses of Horticultural Commodities to Insecticidal CA

Keeping fresh horticultural products in very low O_2 and/or very high CO_2 atmospheres may have beneficial effects, such as reducing respiration rate, inhibiting ethylene production and action, retarding colour change and softening, maintaining composition and nutritional value, reducing the incidence of some physiological disorders, and inhibiting decay. The specific beneficial effects observed depend on the commodity, maturity stage, storage temperature, atmospheric composition, and time of exposure.

As O_2 level is reduced to 1–3% or CO_2 level raised to 5–20%, respiration rate usually decreases. However, if O_2 concentration is reduced to near 0% or CO_2 level increased to 50–90%, respiration rate may be similar to or even higher than that of the commodities kept in air since anaerobic respiration could occur (Ke *et al.*, 1991b). For most commodities, decreasing O_2 level to 1% or below and/or enhancing CO_2 level to 50% or above greatly inhibits ethylene production and action (Ke *et al.*, 1990, 1991b,c). Kubo *et al.* (1990) investigated the effects of 20% O_2 + 60% CO_2 on respiration and ethylene production rates of 18 kinds of fruits and vegetables. They found this atmosphere to decrease respiration rates in climacteric fruits and broccoli, but the effect on respiration rates of non-climacteric commodities was small. They proposed that the respiratory responses of horticultural crops to high CO_2 might be mediated by the effects of CO_2 on the action and/or synthesis of ethylene.

Insecticidal low O_2 and/or high CO_2 atmospheres retard skin yellowing in apples and 'Bartlett' pears and skin darkening in plums and cherries (Ke *et al.*, 1990, 1991b; Ke and Kader, 1992b). Exposure to insecticidal CA also retards flesh softening in apples, Bartlett pears, plums, and strawberries (Ke *et al.*, 1990, 1991a,b). For Asian pears, oranges, peaches, and nectarines, insecticidal CA atmospheres do not influence their normal changes in colour and flesh firmness (Ke and Kader, 1990, 1992b; Ke *et al.*, 1991b,c).

Short-term exposures to insecticidal CA usually do not significantly influence soluble solids content. In some commodities, such as peach and apple, loss of organic acids may be retarded (Ke *et al.*, 1991b,c). Insecticidal CA treatments do not significantly influence vitamin C content in strawberries and oranges (Ke and Kader, 1990; Ke *et al.*, 1991a). For long-term storage of apples and pears, however, reducing O_2 level to 1–3%

with or without the addition of 3–5% CO_2 maintains higher levels of soluble solids content, titratable acidity, and amino acids (Mellenthin *et al.*, 1980; Chen and Mellenthin, 1982). Through these beneficial effects, dessert quality is maintained.

Shaw *et al.* (1991) stored 'Hamlin', 'Pineapple', and 'Valencia' oranges in CA of 100% N_2, 100% CO_2, or 0.1–0.7% acetaldehyde in air, for 8 to 24 h. These treatments caused two- to threefold increases in acetaldehyde, ethyl acetate, ethyl butyrate, and ethanol within 24 h after treatment, but no consistent flavour changes in juice were noted for up to 8 days after treatment. Pesis *et al.* (1991) found that treating feijoa fruits with 98% N_2 for 24 h enhanced their flavour quality.

Insecticidal CA treatments reduce severity of internal breakdown (chilling injury) in peaches (Ke *et al.*, 1991c) and some browning and scald disorders in apples and pears (Mellenthin *et al.*, 1980; Lau, 1983). Lowering O_2 level to below 1% or increasing CO_2 level to 10–30% substantially reduces decay incidence in strawberries, blueberries, and cherries (Patterson, 1982; Smittle and Miller, 1988; Ke *et al.*, 1991a). These treatments may have a good potential for controlling postharvest decay, especially for the soft fruits that usually have very short storage life under normal cold storage in air. The effects of CA on decay may include the direct effect on inhibiting pathogen growth and the indirect effect on delaying ripening and senescence of the commodities and consequently maintaining their resistance to pathogen attack (El-Goorani and Sommer, 1981; Barkai-Golan, 1990).

Tolerance of Horticultural Commodities to Insecticidal CA

The possible detrimental effects of insecticidal CA on fresh commodities include low O_2 and/or high CO_2 injury and off-flavour development. The period of tolerance of fruits and vegetables to insecticidal CA conditions is, therefore, limited to the time before the onset of these effects. The tolerances of some fruits to insecticidal low O_2 and/or high CO_2 atmospheres are summarized in Table 11.3. Delate and Brecht (1989) found that sweet potato roots tolerated exposure to 2% O_2 + 40% CO_2 at 25°C for up to 10 days.

For a few commodities such as Asian pear, nectarine, and papaya, low O_2 injury could be observed after exposure to 0–0.25% O_2 for 3–14 days (Yahia *et al.*, 1989; Ke and Kader, 1990, 1992b; Ke *et al.*, 1991b). For most other commodities, low O_2 injury is not observed during short-term low O_2 treatments. Most commodities could tolerate CO_2 levels in the 10–30% range for only a few days. The exceptions include cherry, strawberry, and blueberry that can be kept in these CO_2 atmospheres for more than 10 days without detrimental effects (Patterson, 1982; Smittle and

Table 11.3. Examples of tolerance limits of selected fruits to insecticidal controlled atmosphere (CA) treatments at indicated temperatures.

Commodity	Temp. (°C)	CA composition O_2%	CO_2%	Tolerance (days)	References
Apple	0	0.02		12	Chen *et al.*, 1985;
	0	0.25		23	Chu, 1992;
	0	0.5		100	Dewey and Bourne, 1982;
	0	1	1–3	220	Ke *et al.*, 1991b;
	0	2–3	1–3	300	Little *et al.*, 1982;
	1 or 3	1.5–1.7	1.0–1.1	217–231	Lau, 1983;
	5	0.02		9	Morgan and Gaunce, 1975
	5	0.25		11	
	5	0.5		55	
	5	1	1–3	180	
	5	2–3	1–3	210	
	10	0.02		5	
	10	0.25		7	
	22		90	1	
Blueberry	5	0		14–21	Smittle and Miller, 1988;
	5	5	10–20	42	Prange and Lidster, 1992
	21		54	2	
Cherry	0	0.02		25	Chen *et al*, 1981;
	0	0.25		35–44	Patterson, 1982;
	0	1		56	Ke and Kader, 1992b
	5	0.02		21	
	5	0.25		38	
	0		20–40	42	
	0		60	7	
	0		80–100	4–7	
Mango	20	0.1–0.2		5	Yahia *et al.*, 1989
	20	2	50	5	
	20	0.5	70–80	4	
Nectarine	0	0.02		14	Mitchell *et al.*, 1984;
	0	0.25		28	Smilanick and Fouse, 1989;
	5	0.5		3–6	Ke and Kader, 1992b
	10	0.02		10	
	10	0.25		13	
	15	0.5		2–3	
	25	0.5		2–3	
	0	4.2	80	10	
	0	1	99	6	

Table 11.3. *contd*

Commodity	Temp. (°C)	CA composition O$_2$%	CO$_2$%	Tolerance (days)	References
Orange	5	0.02		16	Ke and Kader, 1990
	5	0.25		23	
	10	0.02		15	
	10	0.25		20	
	5	8.4	60	5	
Papaya	20	0.2–0.4		3	Yahia *et al.*, 1989, 1992
Peach	0	0.02		40	Mitchell *et al.*, 1984;
	0	0.25		>40	Ke *et al.*, 1991c
	5	0.02		14	
	5	0.25		14	
	0	4.2	80	6	
	0	1	99	3	
Pear (Asian)	0 or 5	0.02		14	Ke *et al.*, 1991b
	0 or 5	0.25		14	
Pear (European)	0	0.02		>10	Mellenthin *et al.*, 1980;
	0	0.25		40	Chen and Mellenthin, 1982;
	0	0.5		60	Yoshida *et al.*, 1986;
	0	1	1–3	90	Ke *et al.*, 1990
	5	0.25		18	
	5	1		25	
	10	0.25		10	
	10	1		>10	
	25	0.25		3–5	
	0–10	10–16	20–50	6	
	5–10	4.2	80	4	
Plum	5	0.02		32	Ke *et al.*, 1991b;
	5	0.25		41	Mitchell *et al.*, 1984
	10	0.02		9	
	10	0.25		14	
	0	1	99	5	
	5	4.2	80	6	
Strawberry	0	0.02		6	Ke *et al.*, 1991a
	0	0.25		10	
	5	0.25		8	
	5	1		>10	
	0–5		20	10–15	
	0–5	4.2–10	50–80	8	

Miller, 1988; Ke *et al.*, 1991a). With CO_2 levels at 50–100%, however, injury could be observed in all commodities after 3–8 days of treatment (Mitchell *et al.*, 1984; Yahia *et al.*, 1989; Ke and Kader, 1990; Ke *et al.*, 1990, 1991a). Symptoms of low O_2 and/or high CO_2 injury may include internal and/or external browning and skin pitting (Ke and Kader 1990, 1992b; Ke *et al.*, 1991a,b). The injured fruits may fail to ripen after removal to air and show increased susceptibility to decay.

Another very common detrimental effect on exposing fruits and vegetables to insecticidal CA is off-flavour development. This is generally thought to be caused by the accumulation of fermentation products such as ethanol, acetaldehyde, ethyl acetate, and probably some other volatiles under the very low O_2 and/or very high CO_2 conditions. At the early stage of off-flavour development, slight alcoholic flavour could be detected; at a later stage, unpleasant off-flavour occurs. Ethanol and acetaldehyde contents greatly increase during off-flavour development, with ethanol as the predominant volatile (Nichols and Patterson, 1987; Patterson and Nichols, 1988; Ke and Kader, 1990, 1992b; Ke *et al.*, 1991a,b).

For many commodities, such as apple, Bartlett pear, plum, cherry, orange, and strawberry, off-flavour development is the only detrimental effect observed following short-term exposure to O_2 level at or below 1% (Ke and Kader, 1990, 1992b; Ke *et al.*, 1990, 1991a,b). The major factors that influence off-flavour development include the O_2 level used, storage temperature, respiration rate, resistance to gas diffusion, soluble solids content, and ethanol accumulation rate of the commodity under a low O_2 treatment (Ke *et al.*, 1991b; Ke and Kader, 1992b).

Decreasing O_2 level to near 0% causes faster off-flavour development. Increasing storage temperature dramatically enhances off-flavour. If the relative tolerance of fruits to 0.25% O_2 at 0°C is used as a reference point (100%), then increasing O_2 concentration to 1.0% at the same temperature usually increases the tolerance threefold; on the other hand, raising the temperature to 10°C at the same O_2 concentration decreases the tolerance to about 35% (Fig. 11.1). A commodity with higher respiration rate and greater resistance to gas diffusion is less tolerant to insecticidal low O_2 atmospheres (Ke and Kader, 1992b). On the other hand, higher soluble solids content requires more ethanol to cause off-flavour, probably due to the interaction between soluble solids and ethanol in determining flavour (Ke *et al.*, 1991b; Ke and Kader, 1992b). For example, Asian pear, 'Yellow Newtown' apple, and plum had 10.7%, 13.4%, and 15.3% average soluble solids contents and their threshold ethanol concentrations for off-flavour detection were about 300, 1000, and 3000 $\mu l\ l^{-1}$, respectively (Fig. 11.2). Flavour and ethanol content appear to have a logarithmic relationship and the threshold ethanol content for off-flavour detection (E_o, $\mu l\ l^{-1}$) could be predicted by soluble solids content (S, %) of the commodity as follows: $E_o = 10^{0.228\,S}$ (Ke *et al.*, 1991b). It should be

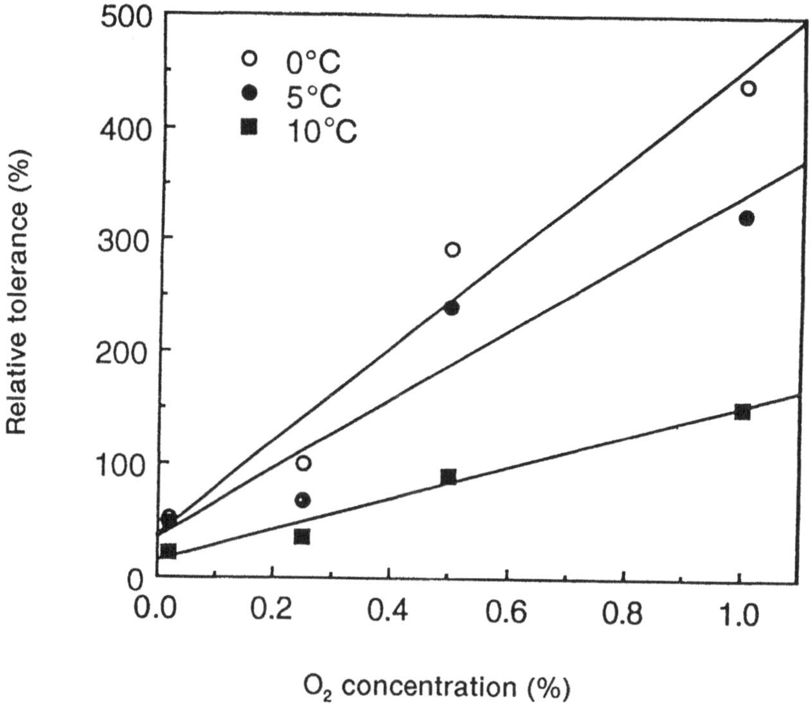

Fig. 11.1 Relative tolerance of fruits to O_2 concentrations at or below 1% as influenced by temperature. Tolerance to 0.25% O_2 at 0°C is used as a reference point (100%).

pointed out that tolerance to low O_2 is not determined by a single factor and all the related factors should be considered. More mature fruits may have higher respiration rates and resistance to gas diffusion and they are consequently less tolerant to low O_2. Over-ripe commodities should not be treated with low O_2 and/or very high CO_2 atmospheres due to their low tolerance to these CA conditions.

By using soluble solids content (%) and average ethanol accumulation rate per day (V_E, μl l^{-1} day^{-1}), a model has been developed for predicting fruit tolerance limit (T_l, day) to insecticidal low O_2 atmospheres: $T_l = (10^{0.228\,S})/V_E$ (Ke *et al.*, 1991b). Using storage temperature (T, °C), O_2 concentration (C, %), respiration rate (R, ml CO_2 kg^{-1} h^{-1}), resistance to gas diffusion (r, % (ml kg^{-1} $h)^{-1}$), and soluble solids content of the commodity as variables, the following multiple regression model was developed (Ke and Kader, 1992b) for quick prediction of fruit tolerance to insecticidal low O_2 atmospheres:

$$T_l = -0.484\,T + 64.6\,C - 0.722\,R - 4.20\,r + 3.00\,S - 5.90\,T \times C - 18.2 \quad [11.1]$$

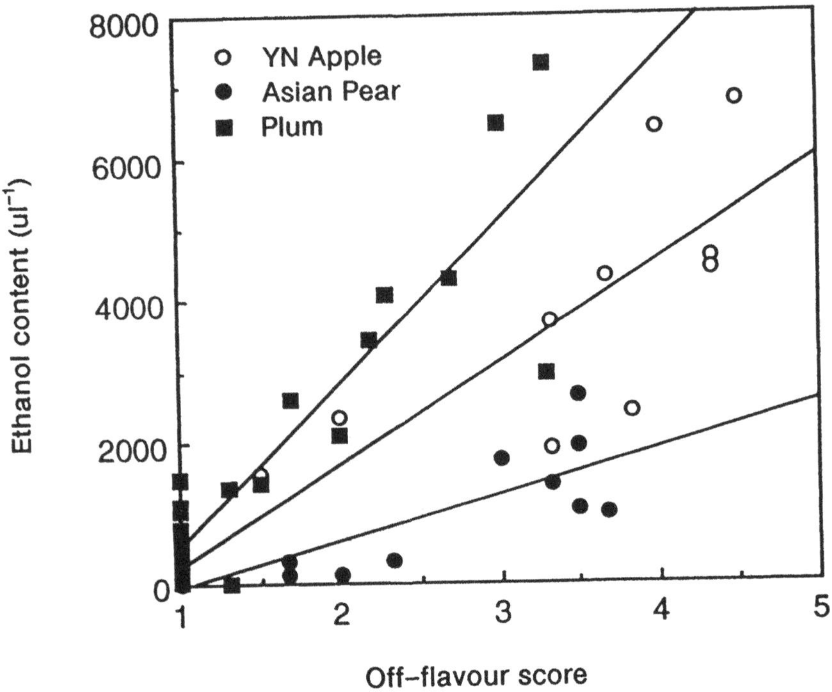

Fig. 11.2 Influence of ethanol concentration on off-flavour severity in 'Yellow Newtown' apple, 'Asian' pear, and 'Angeleno' plum. Off-flavour was scored using a scale of 1 to 5 where 1 = none, 2 = slight, 3 = moderate, 4 = severe, and 5 = extreme off-flavour.

Equation [11.1] indicates that relatively high temperature, respiration rate, and resistance to gas diffusion reduce fruit tolerance to low O_2 atmospheres; relatively high O_2 concentration and *S* increase fruit tolerance. The interaction between temperature and O_2 concentration is also considered in this model.

A correlation coefficient of 0.902 was obtained for Equation [11.1] at $P<0.0001$. This model could reasonably predict T_1 in most cases and the average difference between observed tolerance from experiment (T_{le}) and predicted tolerance is about 4 days. In Equation [11.1], *T* and *C* are predetermined or specified, *R* and *r* are measured after 3 days of low O_2 treatment, and *S* is commodity specific, that is measured after the fruits are allowed to ripen.

It should be noted that this regression model is limited to *T* of 0, 5, and 10°C and *C* of 0.25% and 0.02% O_2. For other temperatures and O_2 concentrations, this model should be used with care or modified as needed.

Since *T* influences both *R* and *r*, variables *R* and *r* are not completely independent in Equation [11.1]. The interaction between T and C also reduces the accuracy of using this model for predicting T_l. To improve prediction accuracy, the following individual models for each temperature (0, 5, and 10°C) were developed (Ke and Kader, 1992b):

For 0°C

$$T_l = 54.2\,C - 12.2\,R - 7.23\,r + 6.56\,S - 39.8 \qquad [11.2]$$

For Equation [11.2], correlation coefficient = 0.984, $P = 0.013$, average difference between T_{le} and T_l = 1.6 days.

For 5°C

$$T_l = 14.5\,C - 9.90\,R - 1.07\,r + 7.68\,S - 56.0 \qquad [11.3]$$

For Equation [11.3], correlation coefficient = 0.986, $P = 0.011$, average difference between T_{le} and T_l = 1.6 days.

For 10°C

$$T_l = 11.4\,C + 0.352\,R - 1.88\,r + 1.30\,S - 10.4 \qquad [11.4]$$

For Equation [11.4], correlation coefficient = 0.963, $P = 0.046$, average difference between T_{le} and T_l = 0.7 days. In Equation [11.4], the regression coefficient of 0.352 is small compared to those of C, r, and S and statistically not significant ($P = 0.224$). Since the influence of R on T_l should be negative according to Equation [11.1], the involvement of R data in the model did not appear to be appropriate and it was removed to obtain the following model:

$$T_l = 10.3\,C - 2.60\,r + 1.06\,S - 4.35 \qquad [11.5]$$

For Equation [11.5], correlation coefficient = 0.934, $P = 0.029$, average difference between T_{le} and T_l = 0.9 days.

By predicting the tolerances of fruits and vegetables to insecticidal CA conditions, optimum temperature/O_2 concentration/time combinations could be selected so that insects of concern may be controlled without detrimental effects on the fresh commodities.

References

Barkai-Golan, R. (1990) Postharvest disease suppression by atmospheric modifications. In: Calderon, M. and Barkai-Golan, R. (eds), *Food Preservation by Modified Atmospheres*. CRC Press, Boca Raton, pp. 237–264.

Blanpied, G.D. (1990) Controlled atmosphere storage of apple and pears. In: Calderon, M. and Barkai-Golan, R. (eds), *Food Preservation by Modified Atmospheres*. CRC Press, Boca Raton, pp. 265–299.

Boersig, M.R., Kader, A.A. and Romani, R.J. (1988) Aerobic–anaerobic respiratory transition in pear fruit and cultured pear fruit cells. *Journal of the American Society for Horticultural Science* 113, 869–873.

Burton, W.G. (1978) Biochemical and physiological effects of modified atmospheres and their role in quality maintenance. In: Hultin, H.O. and Milner, M. (eds), *Postharvest Biology and Biotechnology*. Food and Nutrition Press, Westport, pp. 97–110.

Chen, P.M. and Mellenthin, W.M. (1982) Storage behavior of 'd'Anjou' pears in low oxygen and air. In: Richardson, D.G. and Meheriuk, M. (eds), *Controlled Atmospheres for Storage and Transport of Perishable Agricultural Commodities*. Timber Press, Beaverton, pp. 139–148.

Chen, P.M., Mellenthin, W.M., Kelly, S.B. and Facteau, T.J. (1981) Effects of low oxygen and temperature on quality retention of 'Bing' cherries during prolonged storage. *Journal of the American Society for Horticultural Science* 106, 533–535.

Chen, P.M., Olsen, K.L. and Meheriuk, M. (1985) Effects of low-oxygen atmosphere on storage scale and quality preservation of 'Delicious' apples. *Journal of the American Society for Horticultural Science* 110, 16–20.

Chu, C.L. (1992) Postharvest control of San Jose scale on apples by controlled atmosphere storage. *Postharvest Biology and Technology* 1, 361–369.

Delate, K.M. and Brecht, J.K. (1989) Quality of tropical sweet potatoes exposed to controlled-atmosphere treatments for postharvest insect control. *Journal of the American Society for Horticultural Science* 114, 963–968.

Dewey, D.H. and Bourne, M.L. (1982) Low oxygen CA storage of McIntosh apples. In: Richardson, D.G. and Meheriuk, M. (eds), *Controlled Atmospheres for Storage and Transport of Perishable Agricultural Commodities*. Timber Press, Beaverton, pp. 101–107.

El-Goorani, M.A. and Sommer, N.F. (1981) Effects of modified atmospheres on postharvest pathogens of fruits and vegetables. *Horticultural Reviews* 3, 412–461.

Hatton, T.T., Jr and Spalding, D.H. (1990) Controlled atmosphere storage of some tropical fruits. In: Calderon, M. and Barkai-Golan, R. (eds), *Food Preservation by Modified Atmospheres*. CRC Press, Boca Raton, pp. 301–313.

Herner, R.C. (1987) High CO2 effects on plant organs. In: Weichmann, J. (ed.), *Postharvest Physiology of Vegetables*. Marcel Dekker, Inc., New York, pp. 239–253.

Kader, A.A. (1986) Biochemical and physiological basis for effects of controlled and modified atmospheres on fruits and vegetables. *Food Technology* 40, 5, 99–100, 102–104.

Kader, A.A., Zagory, D. and Kerbel, E.L. (1989) Modified atmosphere packaging of fruits and vegetables. *CRC Critical Reviews of Food Science and Nutrition* 28, 1–30.

Ke, D. and Kader, A.A. (1990) Tolerance of 'Valencia' oranges to controlled atmospheres as determined by physiological responses and quality attributes. *Journal of the American Society for Horticultural Science* 115, 779–783.

Ke, D. and Kader, A.A. (1992a) Potential of controlled atmospheres for postharvest insect disinfestation of fruits and vegetables. *Postharvest News and Information* 3, 31N–37N.

Ke, D. and Kader, A.A. (1992b) External and internal factors influence fruit tolerance to low-oxygen atmospheres. *Journal of the American Society for Horticultural Science* 117, 913–918.

Ke, D., van Gorsel, H. and Kader, A.A. (1990) Physiological and quality responses of 'Bartlett' pears to reduced-O_2 and enhanced-CO_2 atmospheres and storage temperatures. *Journal of the American Society for Horticultural Science* 115, 435–439.

Ke, D., Goldstein, L., O'Mahony, M. and Kader, A.A. (1991a) Effects of short-term exposure to low O_2 and high CO_2 atmospheres on quality attributes of strawberries. *Journal of Food Science* 56, 50–54.

Ke, D., Rodriguez-Sinobas, L. and Kader, A.A. (1991b) Physiology and prediction of fruit tolerance to low oxygen atmospheres. *Journal of the American Society for Horticultural Science* 116, 253–260.

Ke, D., Rodriguez-Sinobas, L. and Kader, A.A. (1991c) Physiological responses and quality attributes of peaches kept in low oxygen atmospheres. *Scientia Horticulturae* 47, 295–303.

Kubo, Y., Inaba, A. and Nakamura, R. (1990) Respiration and C_2H_4 production in various harvested crops held in CO_2-enriched atmospheres. *Journal of the American Society for Horticultural Science* 115, 975–978.

Lau, O.L. (1983) Effects of storage procedures and low oxygen and high carbon dioxide atmospheres on storage quality of 'Spartan' apples. *Journal of the American Society for Horticultural Science* 108, 953–957.

Leshuk, J.A. and Saltveit, M.E., Jr (1990) Controlled atmosphere storage requirements and recommendations for vegetables. In: Calderon, M. and Barkai-Golan (eds) *Food Preservation by Modified Atmospheres.* CRC Press, Boca Raton, pp. 315–352.

Little, C.R. and Peggie, I.D. (1987) Storage injury of pome fruit caused by stress levels of oxygen, carbon dioxide, temperature, and ethylene. *HortScience* 22, 783–790.

Little, C.R., Fragher, J.D. and Taylor, H.J. (1982) Effects of initial oxygen stress treatments in low oxygen modified atmosphere storage of 'Granny Smith' apples. *Journal of the American Society for Horticultural Science* 107, 320–323.

Lougheed, E.C. (1987) Interactions of oxygen, carbon dioxide, temperature, and ethylene that may induce injuries in vegetables. *HortScience* 22, 791–794.

Mellenthin, W.M., Chen, P.M. and Kelly, S.B. (1980) Low oxygen effects on dessert quality, scale prevention, and nitrogen metabolism of 'd'Anjou' pear fruit during long-term storage. *Journal of the American Society for Horticultural Science* 105, 522–527.

Mitchell, F.G., Kader, A.A., Crisosto, G. and Mayer, G. (1984) *The Tolerance of Stone Fruits to Elevated CO_2 and O_2 Levels.* Report to the California Tree Fruit Agreement, Sacramento, California, USA, 11pp.

Morgan, C.V.G. and Gaunce, A.P. (1975) Carbon dioxide as a fumigant against the San Jose scale (Homoptera: Diaspididae) on harvested apples. *Canadian Entomologist* 107, 935–936.

Nichols, W.C. and Patterson, M.E. (1987) Ethanol accumulation and poststorage quality of 'Delicious' apples during short-term, low-O_2 CA storage. *HortScience* 22, 89–92.

Patterson, M.E. (1982) CA storage of cherries. In: Richardson, D.G. and Meheriuk, M. (eds), *Controlled Atmospheres for Storage and Transport of Perishable Agricultural Commodities.* Timber Press, Beaverton, pp. 149–154.

Patterson, M.E. and Nichols, W.C. (1988) Metabolic responses of 'Delicious' apples to carbon dioxide in anoxic and low oxygen environments. *HortScience* 23, 866–868.

Pesis, E., Zauberman, G. and Avissar, I. (1991) Induction of certain aroma volatiles in feijoa fruit by postharvest application of acetaldehyde or anaerobic conditions. *Journal of the Science of Food and Agriculture* 54, 329–337.

Prange, R.K. and Lidster, P.D. (1992) Controlled-atmosphere effects on blueberry maggot and lowbush blueberry fruit. *HortScience* 27, 1094–1096.

Shaw, P.E., Moshonas, M.G. and Pesis, E. (1991) Changes during storage of oranges pretreated with nitrogen, carbon dioxide and acetaldehyde in air. *Journal of Food Science* 56, 469–474.

Smilanick, J.L. and Fouse, D.C. (1989) Quality of nectarines stored in insecticidal low-O_2 atmospheres at 5 and 15°C. *Journal of the American Society for Horticultural Science* 114, 431–436.

Smittle, D.A. and Miller, W.R. (1988) Rabbiteye blueberry storage life and fruit quality in controlled atmospheres and air storage. *Journal of the American Society for Horticultural Science* 113, 723–728.

Wang, C.Y. (1990) Physiological and biochemical effects of controlled atmosphere on fruits and vegetables. In: Calderon, M. and Barkai-Golan, R. (eds), *Food Preservation by Modified Atmospheres*. CRC Press, Boca Raton, pp. 197–223.

Weichmann, J. (1986) The effects of controlled atmosphere storage on the sensory and nutritional quality of fruits and vegetables. *Horticultural Reviews* 8, 101–127.

Yahia, E.M., Medina, F. and Rivera, M. (1989) The tolerance of mango and papaya to atmospheres containing very high levels of CO_2 and/or very low levels of O_2 as a possible insect control treatment. In: Fellman, J.K. (ed.), *Proceedings of the 5th International Controlled Atmosphere Research Conference, Wenatchee, Washington, USA, 14–16 June 1989, Vol. 2*, pp. 77–89.

Yahia, E.M., Rivera, M. and Hernandez, O. (1992) Responses of papaya to short-term insecticidal oxygen atmosphere. *Journal of the American Society for Horticultural Science* 117, 96–99.

Yoshida, T., Borgic, D.M., Chen, P.M. and Mielke, E.A. (1986) Changes in ethylene, acids and brown-core development of 'Bartlett' pears in low oxygen storage. *HortScience* 21, 472–474.

Zagory, D. and Kader, A.A. (1989) Quality maintenance in fresh fruits and vegetables by controlled atmospheres. In: Jen, J.J. (ed.), *Quality Factors of Fruits and Vegetables: Chemistry and Technology*. American Chemical Society, Washington, DC, pp. 174–188.

Modelling Heat and Cold Transfer 12

CHARLES F. HAYES
Department of Physics and Astronomy, University of Hawaii, Honolulu, HI 96822, USA.

The use of thermal treatments to control insects in fruit has necessitated the development of mathematical models to predict fruit temperatures during heating. Combining such mathematical models with theoretical models for insect mortality and heat-induced damage to the fruit provides a powerful tool to evaluate heat treatment protocols. Such tools can reduce the substantial expenses in both time and resources involved in physically testing each proposed method of treatment. This chapter focuses on the derivation of heat transfer equations, the assumptions used to obtain solutions to the equations, a review of the physical parameters that must be measured to use the solutions to the equations, and recent developments in heat transfer modelling theory.

Heat Diffusion Equation

Inside a fruit, heat travels by conduction from regions of higher to lower temperatures. Consider a point located in the interior of a fruit and on a smoothly curved or plane mathematical surface, S. The thermal energy per area per time conducted across the surface is called the heat flux, J, measured in units such as calories per square centimetre per second. Let $T(x,y,z,t)$ be the temperature as a function of position and time in the fruit. The thermal conductivity, K, is defined as the constant of proportionality between the temperature gradient and the heat flux

$$J = -K\nabla T \qquad [12.1]$$

where the negative sign has been introduced since heat flows in the direction of decreasing temperature.

From: Paull, R.E. and Armstrong, J.W. (eds) (1994) *Insect Pests and Fresh Horticultural Products: Treatments and Responses.* CAB INTERNATIONAL, Wallingford, UK.

The heat capacity per volume, C, is the amount of heat required to raise a unit volume of material by one degree in temperature. If surface, S, encloses a volume, V, such as the interior of a fruit, the thermal energy passing through S must equal both the change in thermal energy of V as well at the rate of heat flow through the S determined by the flux

$$\frac{\partial}{\partial t} \int CT\mathrm{d}V = -\int \vec{J} \cdot \mathrm{d}\vec{A} \qquad [12.2]$$

which can be changed using the divergence theorem to

$$\frac{\partial}{\partial t} \int CT\mathrm{d}V = -\int \nabla \cdot \vec{J}\mathrm{d}V \qquad [12.3]$$

and by Equation [12.1]

$$\frac{\partial}{\partial t} \int CT\mathrm{d}V = +\int K\nabla \cdot \nabla T\mathrm{d}V \qquad [12.4]$$

$$= \int K\nabla^2 T\mathrm{d}V \qquad [12.5]$$

So we obtain the heat flow equation

$$C\frac{\partial T}{\partial t} = K\nabla^2 T \qquad [12.6]$$

Defining K/C as k, the thermal diffusivity

$$\frac{\partial T}{\partial t} = k\nabla^2 T \qquad [12.7]$$

It is this partial differential equation that must be solved to determine the temperature as a function of time and position in the fruit. A common way of solving the equation is by describing the temperature as a Fourier series

$$\frac{1}{2}a_0 + \sum_{n=1}^{\infty} (a_n \cos nx + b_n \sin nx) \qquad [12.8]$$

where

$$a_n = \frac{1}{d}\int_{-d}^{d} T(x)\cos\frac{n\pi x}{d}\,\mathrm{d}x \qquad [12.9]$$

$n=0,1,2 \ldots$ and

$$b_n = \frac{1}{d}\int_{-d}^{d} T(x)\sin\frac{n\pi x}{d}\,\mathrm{d}x \qquad [12.10]$$

$n=1,2,3 \ldots$ for T between $x=-d$ and $x=+d$. For example, suppose the temperature is a constant T_o, between $x=0$ and $x=L$ and $-T_o$ between $x=-L$ and $x=0$. Then the a_ns are zero since $T(x)$ is an odd function and the b_ns are given by

$$b_n = \frac{1}{L}\int_{-L}^{L} T_o \sin\frac{n\pi x}{L}\,dx \qquad [12.11]$$

$$= \frac{2T_o}{L}\int_{0}^{L} \sin\frac{n\pi x}{L}\,dx \qquad [12.12]$$

$$= \frac{4T_o}{(2m+1)\pi}\sin[\frac{\pi(2m+1)x}{2}] \qquad [12.13]$$

where $m=0,1,2,\ldots$

$$T(x) = \sum_{m=0}^{\infty} \frac{4T_o}{(2m+1)\pi}\sin[\frac{\pi(2m+1)x}{2}] \qquad [12.14]$$

For a spherical fruit of radius R, assuming radial symmetry, Equation [12.7] becomes

$$\frac{\partial T}{\partial t} = k[\frac{1}{r^2}\frac{\partial}{\partial}r(r^2\frac{\partial T}{\partial r})] \qquad [12.15]$$

Let us simplify this equation by the transformation

$$U = Tr$$

Then Equation [12.15] becomes

$$\frac{\partial U}{\partial t} = k\frac{\partial^2 U}{\partial r^2} \qquad [12.16]$$

Analytical Solution

Consider the case of a fruit that is at room temperature. At time equal zero, the fruit is placed in a water bath at an elevated temperature. Assume that the surface of the fruit, $r=R$, is held at the temperature of the water. To simplify the equations, zero is used for the room temperature. The temperature difference between the fruit and the water is T_o. Once the calculation is completed, the original temperature of the fruit can be added to restore the usual temperature scale. With these changes

$T=0$ at $t=0$ for $0 \le r < R$

$T=T_o$ at $r=R$ for all times

Using a Fourier series the solution of Equation [12.16] becomes

$$U = T_o r + \sum_{n=1}^{\infty} a_n \sin(\frac{n\pi r}{a})e^{-kn^2\pi^2 t/R^2} \qquad [12.17]$$

where

$$a_n = \frac{(-1)^n 2RT_o}{n\pi} \qquad [12.18]$$

So the temperature of the fruit as a function of time and position is

$$T(r, t) = T_o + \frac{2RT_o}{r\pi} \sum_{n=1}^{\infty} \frac{(-1)^n}{n} \sin\left(\frac{n\pi r}{R}\right) e^{-kn\pi^2 t/R^2} \qquad [12.19]$$

(Carslaw and Jaeger, 1959).

For apples, grapefruit, and oranges, the thermal diffusivity is 0.0049 $ft^2\,h^{-1}$ (4.55 $cm^2\,h^{-1}$), 0.0047 $ft^2\,h^{-1}$ (4.37 $cm^2\,h^{-1}$), and 0.0049 $ft^2\,h^{-1}$ (4.55 $cm^2\,h^{-1}$) respectively (Rha, 1975). Taking $R \sim 6$ cm, for times greater than 5 min the exponential assures less than a 1% error if terms after $n = 10$ are deleted.

Equation [12.19] may be used for homogeneous fruit of spherical symmetry that are initially of uniform temperature when immersed in a well-stirred bath of constant temperature. T_o equals the difference between the bath temperature and the initial fruit temperature. If the initial temperature of the fruit is not uniform but a function of radius, $f(r)$, Equation [12.19] may be generalized to

$$T(r, t) = T_o + \frac{2RT_o}{r\pi} \sum_{n=1}^{\infty} \frac{(-1)^n}{n} \sin\left(\frac{n\pi r}{R}\right) e^{-kn^2\pi^2 t/R^2} \qquad [12.20]$$

$$+ \frac{2}{Rr} \sum_{n=1}^{\infty} \left[\int_0^R r' f(r') \sin\left(\frac{n\pi r'}{R}\right) dr'\right] \cdot \sin\left(\frac{n\pi r}{R}\right) e^{-kn^2\pi^2 t/R^2}$$

To program Equation [12.20] for a computer, it is convenient to use an approximation for the integral. Simpson's formula (Spiegel, 1968), also called the parabolic formula, for even n is

$$\int_a^b f(x)dx = \frac{h}{3}(Y_o + 4Y_1 + 2Y_2 + 4Y_3 + \ldots + 2Y_{n-2} + 4Y_{n-1} + Y_n) \qquad [12.21]$$

where $Y_o = f(x_o)$, $y_1 = f(x_1)$, and the interval $x = a$ to $x = b$ is evenly divided into n segments.

Equations [12.20] and [12.21] have been used for papayas initially warmed with microwaves and subsequently immersed in water (Hayes *et al.*, 1984). The focusing of microwaves by the fruit surface coupled with the absorption length of the microwaves resulted in the centre of the papaya having the highest pre-immersion temperature. The pre-immersion temperature distribution is the $f(r')$ of Equation [12.20].

Heat Transfer in Air

Thus far the temperature of the fruit surface is taken to be that of the immersion bath, a good approximation for a well stirred fluid. If the fruit is heated by air, a new physical parameter, H, must be introduced called the surface heat transfer coefficient or the convective coefficient. This proportionality constant couples the rate of heat flow through the surface per surface area, A, to the difference between the surface temperature, T_s, and that of the ambient temperature T_A

$$\frac{1}{A}\frac{dQ}{dt} = H(T_s - T_A) \qquad [12.22]$$

Since the heat flow through the surface is also equal to the heat conducted at the surface

$$\frac{1}{A}\frac{dQ}{dt} = k\frac{\partial T}{\partial r}\Big|_{r=R} \qquad [12.23]$$

$$\frac{\partial T}{\partial r}\Big|_{r=R} = h(T_s - T_A) \qquad [12.24]$$

where we have taken $h = H/k$.

If the fruit's temperature is initially uniform, and if the fruit is placed in air of an elevated temperature, the fruit's subsequent temperature distribution as a function of time is

$$T(r, t) = T_o\,[1 - \sum_{n=1}^{\infty} G_n\, e^{-k\alpha_n^2 t}] \qquad [12.25]$$

(Hayes and Young, 1989).

where

$$G_n = 2h[R^2\alpha_n^2 + (Rh-1)^2] \cdot \frac{\sin(R\alpha_n)\sin(r\alpha_n)}{r\alpha_n^2[R^2\alpha_n^2 + Rh(Rh-1)]} \qquad [12.26]$$

where α_n are the roots of

$$R\alpha\cot(R\alpha) + Rh - 1 = 0 \qquad [12.27]$$

T_o is the difference between the ambient temperature and the initial fruit temperature, i.e. Equation [12.25] assumes the initial fruit temperature is zero.

If h is known, Equation [12.25] may be used to find the temperature distribution of the fruit at any time t after it is placed in the heated air. Conversely, by measuring the temperatures, h can be determined. This method was used successfully to determine h for papayas (Hayes and Young, 1989). Temperature measurements were taken at the centre of the fruit as a function of time under a variety of air velocities to determine h.

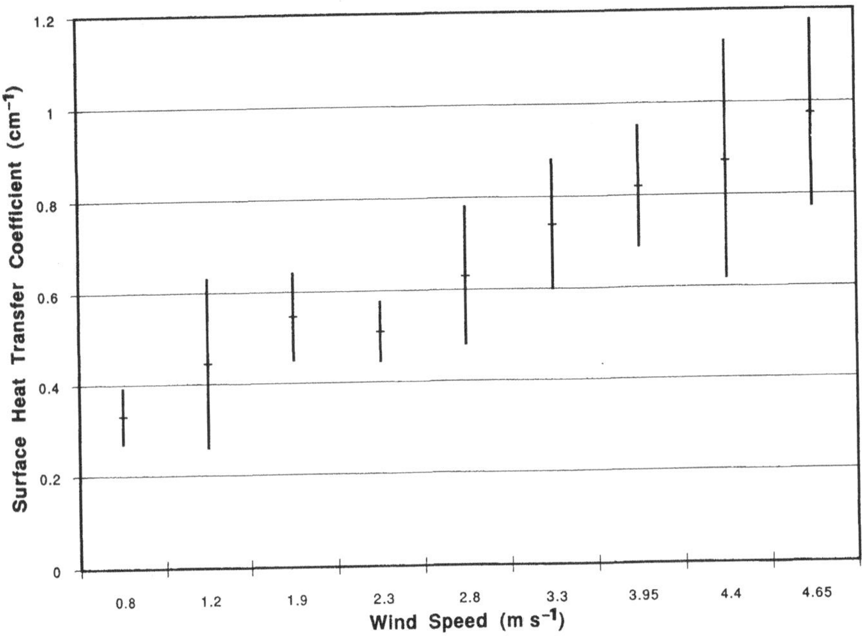

Fig. 12.1 Relationship between surface heat transfer coefficient, h (cm^{-1}), and wind speed (m s^{-1}) for papaya.

Papayas originally at room temperature (24°C ± 1°C) were placed in air at a higher constant temperature that was circulated at a rate of 0.8 to 4.6 m s^{-1} air velocity. The temperature profiles versus time were fitted for best h (Fig. 12.1) using Equation [12.25] as a function of air velocity. The data show a linear relationship between h and air velocity:

$$h = \frac{0.22}{\text{cm}} + \frac{0.154 \cdot v}{\text{cm}^2\ \text{s}^{-1}} \qquad [12.28]$$

velocity = v, $\chi^2 = 0.16$, $P = 0.99$. Differences in fruit orientation, regardless of whether the blossom or the stem ends faced the wind, did not significantly alter the derived h values. When the stem ends faced the wind $h = (0.535 \pm 0.130)$ cm^{-1} and when the blossom end faced the wind $h = (0.517 \pm 0.116)$ cm^{-1}.

Papaya fruit shipped from Hawaii to the mainland United States must first be treated for possible fruit fly infestation. There are two commercial heat treatments used in the Hawaiian papaya industry to disinfest fruit fly eggs and larvae from the fruit: (i) the vapour heat treatment (Couey, 1989) in which fruit is heated by hot air saturated with water vapour; and (ii) the forced hot-air treatment (Armstrong *et al.*, 1989) in which fruit is heated by air not saturated with water vapour. It may be possible to use hot water of varying temperature to reproduce the rate of heating of the air treatments to kill the insect throughout the fruit. The advantage would be that it is more efficient to heat water than air, and water would provide more uniform heating of fruit. In air heating, the heat transport at the fruit surface depends strongly on the air speed as shown above. In water heating, differential heating rates are not such a problem since the fruit surface assumes the temperature of the surrounding water even at modest rates of fluid motion.

The temperatures of fruit heated with air saturated with water vapour or at lower humidities are expected to be approximately equal. Therefore, the thermal stress for fruit heated by either method should be the same. However, respiration of the fruit during heating may be hampered during water treatments. Jones (1939) compared fruit heated by air of low humidity with fruit heated by nearly saturated air and observed greater fruit damage in the saturated air treatment. Jones hypothesized that water condensing on the fruit surface prevented respiration and contributed to the damage. One could argue that additional heating from condensation caused the damaged fruit to obtain a higher temperature causing the difference in observed damage. However, a programmed increase in water temperature would give the same internal fruit temperature profile as for heated air.

This protocol was tested with a thermally insulated double-walled chamber. The chamber has a space of 5 cm between the inner and outer walls. A temperature-controlled circulation pump forced the water through the space between the inner and outer walls with baffles ensuring even flow over each of the six sides of the chamber. The temperature variation of this unit is typically less than 0.1°C over a time period of several hours. A fan circulated the air through the chamber with a speed <1.5 m s^{-1}. Humidity typically rose from 20 to 75% RH because of evaporation from the fruit. Comparisons were made with fruit immersed in a bath in which the water is circulated through the tank and two heaters. Temperatures were sensed by thermistors to allow constant control of the power to the heaters so that proper temperature was maintained.

The surface of water-heated fruit immediately assumes the bulk temperature of the surrounding water if the bath is well stirred. However, the surface temperature of fruit heated by air does not equal the bulk air temperature immediately after treatment has begun. Rather, the surface temperature gradually rises, as shown by the lower curve in Fig. 12.2 for 49°C air. The temperature at any point inside a fruit is determined by the surface temperature of the fruit and the fruit's thermal diffusivity. If water in which fruit is placed is gradually heated, reproducing the surface

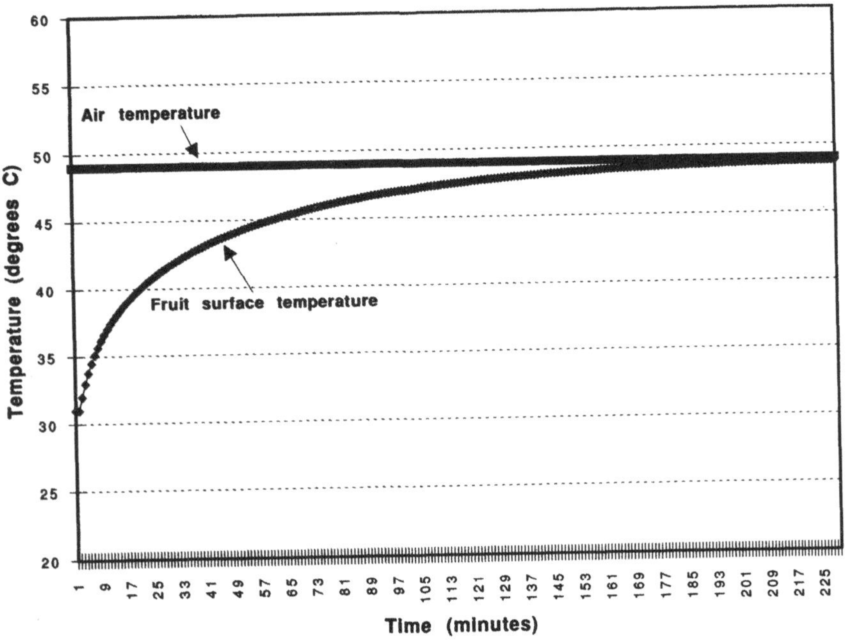

Fig. 12.2 Surface temperature of fruit heated by air at 49°C as a function of time.

temperature curve expected with air treatment, the surface temperature of the water-heated fruit will be the same as that of the air-treated fruit at any given time. Therefore, at any time the internal temperature at a given position in a fruit heated by water should be equal to that in a fruit heated by air. Thus, comparisons can be made between the effect of heat stress from treatment by air and of equal heat stress from treatment by water.

In experimental results represented in Fig. 12.3, water-heated fruit reached the value of 47.2°C about 20 min before the air-heated fruit. None of the control or air-heated fruit was damaged but all of the water-heated fruit were damaged despite the shorter immersion time. The force required to deform the surface of the water-heated fruit was approximately twice that of the control.

These results support Jones' (1939) hypothesis that fruit respiration is inhibited by the water condensation on the fruit surface in a saturated environment. This inhibition predisposes the fruit to injury. The implication with respect to the double immersion, vapour heat and forced hot air treatments is that the latter has the smallest probability of causing damage to the fruit.

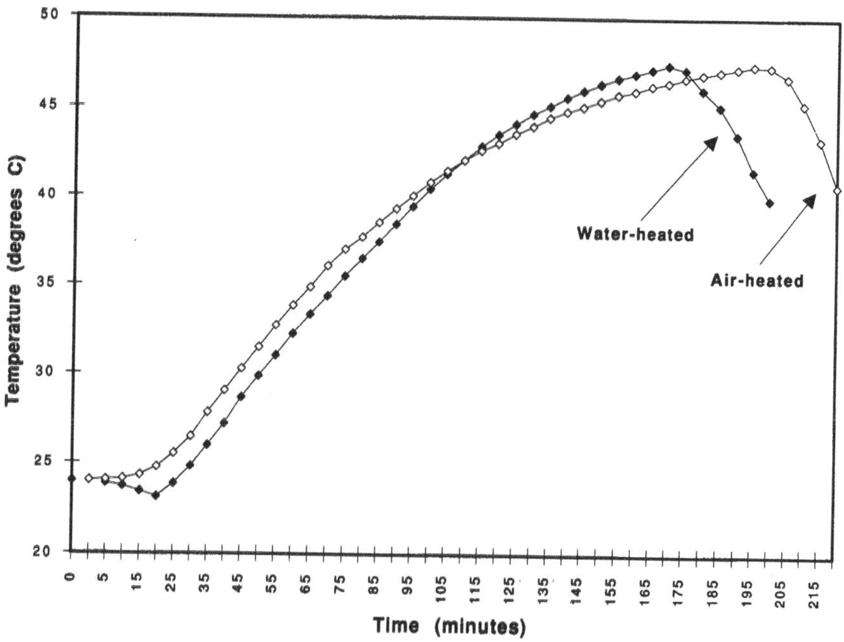

Fig. 12.3 Comparison of central fruit temperature for fruit heated by air and fruit heated using water whose temperature is gradually increased to give the same pattern of heating as expected for fruit heated by air.

Finite Difference Equations and Phenomenological Approaches

Laidlaw and Hayes (1990) used the model of Baird and Gaffney (1977) to develop a set of finite difference equations for heat flow. These equations allow numerical rather than analytical solutions for temperature as a function of position and time during heating. The fruit is assumed to be spherical with the fruit divided mathematically into a grid of cells of length Δr. Time is also divided into a series of steps, each of duration Δt. The temperature at one time step is obtained from the temperature in the previous time step with an orderly progression during heat treatment. Each cell may have different thermal properties. There are three fundamental equations: the central fruit temperature, the bulk of the fruit, and the surface of the fruit. Comparison of solutions from these equations with previously presented analytical solutions confirmed their accuracy to within 1% of predicted values (Laidlaw and Hayes, 1990). These equations are:

1. Centre

$$T_i^{n+1} = \frac{6\alpha_i \Delta t}{(\Delta r)^2} T_{i+1}^n + \left(1 - \frac{6\alpha_i \Delta t}{(\Delta r)^2}\right) T_i^n \qquad [12.29]$$

2. Bulk

$$T_i^{n+1} = \{(1-\frac{\alpha_i \Delta t}{\Delta r})^2 \cdot [(1-\frac{\Delta r}{2r_i})^2 + (1+\frac{\Delta r}{2r_i})^2]\}T_i^n$$
$$+\{(\frac{\alpha_i \Delta t}{(\Delta r)})^2 \cdot [(1-\frac{\Delta r}{2r_i})^2 \, T_{i-1}^n + (1+\frac{\Delta r}{2r_i})^2]\}T_{i+1}^n \qquad [12.30]$$

3. Surface

$$T_i^{n+1} = \{[(\frac{3\alpha_i \Delta t}{\Delta r})\ (R_\infty - \frac{\Delta r}{2})^2] + [R_\infty^3 - (R_\infty - \frac{\Delta r}{2})^3]\}\ T_{i-1}^n$$
$$+ \{[\frac{3R_\infty^2 h \alpha_i \Delta t}{\kappa_i}] + [R_\infty^3 - (R_\infty - \frac{\Delta r}{2})^3]\}\ T_{bath}^n \qquad [12.31]$$
$$+ \{(1 - [\frac{3\alpha_i \Delta t}{\Delta r}] + [R_\infty^3 - (R_\infty - \frac{\Delta r}{2})^3]) \cdot [\frac{(R_\infty - \frac{\Delta r}{2})^2}{\Delta r} + \frac{R_\infty^2 h}{\kappa_i}]\}\ T_i^n$$

Analytical solutions are valid for fruit approximating regular shape. Generally, the approximation is that the fruit is a sphere or cylinder. The use of numerical methods by the finite difference equations allows irregularly shaped fruits to be handled (Eschleman and Gaffney, 1976).

Hansen (1992), rather than using equations derived from the diffusive heat flow equations, has examined a variety of mathematical models to predict fruit temperature as a function of time. Linearized growth model as well as exponential and logistic growth models for grapefruit, carambola, tangerine, lychee, and small and large sweet potatoes were examined. The commodities were treated using either hot water or hot air environments. The overall best fit was found to be the logistic growth model

$$T_t = \frac{T_F}{1 + b'e^{-bt}} \qquad [12.32]$$

Gaffney and Armstrong (1990) developed a computer-controlled hot forced-air facility that controlled both the temperature and humidity. Finite difference heat and mass transfer models previously used for citrus and other fruits were modified to accommodate stacks of papayas. As air passes through the layers of fruit there is a decrease of air temperature. This decrease is minimized by providing a high air flow rate. The relative humidity of the air was controlled to avoid condensation on the fruit by keeping the dew point temperature of the air below the coldest fruit surface temperature.

Although it was used to analyse the cooling of tomatoes, Pan and Bhowmik (1991) developed a finite element technique to determine heat flow. An ANSYS program from Swanson Analysis Systems Incorporated was used in developing the model. A tomato cross-section was divided into small volume elements, each with the ability to conduct heat in two dimensions. Predictions of the model agreed within 1°C for tomatoes with initial temperature of 20°C being cooled with an air temperature of 12°C.

Chau and Gaffney (1990) introduced a finite difference numerical model which includes both respiration and transpiration. Most previous models, with the exception of Hayakawa and Succar (1982) and Chau *et al.* (1984), did not include the development of heat internally or cooling due to transpiration. Comparisons between this model and previous models were made with a variety of surface heat transfer coefficients. For a spherically symmetrical fruit divided into equivalent numbers of spherical shells the proposed model resulted in more accurate temperature values than previous models.

The last ten years have seen the development of sophisticated mathematical models of heat flow in fruit. Major challenges lie ahead for combining these models with models for accurate prediction of insect mortality and phytotoxicity.

References

Armstrong, J.W., Hansen, J.D., Hu, B.K.S. and Brown, S.A. (1989) High-temperature forced-air quarantine treatment for papayas infested with tephritidae fruit flies (Diptera: Tephritidae). *Journal of Economic Entomology* 82, 1667–1674.

Baird, C.D. and Gaffney, J.J. (1977) A numerical procedure for calculating heat transfer in bulk loads of fruit or vegetables. *American Society of Heating, Refrigeration and Air-Conditioning, Engineering Transactions* 82, 525.

Carslaw, H.S. and Jaeger, J.C. (1959) *Conduction of Heat in Solids.* Clarendon Press, Oxford.

Chau, K.V. and Gaffney, J.J. (1990) A finite-difference model for heat and mass transfer in products with internal heat generation and transpiration. *Journal of Food Science* 55, 484.

Chau, K.V., Gaffney, J.J. and Bellagha, S. (1984) Simulation of heat and mass transfer in products with internal heat generation and transpiration. Paper no. 84-6513. Presented at the 1984 winter meeting of the American Society of Agricultural Engineers, 11–14 December.

Couey, H.M. (1989) Heat treatment for control of postharvest diseases and insect pests of fruits. *HortScience* 24, 198–202.

Eschleman, W.D. and Gaffney, J.J. (1976) A numerical simulation of transient heat flow in irregular shaped foods. Paper no. 76-6504. Presented at the 1976 winter meeting of the American Society of Agricultural Engineers, 14–17 December.

Gaffney, J.J. and Armstrong, J.W. (1990) High temperature forced air research facility for heating fruits for insect quarantine treatments. *Journal of Economic Entomology* 83, 1959–1964.

Hansen, J.D. (1992) Heating curve models of quarantine treatments against insect pests. *Journal of Economic Entomology* 85, 1846–1854.

Hayakawa, K.I. and Succar, J. (1982) Heat transfer and moisture loss of spherical fresh produce. *Journal of Food Science* 47, 596.

Hayes, C.F. and Young, H. (1989) Extension of model to predict survival from heat treatment of papaya infested with oriental fruit flies (Diptera: Tephritidae). *Journal of Economic Entomology* 82, 1157–1160.

Hayes, C.F., Chingon, H.T.G., Ha, F.A.N. and Wang, W.J. (1984) Temperature control as an alternative to ethylene dibromide fumigation for the control of fruit flies (Diptera: Tephritidae) in papaya. *Journal of Economic Entomology* 77, 683–686.

Jones, W.W. (1939) The influence of relative humidity on the respiration of papaya at high temperatures. *Proceedings of the American Society for Horticultural Science* 37, 109–124.

Laidlaw, W.G. and Hayes, C.F. (1990) Calculation of ethylene-forming enzyme activity in papayas treated for control of oriental fruit flies (Diptera: Tephritidae). *Journal of Economic Entomology* 83(5), 1944–1948.

Pan, J.C. and Bhowmik, S.R. (1991) The finite element analysis of transient heat transfer in fresh tomatoes during cooling. *Transactions of the ASAE* 34(3), 972–976.

Rha, C. (1975) *Theory, Determination and Control of Physical Properties of Food Materials*. Reidel, Boston.

Spiegel, M.R. (1968) *Mathematical Handbook of Formulas and Tables*. McGraw-Hill, New York.

Heat Treatment Facilities 13

Michael Williamson and Paul Winkelman
Department of Agricultural Engineering, University of Hawaii, 3050 Maile Way, Honolulu, HI 96822, USA.

Introduction

Many designs are possible for disinfestation chambers with all successful designs based on properly established entomological and physiological information relating to the disinfestation objective (Couey, 1989; Paull, 1990).

The design of heat treatment facilities for commercial application involves the transfer of laboratory results to commercial 'real world' situations. These laboratory experiments must establish the entomological and physiological changes occurring within the fresh horticultural produce as its internal temperature is raised. The facility can only be properly designed upon the establishment of working parameters which include the following:

1. The efficacy temperature, T_e, and elapsed time, t_e, causing 100% mortality of the target insects. There may be an efficacy temperature at which the insect is killed immediately (t_e is close to 0) or a lower temperature value at which the fruit is held for a longer period of time causing 100% mortality.
2. The damage temperature, T_d, and elapsed time, t_d, causing unacceptable damage to the fruit. There may be a wide range of temperature and times (T_d and t_d) which damage the commodity.
3. The location of insect eggs and larvae in the commodity determines if the surface, subsurface layers or the entire fruit have to be heated to the efficacy temperature. This information determines the size and type of probe necessary to monitor heating.

From: Paull, R.E. and Armstrong, J.W. (eds) (1994) *Insect Pests and Fresh Horticultural Products: Treatments and Responses.* CAB INTERNATIONAL, Wallingford, UK.

4. The weight a commodity is able to support, over the entire treatment temperature range, without damage, determines the maximum depth of the commodity in a single container.
5. The relative humidity of the heating air influences the rate of commodity water loss. Scalding may result if water vapour from humid air condenses on the surface.
6. The rate of heating of the commodity affects the treatment time. Improper heating rates may have adverse effects on commodity quality.

The values of these parameters depend upon commodity maturity, size and season when harvested.

Design Considerations

The disinfestation treatment chamber must be capable of treating each and every commodity within a batch to the required quarantine standards. For example, Hawaiian papaya must be heated for a period of at least 4 h until the fruit centre temperature reaches 47.2°C to meet US mainland quarantine standards (Animal and Plant Health Inspection Service, 1990). The relative humidity of the air and the type of fruit container used are left to the discretion of the packer. Papayas destined for Japan must all reach a fruit centre temperature of 47.2°C with the standard deviation of the elapsed time not to exceed 9 min for the monitored fruit. In addition, the relative humidity within the chamber must exceed 90% for the final hour of treatment (Japan MAF, 1985). Regardless of the market, all fruit must be unloaded into a fly-free zone for packaging (USDA-APHIS, 1990).

The certification standard developed for papaya further requires that, during normal commercial runs, the monitored fruit be located in designated 'cold spots' within the chamber (Williamson *et al.*, 1991). A 'cold spot' is the location within a stack, bin or even the chamber as a whole, that consistently contains the slowest heating fruit. These fruit would therefore be the last fruit in the bins or stacks to reach efficacy temperature. 'Cold spots' allow the chamber to be operated with only a few fruit temperature monitoring probes. When these few probes reach efficacy temperature, all other fruit would be expected to have reached or exceeded this temperature. Using the largest fruit in the 'cold spot' further increases the confidence of the test. If the probed fruit is not placed in the 'cold spots', then there is a greater likelihood that some unprobed fruit will not have reached efficacy temperature before all the probed fruit.

Although a disinfestation treatment may exist, it may not be applicable to commercial use. Treatment viability is determined by the temperature difference window, T_w, between the required insect efficacy temperature, T_e, and the fruit temperature at which commodity damage occurs, T_d. This can be stated as:

$$T_d - T_e = T_w$$

It is necessary that T_w be greater than or equal to 0. Small values of T_w may allow for a possible treatment which may not be practical due to economic or technical limitations.

Capital investment costs include chamber housing, fan(s) and motor(s), heating equipment, loading equipment, computer and related monitoring and control equipment, humidifier, fruit containers, and installation. Higher quality equipment may prove economical if it greatly reduces the likelihood of breakdowns and resulting fruit losses. Labour, heating fuel and electricity account for the majority of operating costs. The cost of handling, loading and unloading the chamber depends upon the container configuration. For example, baskets or trays will require more labour to fill than bulk loaded bins. Automated handling reduces labour requirements.

The capacity of the chamber is critical as it becomes more difficult to achieve homogeneous heating throughout the fruit mass as the amount of fruit increases. Using two or more smaller chambers ensures redundancy should one fail. Labour allocation for loading/unloading and inspection services can be made more efficient with smaller chambers. Maximum heat loads, occurring immediately after starting a treatment, can follow a staggered schedule.

Fruit handling

The quarantine standards of the importing countries may dictate the type of container used to hold the commodity during treatment. For example, papaya destined for the Japan market must be treated in interlocking plastic containers with solid sides and perforated floors called lugs. Each lug contains approximately 20 fruits in a single layer. A typical stack consists of 72 lugs loaded on a pallet. The cost of the lugs and the additional labour required to load the fruit makes this method more expensive. However, heat distribution is more uniform due to the relatively unrestricted airflow around each fruit. Labour and capital costs can be reduced when the treatment can be done in bulk bins, as for the US mainland. The use of bins increases the volume of fruit per unit volume of the chamber but reduces the uniformity of airflow and increases the variation of exposure of each fruit to the heated air (Winkelman and Williamson, 1990).

A well-designed chamber should accommodate any of the commonly used commodity containers without modification. This allows one chamber to treat commodities in any container dependent upon the market. The target market requirement for a particular container determines the volume of fruit which can be handled in a given batch.

The commodity should be graded by size and maturity prior to treatment. A wide variety of sizes within a single batch seriously affects heating uniformity and consequently quality. A batch of similar size and maturity heats more uniformly. The shape of a fruit determines its surface area to volume ratio. Fruits with higher ratios heat more rapidly. As a fruit matures, its mechanical properties change and it becomes more susceptible to compression damage (Williamson and Winkelman, 1992).

Papayas, in bins or stacked lugs, exhibit a high resistance to airflow and the air can easily bypass the fruit through existing leaks. The fruit stacks should be properly sealed within the duct or plenum to force all the air through the fruit. These seals must also allow the fruit stacks to slide past during loading and unloading. Inflatable bags fitted around the stacks and solid support tracks for the stack bases are currently used.

A mechanical method of loading and unloading commodity containers from the treatment chamber is desirable. Loading and unloading, however, must not damage the air seals. In addition, the temperature probe in every bin or stack must be physically connected to the data interface and not be damaged during loading or unloading. Since untreated fruit is loaded from a fly-infested zone and treated fruit unloaded into a fly-free zone, separate loading and unloading doors are necessary. These doors must be opened in sequence so that both doors are never opened at the same time. A door status alarm device is required to satisfy quarantine inspectors.

Airflow considerations

The heated air used for disinfestation is circulated through the fruit mass and returned for reheating. The design of the air circulation system depends upon the intended commodity. The pressure drop across the fruit mass, as a function of airflow, must be established for each potential commodity. Small fruit, like litchis, have a higher resistance to airflow than large fruit, like papaya or mango. The resistance to airflow of different commodities can be readily determined by placing a sample of the commodity in a ducted airflow. Piezometer rings, or static pressure tubes, are placed upstream and downstream of the fruit to measure the pressure drop as the air flows through the fruit. As the airflow increases, the pressure drop increases. Airflow is typically measured using a standardized nozzle located within the air supply system. The volumetric flowrate can be calculated from the pressure drop measured across the nozzle. Thus, a curve of the volumetric flowrate to pressure drop per unit area of cross-section can be derived for the commodity (Fig. 13.1).

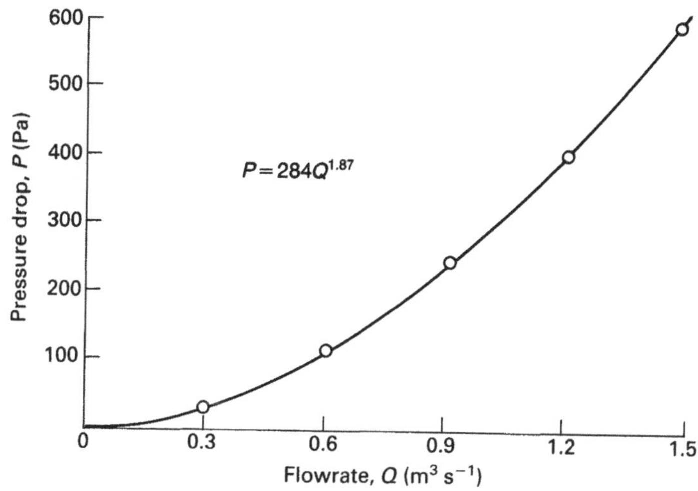

Fig. 13.1 Pressure drop vs flowrate for a sample of papayas 1.2 m deep and a cross-sectional airflow area of 0.81 m^2.

Doubling the area perpendicular to the airflow while maintaining a constant depth would require doubling the airflow to maintain the same pressure drop across the column. Doubling the commodity depth while maintaining a constant airflow and flow area would double the pressure drop. A different fan would be needed to accommodate each of the above changes in fruit resistance, fruit depth, flow area, and airflow for a treatment facility and must be known before a suitable fan can be selected.

There are only two practical ways to pass air through the fruit mass regardless of the container used: vertical and horizontal, vertical airflow being the most common. Fruit containers used in chambers with vertical airflow must have perforated floors to allow the heating air to pass vertically through the fruit. In lugs, the plastic floor has to support only one layer of fruit stack. In bins, the floor supports a column of fruit and the perforations greatly reduce the strength of the floor. A suitable material, such as stainless steel, must be selected with an elastic-type coating to reduce damage to the bottom layer of fruit. The most important design consideration of the floor is the percentage open area, defined as:

$$\%\ \text{Open area} = \frac{(\text{Number of holes})\ (\text{Area per hole})}{(\text{Total area of floor})}$$

A small percentage of open area increases resistance to airflow. Increasing the open area reduces both the load the floor can support and the area available to support the commodity. A reduced support area increases the likelihood of mechanical damage to the commodity surface. Currently, floors with 50–60% open area are in commercial use.

In horizontal airflow, the fruit container must have perforated sides. The floors are usually perforated to provide drainage during cool down with water at the end of a run. Horizontal airflow can be used with flat trays having low sides which allow free passage of air across the layers of fruit.

Airflow reversal systems greatly affect the commodity heating and influence the location of the 'cold spot'. Heat absorption by fruit on the layer nearest the incoming air lowers the temperature of the air passing through subsequent layers. Fruit furthest downstream will tend to heat most slowly. Although increasing the airflow or decreasing fruit depth will reduce the lag, the 'cold spot' is normally located in this downstream layer. This is particularly true if the last layer is totally open to the exiting air. If the airflow is reversed, the upstream layer becomes the downstream layer and vice versa, and the location of the 'cold spot' becomes very unpredictable. The frequency and number of airflow direction changes must be selected by the operator prior to any certification. If the chamber is approved and consistent cold spots are identified, then the reversal schedule becomes part of the certified operating procedure and must be strictly adhered to for all subsequent runs. Any change in the reversal pattern would require additional testing and re-certification.

The last airflow direction change should take place well before the treatment is over so as to establish a stable thermal distribution. Changing the direction too close to the completion of a treatment may shift the 'cold spot' to an unmonitored location.

Once the required airflow and total static pressure losses within the chamber are known, the fan can be specified. The selected fan can be either an axial or centrifugal type. Axial fans are typically used for high volume/low pressure applications, such as ventilation. Performance characteristics vary according to the number and pitch of the blades, the fan and centre hub diameter, and the speed of rotation. Air enters and leaves the fan in the same (axial) direction. Reversible axial fans are available with the same performance regardless of direction of rotation. Centrifugal fans, also known as blowers, are non-reversible and perform better at higher pressures but have lower flowrates than axial fans. Fan performance varies according to blade design which may be backward inclined, radial, or forward inclined. For most centrifugal fans, the air direction changes 90° between entry and exit.

If airflow reversibility is required, a reversible axial fan may be installed or flow dampers used to redirect airflow. If variable airflow is required for different commodities, the fan can be driven by a variable speed motor (AC or DC), a hydraulic motor or a belt drive with variable sheave.

Heat balance

Heat load demand changes as the treatment progresses. The greatest load demanded by the fruit occurs when the temperature difference between the air and the commodity is the largest. The maximum rate of heat gain of a fruit occurs at the greatest slope of the temperature curve. For papaya, a typical value is approximately 0.13°C min^{-1}, 90 min into the run (Fig. 13.2). Increasing the air temperature or airflow increases the rate of temperature rise. Near the end of the run, the fruit temperature rise can decline to 0.025°C min^{-1}. Fruit specific heat, weight and maximum acceptable rate of temperature rise determine the heat load.

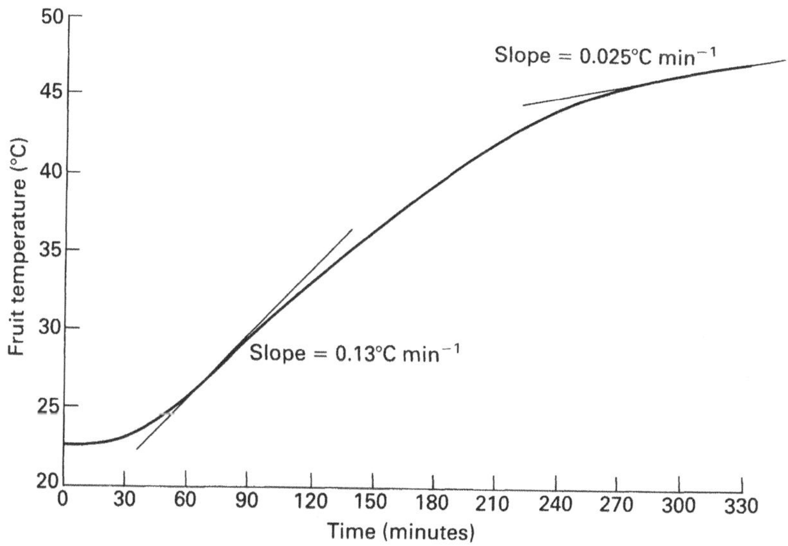

Fig. 13.2 Typical heating curve of a papaya seed cavity using hot air.

Significant heat losses occur from the chamber, the rate being determined by the difference between the interior and exterior air temperatures. Heat loss through the chamber walls depends on the exposed wall area, the wall insulative value, the interior and exterior air speed and temperature difference.

Although wall insulation reduces heat loss, care must be taken to avoid over-insulating. A significant amount of heat is supplied by the fan through frictional losses. Most of this heat is added to the chamber air even if the fan motor is mounted outside. For example, a 10 kW fan can supply enough heat in 1 h to raise the temperature of 1000 kg of fruit 8.6°C. The insulation must be low enough to allow the dissipation of excess heat from the fan when the fruit heat demand is low. Over insulating will cause the air temperature to rise above acceptable limits, causing some of the fruit to overheat.

Electric resistive heaters are easy to control electronically but require an expensive form of energy. Propane is a more economical alternative. Direct heating of the air using a gas burner is not recommended due to ethylene production resulting in premature fruit ripening (Abeles *et al.*, 1992). Heating water or generating steam with propane requires pumping the hot water or steam through a heat exchanger. Steam introduced directly into the airstream interferes with humidity control and the high heat content makes temperature control difficult.

Computer monitoring and control

The computer and data acquisition units must be capable of accepting and handling the incoming data. Readings are typically taken every 1–5 min. The data acquisition units are needed to convert the analog signals from the measuring devices to digital signals and convert computer outputs to control chamber parameters.

Measuring devices

Temperatures are typically measured using resistive thermal detectors (RTDs), thermistors or thermocouples. The resistance of RTDs and thermistors changes as a function of the temperature surrounding the element in the tip of the sensor. This allows the absolute temperature surrounding the probe tip to be determined by measuring the resistance. RTDs have platinum elements making them very stable. The resistance of an RTD changes linearly with temperature, typically at the rate of 0.385 Ω $°C^{-1}$ over the range of 0–100°C, whereas a thermistor is non-linear. To measure the resistance of a RTD or thermistor, a small current must be supplied that produces a slight self-heating effect in the sensor, raising its temperature. This effect is negligible when measuring temperature in a heat sink as large as a papaya. Thermistors are less stable and considerably less expensive than RTDs.

Thermocouples consist of two wires made of dissimilar metals. Temperature measurement is based on the Seebeck effect. If a loop is made of these two wires and one junction is held at a different temperature than the other, a small current will flow. The voltage potential varies from 0.3 to 6.7 mV for a 100°C difference, depending on thermocouple type. Since only relative temperature differences can be measured, a reference temperature is required to measure the absolute value. Another temperature-measuring device, such as a thermistor or RTD, is normally provided within the data acquisition unit as a reference. The advantage of thermocouples is their low cost and ease of repair, though they are not as stable as RTDs.

Sensors are needed to measure the temperature of the air heating the commodity. Air temperature sensors are seldom moved, except for the occasional recalibration. They suffer little abuse, although it may be necessary that they withstand drenching during hydrocooling.

Papaya must have the temperature-sensing device located in the seed cavity to meet quarantine requirements. In this case, the sensing element is placed in a stainless steel sheath long enough to reach the fruit centre. Metal exposed to the heating air is not recommended as it may conduct heat to the sensing element. It is desirable that the commodity probe be capable of reading the temperature to within 0.1°C. Such accuracies are not normally attained without periodic calibration.

Temperature probes can be used to measure the relative humidity from the dry bulb and wet bulb temperatures. The dry bulb temperature is determined by placing a probe directly in the airstream. To measure the wet bulb temperature, a moist wick is placed over the sensing tip of a second probe held in the same airstream. More expensive sensors measure the electrical properties of a material which change with humidity, such as a hygroscopic capacitor, or the dew point of the air.

Chamber control

Good air temperature control is crucial to ensure that the commodity is treated properly and predictably without suffering heat damage. The purpose of a control system is to keep the difference between the actual and desired air temperatures within an acceptable window of error. A simple on/off control is unable adequately to maintain a proper temperature, particularly at the end of the run. More sophisticated control, where the heat source may be partially on, is needed. A PID equation is a common method of obtaining this type of control (Stoecker, 1989). The PID equation (Proportional, Integral, Derivative) produces a value which is proportional to either the heat or change in heat required. The computer passes this value to the data interface, which is then relayed to the appropriate control mechanism. Example A, below, shows how these values can be readily obtained from the air temperature data.

A proportional valve is a typical control mechanism. The valve regulates, for instance, the amount of steam passing through the exchanger and therefore the heating rate. Hot water pumped through a heat exchanger can be controlled using several methods. As with steam, a proportional valve may be used. In this case, the water pump will run at a constant speed with varying pressure and must be capable of operating even when the valve is completely closed. Another method is to eliminate the valve and vary the speed of the pump with a computer-controlled, variable speed motor.

As the air temperature rises, moisture must be added to maintain a constant relative humidity. Water can be sprayed directly into the chamber. As vaporizing water requires a significant amount of heat, it should be sprayed against a heat source, such as a heat exchanger, to increase vaporization. This will decrease the air temperature, due to the latent heat of vaporization, while increasing the relative humidity. However, it is difficult to obtain high humidities with this method. Introducing steam into the airflow is a better way to add moisture. Saturated steam is preferable to superheated steam which may add too much heat and make air temperature control difficult.

The required capacity of the humidity control system depends largely on the amount of air leakage from the chamber to the outside and the size of the chamber. If the chamber is perfectly sealed, the only water needed would be to maintain humidity as the air temperature is raised with perhaps additional moisture should the initial relative humidity be too low.

Operating protocols

The operating protocol controls the factors that affect the heating rate and maximum temperature of the fruit. This, in turn, affects the efficacy of the disinfestation and the quality of the fruit after treatment. The two main variables are the heating rate and the relative humidity level. If airflow can be reversed, then the airflow reversal schedule provides a third possible variable.

Air temperature is the major consideration affecting the heating rate. The air temperature must be high enough to ensure that the commodity reaches efficacy temperature, T_e, to kill the target pest. In addition, the air temperature must be low enough to prevent the fruit temperature from rising above its damage temperature, T_d, for a period longer than t_d.

The fastest way to bring the fruit to quarantine temperature is to maintain the air temperature at T_d for the entire duration of the treatment. However, to maintain fruit quality, a gradual temperature increase is recommended for most applications (Couey, 1989). The desired air temperature or set point, T_{set}, is typically selected to be a function of only the elapsed treatment time. A number of time–temperature curves are possible (Fig. 13.3). The first quarantine protocol (Armstrong *et al.*, 1989) for forced hot air treatment of papaya involved a step function (Fig. 13.3a). A relatively smooth heating curve using a number of small steps can be based on a mathematical equation provided that its maximum value is not allowed to exceed T_d. The mathematical function may take several forms, such as a linear or asymptotic function (Fig. 13.3b). An alternative control algorithm (Fig. 13.3c) is to adjust the air temperature according to the temperature of some or all of the probed fruit (coldest,

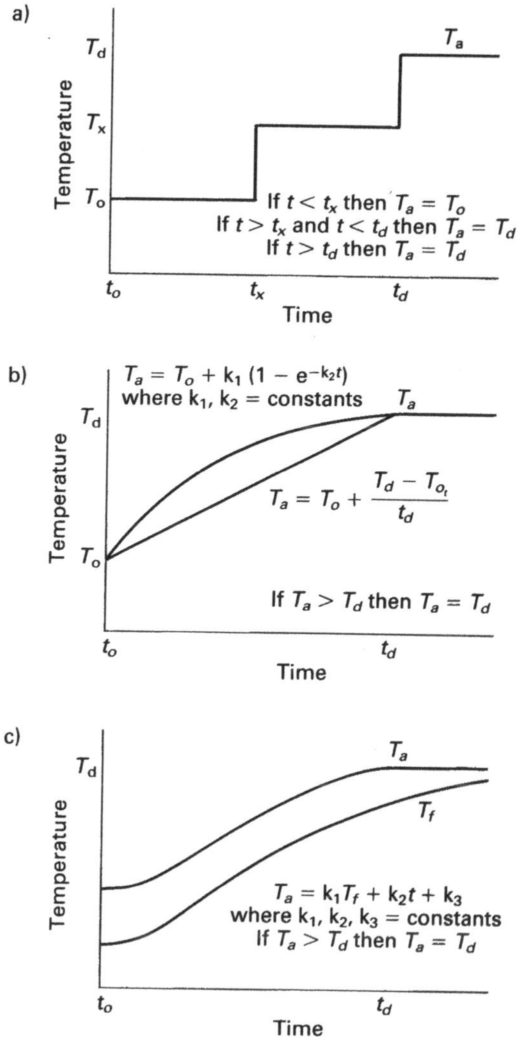

Fig. 13.3 Air temperature curves as: (a) a step function of time; (b) a smooth function of time; and (c) a smooth function of time and fruit temperature.

hottest, average, etc.). Treatments containing larger fruit would require more time for the air temperature to reach a given value. A constant difference can be maintained between the fruit and air temperatures until T_d is reached. However, this difference need not be constant over the treatment period. Air temperature may also be a function of both fruit temperature and time.

The relative humidity of the heating air affects fruit quality. Low relative humidity leads to fruit weight loss and shrivelling. Humidity levels close to saturation risk scalding of fruit if the water vapour condenses on the fruit surface. 'Vapour heat' (VH) disinfestation of papaya requires that the relative humidity be held above 90% for the final hour of treatment. High temperature forced air (HTFA) has no such requirement, although humidity control is still important to preserve quality. A 3.5 h treatment at 50% relative humidity causes a weight loss of about 2% (Paull and Chen, 1989). This weight loss increases the risk of fruit shrivelling occurring earlier in the marketing chain.

Once all the fruit has reached efficacy temperature, it must be cooled to prevent further heat damage. This is typically done using a cold water shower. The chamber must be capable of supplying large amounts of water evenly over the fruit surface. All interior chamber surfaces must be properly treated to prevent rust and corrosion. Any mechanical parts, such as used for loading and unloading, must be resistant to corrosion. To remove the water, the chamber floor must be equipped with drains fitted with either a water trap or valve.

Example A

The air temperature of a chamber is to be maintained by a variable speed pump whose pumping rate varies according to a 0–10 V DC signal from the computer (Table 13.1). The desired air temperature, T_{set}, in °C, varies linearly with the time, t, in minutes, according to Equation [13.1]

$$T_{set} = 35 + 0.075\,t \qquad [13.1]$$

The actual air temperature, T_{act}, measured at 2 min intervals, was 46.2, 46.3, 46.6, 46.85, and 47.2°C at 152, 154, 156, 158, and 160 min, respectively. The error, e, is the difference between the two temperatures ($T_{set} - T_{act}$). The PID equation is used to bring this error as close to zero as possible.

The general form of the PID equation is

$$O(t) = P\,e(t) + I \int_{t_l}^{t_u} e(t)\,\mathrm{d}t + D\,\frac{\mathrm{d}e(t)}{\mathrm{d}t} \qquad [13.2]$$

where $O(t)$ = output signal at time t, t = elapsed time, $e(t)$ = error at time t, P = proportional constant, I = integral constant, D = derivative constant, t_l = lower integral limit, t_u = upper integral limit.

Appropriate integral limits must be determined. Since the integral and derivative are calculated numerically, Equation [13.2] is more useful in the following form

Table 13.1. Tabulated values and calculated computer output to control a variable speed pump using a PID equation.

Reading i	Time t_i (min)	Temperature $T_{set\,i}$ (°C)	$T_{act\,i}$ (°C)	Error e_i (°C)	Integral $\int e_i\,dt$ (min °C)	Derivative de/dt (°C min^{-1})	Output O_i (V DC)
77	152	46.40	46.20	0.20			
					0.45		
78	154	46.55	46.30	0.25			
					0.35		
79	156	46.70	46.60	0.10			
					0.10		
80	158	46.85	46.85	0.00			
					−0.20	−0.10	7.00
81	160	47.00	47.20	−0.20			
							7.70

$$O_i = P(e_i) + I\sum_{n=i-k}^{n=i}(t_n - t_{n-1})\left[\frac{e_n+e_{n-1}}{2}\right] + D\left[\frac{e_i+e_{i-1}}{t_i-t_{i-1}}\right] + O_{i-1} \qquad [13.3]$$

where O_i = output signal at time t_i, t_i = ith time value, e_i = error at time t_i, $k+1$ = number of time intervals included in integral.

The output in this example ranges from 0 to 10 V DC. Notice the addition of the term O_{i-1}. If all error values are zero, the output will be the same as the previous output, rather than 0. The limits of the summation depend on the speed of the process. The previous two to four time periods should be sufficient. The values of the constants P, I, and D must be large enough to produce a relatively quick response, yet small enough to prevent wild oscillations.

Figure 13.4(a) shows the temperature data graphically, from which the error is calculated (Fig. 13.4b). If k =3, then the last four time intervals will be used in the summation and are represented by the four hatched areas of Fig. 13.4(b). The only derivative of interest is that of the last time interval. If P, I, and D are set to 3, 2, and 1, respectively, and the previous computer output, O_{i-1}, was 7.00 V DC, then the new computer output will be 7.70 V DC, thus increasing the pump speed. This pump speed will not change until 2 min later when the new temperature data are taken.

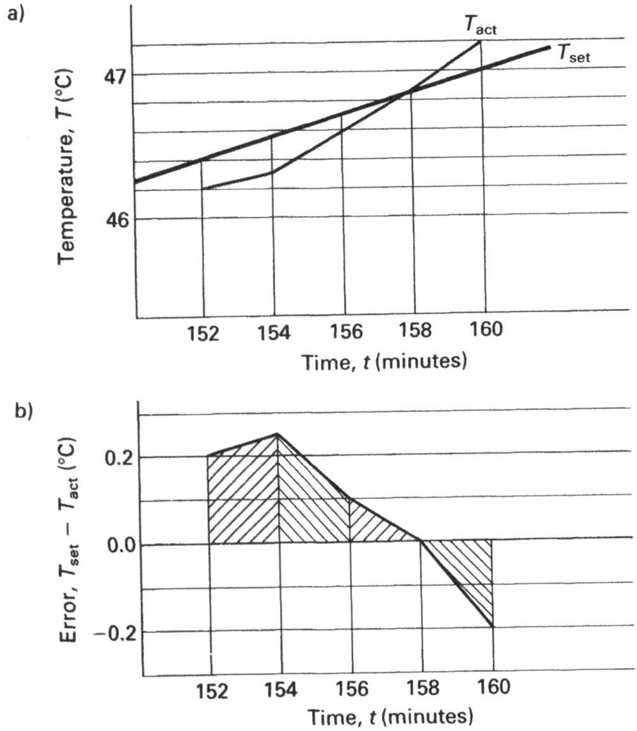

Fig. 13.4 Application of the PID equation: (a) desired and actual air temperatures as a function of time; and (b) error as a function of time.

Example B

The owners of a typical packing plant in Hawaii wish to disinfest papaya of fruit fly before shipment to the US mainland, with future expansion to Japan. Approximately 12 000 kg per week are packed on average during peak harvest season. Most of the fruit is grown on their own property, though 2000 kg are purchased from part-time farmers who bring their fruit to the packing plant at the weekend. Packing plant employees work on staggered shifts ranging from 7 a.m. to 7 p.m., 5 days a week. The suggested solutions are based on an upgraded version of a treatment chamber designed and built by the University of Hawaii and presently in commercial use on the island of Kauai.

The chamber(s) must be capable of handling the treatment load during the peak harvest season, which is 12 000 kg per week: 10 000 kg from the owners' own fields, and 2000 kg from independent growers. On two of the five workdays per week, Saturday and Sunday, 1000 kg from the farmers must be treated, along with 2000 kg of the owners' own fruit, leaving a maximum of 3000 kg per day.

A typical treatment time for papayas is about 6 h. Loading, hydro-cooling, and unloading add a further 2 h, leaving a total of 8 h per treatment. A major restriction is that workers are only available for 12 h each day, allowing the chamber to do only one treatment run per day.

Fruit handling

Bins reduce fruit-loading labour requirements to a minimum and are accepted for the treatment of fruit for shipment to the US mainland. A standard bin, 1.2 m by 1.2 m by 0.6 m deep, can hold approximately 415 kg of fruit when full. Slightly more than seven bins of fruit are needed to treat 3000 kg in a single treatment. Designing the chamber(s) to treat eight bins at a time provides a maximum capacity of 3320 kg when the bins are full.

Japan requires that fruit be treated in a single layer container, such as lugs. Though expensive, lugs have traditionally been custom designed. Hence, they can be chosen to fit in the chamber rather than the other way around. If the chamber needs to treat the same amount of fruit per day for the future Japan market, there must be sufficient vertical space to accommodate the additional volume required for a comparable amount of fruit in lugs. If the fruit is stacked vertically, the lugs should be at least 0.20 m high. Assuming each lug contains twenty 600 g fruit, five layers of six lugs will weigh 360 kg. This stack will be 1.00 m high. Vertical space will be allowed for lugs to be stacked on a pallet to a height of at least 1.2 m.

The loaded fruit stacks (lugs or bins) are normally brought to the chamber by forklift. A stack is placed on the ground, the probed fruit(s) are placed in their appointed location(s) by the inspector, and the fruit is loaded into the chamber. Most treatment chamber designs require that the stacks be pushed in, suggesting that rollers may be useful. These rollers may be idlers, requiring manual pushing/pulling, or motorized. However, rollers may allow the heating air to bypass the fruit. This problem can be eliminated by sliding fruit stacks along a flat surface to which a low friction material is attached, pushed by a chain conveyor. A limit switch automatically positions the stacks during loading.

Fan selection

In order to select an appropriate fan, the required airflow and the resistance of the fruit to this airflow must be determined. Laboratory tests determined that an airflow of approximately 0.005 $m^3\ s^{-1}$ is needed for every kg of fruit. In the proposed treatment chamber, each of the eight bins containing 415 kg of fruit will require an airflow of 2.07 $m^3 s^{-1}$. Since the sample of fruit used to derive the air resistance curve (Fig. 13.1 above) had a smaller cross-sectional flow area (0.81 m^2) than that of the standard bin (1.44 m^2), the design flowrate of 2.07 $m^3 s^{-1}$ must be reduced proportionally. The airflow that applies to Fig. 13.1 is 1.16 $m^3 s^{-1}$ since, blowing 2.07 $m^3 s^{-1}$ of air across an area of 1.44 m^2 is equivalent to blowing 1.16 $m^3 s^{-1}$ of air across a flow area of 0.81 m^2.

The pressure drop across the fruit is then estimated for this flowrate. At a flowrate of 1.16 $m^3 s^{-1}$ and a fruit depth of 1.2 m, based on Fig. 13.1, the pressure drop is approximately 380 Pa. Doubling the fruit depth doubles the pressure drop for a given airflow. Therefore, the pressure drop across a fruit depth of 0.60 m in a standard bin, proportionately adjusted, is 190 Pa.

Alternatively, the curve fitted to the points in Fig. 13.1 can be described by the following equation

$$P = 284\, Q^{1.87} \qquad [13.4]$$

where P = pressure drop across the fruit (Pa), Q = airflow through the fruit ($m^3 s^{-1}$). However, this equation is valid for a flow area of 0.81 m^2 and a depth of 1.2 m. The equation can be generalized by replacing Q with VA, where A is the cross-sectional flow area, and V is the air velocity. To express the pressure per unit depth, simply divide the coefficient by the fruit depth. The resulting equation is

$$P = 160\, V^{1.87}$$

where P_d = pressure drop across the fruit (Pa m^{-1}), V = apparent air velocity through the fruit (m s^{-1}).

Note that the air velocity is only an apparent velocity since the actual area through which the air flows is much less than the 0.81 m^2 of fruit area. The apparent velocity of air through the fruit in the design bin is

$$\frac{2.07\ m^3\ s^{-1}}{1.44\ m^2} = 1.44\ m\ s^{-1}$$

The pressure drop for 1 m of papayas is

$$P = (160)\,(1.44^{1.87})$$
$$= 316\ Pa\ m^{-1}$$

For a 0.60 m depth of papayas, the pressure drop is

$$P = (316\ Pa\ m^{-1})\,(0.60m)$$
$$= 190\ Pa$$

The equation yields the same result as the graphical method.

Since a bin is loaded from the top and hydrocooling water is drained from the bottom, this commodity container is best suited to vertical airflow. The direction of vertical airflow should be selected to facilitate the placement of the probed fruit. The probes are normally placed in the coldest locations in the chamber. If the air flows from the top to the bottom of the fruit stack, the probed fruit must be placed on the bottom, requiring that additional fruit be displaced to allow access to this location. On the other hand, if the air flows from the bottom to the top, the probed fruit can be placed on the top of the bin, the location with the greatest accessibility. Although reversing the airflow effectively reduces the temperature variation among the fruit, it becomes difficult to place the probed fruit properly since the coldest location will vary throughout the run.

The bin/fan set-up determines the type and number of fans required. If one bin is stacked on top of another, the pressure drop will double for the same airflow. However, since the amount of fruit has doubled, the required airflow must also double with no increase in the cross-sectional flow area. Doubling the flowrate will cause the pressure drop to increase to $(190)(2^{1.87})$, or 695 Pa for the proposed bin. A centrifugal fan would be the logical choice. In addition, due to the large depth of fruit, air reversal would probably be needed, requiring the installation of dampers. To eliminate the need of air reversal and/or huge fans, bin stacking should be avoided.

If fan performances are identical, air distribution improves with the number of fans. There is, of course, a practical limit. If more than one fan is used, heating the air evenly may be difficult. A chamber capable of treating four bins at a time is quite manageable with one fan, provided the fan is properly located. Installing an axial fan horizontally between two pairs of bins should produce a uniform airflow through the fruit. However, two fans and two heat sources would allow a larger chamber to be divided into two independently operated and controlled units, each approximately 6.0 m long, 1.4 m wide, and 2.5 m high. This would allow greater flexibility in treatments, as well as redundancy to reduce the likelihood of a complete shutdown of the system. Units can be used simultaneously during peak production season or singly when production is lower.

Assuming all losses occur within the fruit mass, the fan must be capable of delivering 8.3 $m^3 s^{-1}$ at a pressure of 190 Pa for four bins. However, the fan must also overcome the pressure losses associated with the tortuous path of the airflow to and from the fruit. Heat exchangers and the perforated walls of the fruit containers also increase the fan requirements. The difficulty of accounting for all the losses demands that an estimate be made. Assuming these losses are equal to those of the fruit, a fan capable of delivering 8.3 $m^3 s^{-1}$ at a pressure of 380 Pa should be installed. As a second means of estimating fan size, the flowrate can be doubled, i.e. a fan capable of delivering 16.6 $m^3 s^{-1}$ at a pressure of 190 Pa. This second method is only valid if fan efficiency does not drop drastically at the higher pressures. Undersizing the fan may cause the fruit to heat slowly so that fruit in the 'cold spots' significantly lag behind the hotter fruit. On the other hand, if the fan is oversized, its frictional heat may force the air temperature to increase beyond control.

Heat balance

The heat load of the fruit will vary throughout the treatment and is proportional to the fruit temperature curve during heating. The point of maximum heat load required by the fruit occurs at the point of greatest temperature rise, the derivative of the temperature with respect to time. The maximum temperature rise (Fig. 13.2) occurs approximately 90 min into the treatment and is 0.13°C min^{-1}. Assuming the heat capacity of papayas is the same as water (4184 J kg^{-1} °C^{-1}), the maximum heat load is:

$$\frac{(0.13°\text{C min}^{-1})(4184\text{ J kg}^{-1}\text{ °C}^{-1})(3320\text{ kg})}{(60\text{ s min}^{-1})} = 30\ 097\text{ J s}^{-1} \qquad [13.5]$$

$$= 30.1\text{ kW}$$

At the end of the treatment, where the fruit temperature rise is only 0.025°C min^{-1}, the heat load is 5.8 kW. Assuming that the fruit start at a temperature of 25°C and finish at 47.2°C, the average heat load over a 6 h treatment is 14.3 kW.

The chamber losses occur due to the heat transfer through the walls. The general equation is:

$$Q_w = \frac{A(T_i - T_o)}{R}$$

where Q_w = heat loss through the walls (J s^{-1}), A = area of wall (m^2), T_i = temperature inside the chamber (°C), T_o = temperature outside the chamber (°C), R = thermal resistance through wall (m^2 s °C J^{-1}).

If the inside temperature is greater than the outside temperature, then Q_w will be positive, meaning that heat is lost from the inside to the outside. The total wall area, A, of the two-unit chamber 6 m by 2.5 m by 2.8 m is 77.6 m^2. However, since heat lost through the floor is minimal, the area contributing to heat loss is only 60.8 m^2. The thermal resistance through the wall is additive and is made up of three components: inside convection, conduction through the wall, and outside convection. Assuming no insulation is used, the resistance of the steel wall is negligible compared to the convective components and need not be considered. The convective components are dependent on the speed of the air past the wall surface. The convective component inside the chamber is constant provided that airflow through the fan does not vary. Although inside airspeed varies from one location in the chamber to another, assume the average is the same as the apparent velocity through the fruit, 1.44 m s^{-1}. At this speed, the convective resistance is approximately 0.070 m^2 s °C J^{-1}.

Heat loss will be greatest when ambient outside airspeed is high, and ambient outside temperature is low. Assume the chamber air temperature is at its maximum, 48.5°C. A minimum ambient temperature in Hawaii would be about 18°C. An ambient airspeed of 6.7 m s^{-1} will give an *R* value of 0.030 m^2 s °C J^{-1} for the outside wall convection. Thus, the maximum heat loss through the walls is:

$$\frac{(60.8\ \text{m}^2)(48.5 - 18.0°\text{C})}{(0.07 + 0.030)\ \text{m}^2\ \text{s}\ °\text{C}\ \text{J}^{-1}} = 18.5\ \text{kW}$$

Minumum heat loss occurs when ambient airspeed is low, and ambient temperature is high. A maximum ambient temperature in Hawaii is about 32°C. An ambient airspeed of 3.35 m s^{-1} will give an *R* value of 0.044 m^2 s °C J^{-1} for the outside wall convection. Thus, the minimum heat loss through the walls is approximately 8.8 kW. The maximum heat load will occur when the fruit temperature rise is at its greatest and and heat loss through the walls is at its maximum, or 48.6 kW.

A significant amount of heat will come from the fan(s). Approximately 15 kW of fan power will be needed to provide the airflow to all 3320 kg of fruit at the determined pressure drop. Mounting the fan motor(s) outside the chamber reduces the amount of heat added within the chamber, since waste heat from the motor(s) is directly exhausted to the atmosphere. Centrifugal fan motors are typically mounted on the outside of the airstream, whereas those of axial fans are often mounted in the airstream. If hydrocooling is to be done within the chamber, direct drive axial fans require a motor capable of withstanding a wet environment.

The amount of heat generated by the fan is crucial in determining if the air will overheat. The smallest heat load demanded by the fruit is 5.8 kW. The minimum chamber heat loss is 8.8 kW, for a total loss of 14.6 kW. In the worst case scenario, 0.4 kW will be added to the air. As the chamber temperature rises, the heat loss through the chamber walls will also increase, until the temperature reaches 49.3°C, at which point the heat input will equal the heat loss. Hence, it will be difficult, if not impossible, to control the air temperature adequately on hot, calm days. The use of insulation will aggravate the problem. Installing a chamber inside a large warehouse with little air circulation is not desirable as the lower airspeed around the outside of the chamber will decrease the heat dissipated. For still air, the *R* value will increase to 0.106 m^2 s °C J^{-1}, thus decreasing the chamber heat losses. On the other hand, installing the chamber outside may increase the chances of overheating if it is exposed to direct sunshine. Sheltering the chamber with a roof or painting the chamber white will reduce the amount or radiant heat added and allow for better temperature control.

The greatest load is 48.6 kW. Average values can be used to calculate the mean energy consumption per kg of fruit. The average air temperatures are approximately 43°C inside the chamber and 27°C outside. Assume a total R value of 0.10. The average loss through the walls is then 10.9 kW. The average heat consumption of the fruit is 14.3 kW, resulting in a total average power requirement of 25.2 kW. Over the 6 h period, the total energy consumption is 151.2 kW h. Assuming energy costs $0.15 per kW h, the average cost per treatment is $22.68, or 0.7c per kg. The fans contribute 90 kW h, or 60% of the total energy input. The cost of the remaining 61.2 kW h can be reduced by providing the supplemental heat from a different energy source, such as propane. Propane contains about 50 400 kJ kg^{-1} and costs $0.55 kg^{-1}. Assuming an efficiency of 70%, 6.24 kg, or about $3.43 worth, are needed per treatment for supplemental heating, compared to $9.18 for electricity, a saving of $5.75. If 200 treatments are done every year, the annual savings will be $1150.

Computer monitoring and control

A single computer is adequate to control the eight-bin chamber. A printer is required to allow the quarantine inspectors to view the fruit temperatures, as well as provide a hard copy of the completed treatment. The data acquisition interfaces must be capable of monitoring eight fruit temperature probes, and at least two air temperature probes. Duplicate air probes are advisable. If a separate relative humidity meter is not used, additional probe(s) will be required to measure the wet bulb temperature. An analog output channel is necessary to control the heat supply, and possibly another for the relative humidity control. A relative humidity meter must be read using an analog input channel. A digital output will be sufficient to control the relative humidity with an on/off switch. Additional digital outputs can be employed to sound an alarm, turn the fans on and off, as well as commence the hydrocooling.

Treatment chambers in Hawaii certified for shipment to the US mainland normally require one fruit probe per bin. These probes should be RTDs, as accuracy is important to ensure that quarantine standards are met, and can be expected to maintain their calibration over long periods. The RTD sensor should be placed in a stainless steel sheath long enough to reach the centre of the seed cavity, and short enough so that no metal is exposed to the heating air. For Hawaiian papaya, the sheath should be between 90 and 120 mm. The probe wires must be located to prevent any interference with the mechanical loading system. All probes must be periodically recalibrated. Unsheathed thermocouples are sufficient for the measurement of air temperatures.

The relative humidity can be measured using either an electronic meter, provided its sensing tip does not come into direct contact with water, or wet/dry bulb system, provided the wet bulb wick is constantly kept wet.

Supplemental heat can be supplied by a propane heater and heat exchanger. The required flowrate of hot water depends on the exposed area of the heat exchanger, the air velocity across it and the temperature difference between the air and the water. A water flowrate of 100 l min^{-1}, with a temperature drop of 4.8°C, will supply the maximum 33.6 kW of supplemental heat. This temperature drop should not be difficult to achieve with the high air velocities passing the heat exchanger.

The flowrate from the hot water heater may be controlled by using either a proportional valve or a variable speed pump. Proportional valves tend to be very expensive. A variable speed pump is an inexpensive alternative. By driving the pump with a DC motor and using a controller that varies a DC voltage according to a small, PID-determined voltage signal from the computer, good air temperature control can be achieved.

To ship papaya to Japan, relative humidity greater than 90% must be achievable during the treatment. Small water particles are difficult to vaporize at high humidities, so spraying water on a heat exchanger is not reliable, necessitating a steam generating system. Non-pressurized steam vessels, which only produce steam on demand, need no valve but require controlling a high electrical current. Steam from a pressurized vessel releases steam on demand, and can be controlled using a simple valve. A solenoid valve can be computer operated via a digital output activating an inexpensive electronic relay. The steam should enter the chamber slowly to prevent any sudden air temperature rises, but quickly enough to ensure that the humidity rises to the desired value. This can be accomplished using a properly sized nozzle where the steam enters the chamber.

The size of the steam generator is largely a function of air leaks from the chamber. Fruit will tend to lose moisture rather than gain it. Air at 60% RH and 24°C will need 0.027 kg water per kg dry air to reach 48.5°C at 90% RH. If the air has a specific volume of 0.875 m^3 kg^{-1} dry air, and the chamber contains approximately 42 m^3 of dry air, then the amount of water needed per chamber full of air is approximately 1.3 kg. It is difficult to determine the number of air changes per hour. At two changes per hour, the steam generation unit must be capable of producing 2.6 kg h^{-1}. A unit of this size is a sizeable investment. To ensure that high humidity levels can be achieved, the steam generator should be capable of producing at least 3 kg h^{-1}.

A simple on/off method can be used to control the relative humidity in the chamber as it is much less critical than the control of the air temperature. An appropriate nozzle size can be chosen that allows the steam to be injected for 1 or 2 minutes without changing the humidity more than a few percentage points. A window of 5% can maintain the humidity at an acceptable level. For instance, the steam valve can be turned on when the humidity drops below 55%, and turned off when it exceeds 60%.

The complete data acquistion system for the eight-bin chamber must have at least eight RTD inputs and four thermocouple inputs and an analog output, an analog input, and a digital output. Back-up input/output channels reduce the probability of complete treatment failure. Additional channels allow room for expansion, or a more sophisticated control system. A printer must produce a continuous hard copy of each treatment for the quarantine inspector.

Operating protocols

The air temperature protocol must be matched to the tolerance of the fruit and follow an acceptable schedule. The efficacy temperature, T_e, of Hawaiian papaya is 47.2°C (Jang, 1986), with a corresponding time, t_e, of essentially 0. The damage temperatures and corresponding times have not been well established. A maximum air temperature of 48.5°C in the fourth hour of treatment has given good, though not definitive, results for Hawaiian papaya (Williamson and Winkelman, unpublished data).

If the treatments are done using 'dry heat', the relative humidity protocol is left to the discretion of the packer. High levels of relative humidity may be difficult to maintain if leaks are present . Lower levels may result in fruit moisture loss. For the Japanese market, relative humidity must be maintained above 90% for the final hour.

Large amounts of water will be needed to cool the fruit, *ca*. 150 l min^{-1} bin^{-1}. It is important that all fruit are adequately exposed to water. If sufficient water pressure is available, spray nozzles may be suitable. Perforated pipes are best suited for lower pressures. Manual starting of hydrocooling is acceptable, although an automated valve would allow for unattended night-time treatments. The drain and associated piping should contain a water trap or check valve and must be large enough to convey the large volume of water from the chamber. If not, the chamber may become flooded. If water is in short supply, the water can be cooled in a cooling tower and recirculated through the chamber by a pump.

Summary

To treat fruit to quarantine standards, the efficacy temperature must be established, and at least an estimate of the damage temperature. The airflow required to obtain adequate heating, and the corresponding pressure drops determine the size of the fan(s), which should be installed to ensure a uniform airflow through the fruit. The cost of energy and its corresponding control system should be considered before deciding on a heat source. If high humidity is required, the chamber should be essentially airtight, and water vapour should be added in the form of steam. Data interfacing should allow for expansion and redundancy.

References

Abeles, F.W., Morgan, P.W. and Saltveit, M.E. (1992) *Ethylene in Plant Biology*, 2nd edn. Academic Press, New York.

Animal and Plant Health Inspection Service (APHIS) (1990) *Plant Protection and Quarantine Treatment Manual.* United States Department of Agriculture.

Armstrong, J.W., Hansen, J.D., Hu, B.K.S. and Brown, S.A. (1989) High-temperature forced-air quarantine treatment for papayas infested with tephritid fruit flies. *Journal of Economic Entomology* 82, 1667–1674.

Couey, H.M. (1989) Heat treatment for control of postharvest diseases amd insect pests of fruit. *Horticultural Science* 24, 198–202.

Jang, E.B. (1986) Kinetics of thermal death in eggs and third instars of three species of fruit flies (Diptera: Tephritidae). *Journal of Economic Entomology* 79, 700–705.

Japan MAF (1985) Letter to USDA-APHIS on Hawaiian Host Company, Hilo, Hawaii, Certification Procedures.

Paull, R.E. (1990) Postharvest heat treatments and fruit ripening. *Postharvest News and Information* 1, 355–363.

Paull, R.E. and Chen, N.J. (1989) Papaya postharvest studies. *Proceedings 25th Annual Papaya Industry Association Conference*, pp. 62–63.

Stoecker, W.F. (1989) *Design of Thermal Systems*. McGraw-Hill, New York.

Williamson, M. and Winkelman, P. (1992) Certification of fruit fly disinfestation chambers in Hawaii. *Computers in Agricultural Extension Programs*, January.

Williamson, M., Winkelman, P. and Armstrong, J. (1991) Certification of fruit fly disinfestation chambers. Presented at American Society of Agricultural Engineers, December.

Winkelman, P. and Williamson, M. (1990) Advances in commercial dry heat disinfestation of papaya. Presented to the American Society of Agricultural Engineers, June.

IV Individual Product Treatments

Tropical and Subtropical Fruits

14

JOHN W. ARMSTRONG
United States Department of Agriculture,
Agricultural Research Service,
Tropical Fruit and Vegetable Research Laboratory,
PO Box 4459, Hilo, HI 96720, USA.

Introduction

Tephritid fruit flies, especially the Mediterranean fruit fly, *Ceratitis capitata* (Wiedemann), are among the most important quarantine pests of fresh subtropical and tropical fruits on a worldwide basis (Gilmore, 1983). The earliest report on disinfestation of tropical or subtropical fruits for quarantine purposes was a refrigeration treatment for citrus against the Mediterranean fruit fly (Hooper, 1907). The initial reports of other disinfestation treatment technologies against fruit flies include, in chronological sequence of development, vapour heat treatment (Baker, 1952), methyl bromide (MB) fumigation (United States Department of Agriculture, 1940), ethylene dibromide (EDB) fumigation (United States Department of Agriculture, 1951), postharvest insecticide immersions (Saunders and Elder, 1966), irradiation (Balock *et al.*, 1966), hot-water immersion treatment (Armstrong, 1982), and high-temperature forced-air treatment (Armstrong *et al.*, 1989). Except for irradiation, which was approved as a quarantine treatment against fruit flies (Anonymous, 1986), all of these treatment technologies have been used to disinfest subtropical and tropical fruits of quarantine pests.

Descriptions of EDB fumigation schedules are given to provide an historical view rather than for practical application in postharvest quarantine treatments. The US Environmental Protection Agency cancelled the registration of EDB for postharvest fumigation of fruit destined for the domestic market, effective 1 September 1984 (Ruckelshaus, 1984). At the time of this writing, Australia (Anonymous, 1985) and New Zealand (Anonymous, 1992) are preparing to lower the maximum allowable EDB

From: Paull, R.E. and Armstrong, J.W. (eds) (1994) *Insect Pests and Fresh Horticultural Products: Treatments and Responses.* CAB INTERNATIONAL, Wallingford, UK.

residue to 0.1 ppm. As 0.1 ppm is well below the normal residue levels found in EDB-fumigated fruit, this regulation will effectively halt the use of EDB fumigation of fruit destined for their domestic markets. Additionally, there is a possibility that MB use could be limited because it is a possible agent of atmospheric ozone depletion (United Nations Environment Programme, 1992).

Quarantine treatment 'schedules' are developed by a variety of agricultural research agencies worldwide. Regulatory agencies, such as USDA-Animal and Plant Health Inspection Service, Japan Ministry of Agriculture, Forestry and Fisheries, New Zealand Ministry of Agriculture and Fisheries, Australian Quarantine and Inspection Service, and others, are responsible for approving, regulating, and certifying quarantine treatments and quarantine treatment facilities. An approved treatment schedule consists of all the conditions, parameters, and procedures needed to meet a specific quarantine treatment. When quarantine treatments are done in the exporting country, they may be carried out either under the supervision of a regulatory official from the importing country, under the supervision of a regulatory official of the exporting country, or under a 'compliance agreement' that may or may not include regulatory supervision to ensure that the quarantine treatment schedule was followed correctly.

Quarantine Treatment Schedules

A selection of currently approved quarantine treatment schedules, or the results of research on possible quarantine treatment schedules reported in the literature, for some subtropical and tropical fruits are presented. Although not all-inclusive, the selection shows the existence of a broad range of treatments.

Many quarantine treatments are no longer applicable because: (i) the technology has become legally restricted (e.g. ethylene dibromide); (ii) the host/pest relationship has changed; (iii) a new pest that is more tolerant to the applicable treatment has been identified. The pests that quarantine treatments are intended to control and their hosts are as dynamic as the ecosystems in which they live. Failure of a quarantine treatment, system, or protocol can occur for any number of reasons. For example, a quarantine system using fruit selection followed by a two-stage hot-water immersion to disinfest tephritid fruit flies from Hawaii-grown papaya exported to the US mainland (Couey and Hayes, 1986) failed when a genetic deformity at the fruit blossom end allowed fruit flies to survive the treatment (Zee *et al.*, 1989; Liquido, 1990). There are examples of quarantine treatments that have lasted for over 60 years, for 6 years, or for as little as 6 months.

The regulatory agencies requiring each quarantine treatment schedule presented below are identified. However, it is important to note that the geographical areas from which the fruit may be exported are not always identified, and the reader should not assume that any given treatment for a fruit is applicable to all potential origins of export for that fruit. One or more of the origins of potential export may be limited or prohibited by regulatory restrictions due to the presence of insects that can survive the given quarantine treatment schedule or for which there is no available quarantine treatment, or because plant pathogens of quarantine importance are present. Quarantine treatment manuals include lists of countries and fruits for which treatment schedules are applicable or not authorized (i.e. exceptions to the treatment schedules).

All EDB and MB fumigation treatment schedules discussed below are at normal atmospheric pressure (NAP). Regulations pertaining to load factor limits (e.g. up to 50% fruit load in the fumigation chamber) for EDB and MB fumigations were omitted on purpose to save space.

Avocado, **Persea americana** *Mill*

Two quarantine treatments to disinfest Mediterranean fruit fly, oriental fruit fly (*Bactrocera dorsalis* (Hendel)), and melon fly (*B. cucurbitae* (Coquillett)) infesting avocado cultivars from Hawaii are: (i) MB fumigation with 32 g m^{-3} methyl bromide for 4 h at 21°C or above. The treatment is considered marginal as to host tolerance and may damage the fruit (Animal and Plant Health Inspection Service, 1992a); (ii) MB fumigation plus refrigeration that consists of fumigation with 32 g m^{-3} MB at 21°C or above: for 3 h followed by refrigeration at 6.1–8.3°C or 8.9–13.3°C for 3 or 6 days, respectively; or for 2.5 h followed by refrigeration at 1.1–2.8°C, 5.0–8.3°C or 8.9–13.3°C for 4, 6 or 10 days, respectively; or for 2 h followed by refrigeration at 0.6–2.8°C or 3.3–8.3°C for 4 or 11 days, respectively. The treatment is considered marginal at the 3 h fumigation and may damage the fruit (Animal and Plant Health Inspection Service, 1992r).

A quarantine treatment to disinfest fruit flies from avocado cultivars exported from Fiji to New Zealand consists of an EDB fumigation for 2 h with 22 g m^{-3} EDB at 20°C or above; 29 g m^{-3} EDB at 15.5–19.5°C; or 33 g m^{-3} EDB at 10–15°C (New Zealand Ministry of Agriculture and Fisheries, 1992f).

Banana, **Musa** *species and cultivars*

Bananas generally are not a host for Mediterranean fruit fly (Back and Pemberton, 1916; Umeya and Yamamoto, 1971; Armstrong, 1983), melon fly or oriental fruit fly (Armstrong, 1983), or other tephritid species (A. Allwood, R. Drew, and N. Heather, personal communications). The exception is the banana fruit fly, *Bactrocera musae* (Tryon), a major pest of bananas in Queensland, Australia, and Papua New Guinea, which will oviposit in green bananas (Drew *et al.*, 1978).

To disinfest bananas of surface feeders, such as noctuid moths, thrips, or *Copitarsia* species, fumigation is required using 24, 32, 40, 48, or 64 g m^{-3} at 27.7 or above, 21.2–26.1, 15.6–20.6, 10.0–15.0, or 4.4–9.4°C, respectively (Animal and Plant Health Inspection Service, 1992b,k).

New Zealand requires bananas from Australia to be immersed in a dimethoate solution at 500 ppm active ingredient for 30 s at a minimum flesh temperature of 18°C (New Zealand Ministry of Agriculture and Fisheries, 1992a), and bananas from the Philippines to be immersed in a fenthion solution at 100 g active ingredient in 100 l of water (New Zealand Ministry of Agriculture and Fisheries, 1992b) to remove quarantine pests from the fruit. Bananas from either country must be shipped in the unripe condition (New Zealand Ministry of Agriculture and Fisheries, 1992a,b).

Australia requires fumigation with 18 g m^{-3} EDB at 21°C or above to disinfest *Bactrocera* species from the fruit (Australian Quarantine and Inspection Service, 1985, 1988).

Breadfruit, Artocarpus communis *Forst.*

Breadfruit imported into New Zealand from Fiji must be disinfested using the same quarantine treatment schedules applied to bananas imported from Fiji (New Zealand Ministry of Agriculture and Fisheries, 1992f).

Carambola (starfruit), Averrhoa carambola *L.*

Carambola imported into New Zealand from Fiji must be disinfested using the same quarantine treatment schedules applied to bananas imported from Fiji (New Zealand Ministry of Agriculture and Fisheries, 1992f).

A quarantine refrigeration treatment to disinfest Florida-grown carambolas of Caribbean fruit fly, *Anastrepha suspensa* (Loew), was developed by Gould and Sharp (1990). The treatment consists of refrigeration at 1.1°C for 15 days and is used for carambolas exported to other areas in the US where *A. suspensa* does not occur (Gould and Sharp, 1990). Other quarantine treatment methods to disinfest Caribbean fruit fly from carambola have appeared in the literature, including a hot-water immersion treatment at 46.2 ± 0.2°C for 45 min (Hallman and Sharp, 1990), a vapour heat treatment until the fruit centre temperature reached 46°C and was held at that temperature for 35 min or more (Hallman, 1990), and a 47 ± 0.2°C high-temperature forced-air (or forced hot-air) treatment until the fruit centre temperature reached 45.5°C (Sharp and Hallman, 1992).

Cherimoya, Annona cherimola *Mill*

Cherimoya imported into New Zealand from Fiji must be disinfested using the same quarantine treatment schedules applied to bananas imported from Fiji (New Zealand Ministry of Agriculture and Fisheries, 1992f).

Cherimoya must be fumigated with MB to disinfest the fruit of the Chilean false red mite of grapes, *Brevipalpus chilensis* Baker, before entry into the US. The MB fumigation schedules are the same as those applied to disinfest bananas of surface feeders except there is no approved fumigation for temperatures below 50°C (Animal and Plant Health Inspection Service, 1992d). Another treatment to disinfest cherimoya of *B. chilensis* consists of immersing the fruit for 20 s in a soapy water bath of one part soap solution to 3000 parts water followed by a pressure shower rinse to remove all soapy excess, then immersion for 20 s in an undiluted fruit wax coating (Animal and Plant Health Inspection Service, 1992m).

Citrus *species and cultivars*

This large genus includes many economically important species and cultivars that are exported, including grapefruit, *C. paradisi* Macfad.; lemons, *C. limon* (L.) Burm. f.; limes, *C. aurantifolia* (L.) Swingle; mandarins, *C. reticulata* Blanco; pummelo, *C. grandis* Osbeck; and sweet oranges, *C. sinensis* (L.) Osbeck. Accordingly, there is a wide range of quarantine treatment schedules applicable to different species and cultivars for different quarantine pests.

Grapefruit entering the US from some areas must be fumigated using 24–28 g m^{-3} MB, depending on temperature, to disinfest the fruit of citrus blackfly, *Aleurocanthus woglumi* Ashby (Animal and Plant Health Inspection Service, 1992g).

Grapefruit, oranges, and tangerines from countries where Mediterranean fruit fly occurs must be refrigerated at 0, 0.6, 1.1, 1.7, or 2.2°C for 10, 11, 12, 14, or 16 days, respectively, to disinfest the fruit before entry into the US (Animal and Plant Health Inspection Service, 1992q). Similar refrigeration treatments are required for citrus from Mexico to disinfest the fruit of Mexican fruit fly, from countries with species of *Anastrepha* other than Mexican fruit fly, and from Australia to disinfest the fruit of Queensland fruit fly (Animal and Plant Health Inspection Service, 1992q).

Grapefruit and sweet oranges entering the US from Mexico must receive a quarantine treatment for Mexican fruit fly, *A. ludens* (Loew). With the loss of EDB (Ruckelshaus, 1984), the only available quarantine treatment for Mexican fruit fly was a vapour heat treatment originally developed by Hawkins (1932). The vapour heat treatment consists of raising the fruit pulp temperature gradually to 43°C until the centres of the fruit equilibrate at that temperature in 8 h, then the fruit centre temperatures must be held at 43°C for 6 h (Animal and Plant Health Inspection Service, 1992p). Williamson *et al.* (1986) developed a methyl bromide fumigation schedule similar to that used for citrus blackfly that is presently the preferred quarantine treatment (Mangan and Ingle, 1993). However, a recently developed two-stage (40°C air for 2 h followed by 50°C air for 1.5 h) forced hot-air quarantine treatment that raises the fruit centre temperatures to 48°C was proven efficacious in disinfesting Mexican fruit fly from grapefruit and eventually may replace the MB fumigation (Animal and Plant Health Inspection Service, 1992n; Mangan and Ingle,1993). A similar forced hot-air treatment was developed by Shellie *et al.* (1993) to disinfest fruit flies from tangerines.

Japan requires a 13-day refrigeration treatment at 0.5°C for grapefruit from Israel to disinfest the fruit of Mediterranean fruit fly (Japan Ministry of Agriculture, Forestry and Fisheries, 1980). Japan requires similar refrigeration treatment for sweet oranges, lemons and grapefruit from Africa (Japan Ministry of Agriculture, Forestry and Fisheries, 1973) and for oranges from Taiwan (Japan Ministry of Agriculture, Forestry and Fisheries, 1980). For grapefruit from Florida, Japan requires compliance with an area freedom programme through trap-monitoring systems for Caribbean fruit fly and proscribed identical to the grapefruit harvest and export protocol that is re-negotiated each year between the states of Florida and California (Voss, 1990).

New Zealand requires either a refrigeration treatment at 1.0±0.6°C for 16 days or fumigation with 18, 25, or 32 g m^{-3} EDB at 21°C or above, 15.5–20.5°C, or 10.0–15.0°C, respectively, to disinfest citrus from Australia of Mediterranean fruit fly and Queensland fruit fly (New Zealand Ministry of Agriculture and Fisheries, 1992d,h).

Australia requires fumigation of citrus with 18 g m^{-3} EDB for 2 h at 21°C or above, or refrigeration at 0±0.5°C for 14 days to disinfest fruits from countries where *Anastrepha* species, Mediterranean fruit fly, or oriental fruit fly occur (Australian Quarantine and Inspection Service, 1985, 1988). A 48 g m^{-3} MB fumigation for 2 h at 21°C is required for citrus infested with external insect pests, although the treatment may damage the fruit (Australian Quarantine and Inspection Service, 1985, 1988).

Grapes, **Vitis vinifera** *L.*

Grapes entering the US must be disinfested of the grapevine moth, *Lobesia botrana* (Schiff.), and other insect pests using the same MB fumigation schedules applied against surface feeders on bananas (Animal and Plant Health Inspection Service, 1992e,f), and of fruit flies using a cold treatment schedule combined with a MB fumigation (Animal and Plant Health Inspection Service, 1992q,r).

New Zealand requires grapes imported from Australia to receive the same cold treatment schedules used to disinfest citrus of fruit flies (New Zealand Ministry of Agriculture and Fisheries 1992d). Grapes imported into New Zealand from the US must be fumigated with 24, 32, 40, or 48 g m^{-3} MB at 26–30, 21–26, 15–21, or 10–15°C, respectively, to disinfest various insect pests from the fruit (New Zealand Ministry of Agriculture and Fisheries, 1992g).

Guava, **Psidium guajava** *L.*

Guava is a relatively new entry into export markets for fresh tropical fruits so there are few available quarantine treatments.

New Zealand requires the same EDB fumigation for guava imported from Fiji that it does for avocado, carambola, and cherimoya (New Zealand Ministry of Agriculture and Fisheries, 1992f).

Recently, Gould and Sharp (1992) developed a hot-water immersion quarantine treatment for guavas infested with Caribbean fruit fly consisting of immersing fruit in water at 46.1±0.5°C for 35 min with only a slight reduction in quality.

Kiwifruit, **Actinidia chinensis** *Planch.*

Kiwifruit entering the US from Chile or Greece must receive a cold treatment using the same treatment schedule as for loquat from Israel to disinfest Mediterranean fruit fly from the fruit (Animal and Plant Health Inspection Service, 1992q). Kiwifruit entering the US from Australia must also receive a cold treatment, but the treatment schedule required is more severe to disinfest the fruit of Queensland fruit fly (Animal and Plant Health Inspection Service, 1992q). For external and surface feeders, the same MB fumigation schedule used to disinfest external and surface feeders from bananas is required (Animal and Plant Health Inspection Service, 1992h).

Litchis, **Litchi chinensis** *Sonn.*

Japan requires litchis from Taiwan to be vapour heat treated until the fruit centres are 46.2°C and held at that temperature for 20 min, followed by refrigeration until fruit centre temperatures are 2°C and held at that temperature for 42 h to disinfest the fruit of fruit flies (Japan Ministry of Agriculture, Forestry and Fisheries, 1980).

New Zealand requires litchis imported from Australia to receive the same EDB fumigation required for citrus from Australia, and litchis from Cook Islands to be fumigated with the same EDB treatment schedule used for guavas from Fiji (New Zealand Ministry of Agriculture and Fisheries, 1992e,f).

Loquat, Eriobotrya japonica *Lindl.*

Loquat entering the US from Israel must be disinfested of Mediterranean fruit fly using a cold treatment schedule at 0, 0.6, 1.1, 1.7, or 2.2°C or below for 10, 11, 12, 14, or 16 days, respectively (Animal and Plant Health Inspection Service, 1992q).

Mango, Mangifera indica *L.*

Mangoes entering the US from the Caribbean, Central and South America, Mexico or Taiwan must be treated by either hot-water immersion or vapour heat to disinfest the fruit of fruit flies. The hot-water immersion consists of immersing the fruit at 46.5°C for 65–90 min depending on fruit weight (Animal and Plant Health Inspection Service, 1992l,p). The vapour heat treatment for mangoes from Mexico is the same as used for grapefruit, oranges and tangerines, while the vapor heat treatment required for mangoes from Taiwan consists of raising the fruit pulp to 46.6°C and maintaining that pulp temperature for 20 min (Animal and Plant Health Inspection Service, 1992p). Recently, forced hot-air quarantine treatments to disinfest mangoes of Caribbean fruit fly (Sharp, 1992) or West Indian fruit fly, *A. obliqua* (Macquart), (Mangan and Ingle, 1992) were developed in the US. The forced hot-air treatments heated mangoes with 48.0±0.3°C air for 160–220 min until the fruit seed surface temperatures are 46.1°C or above to disinfest the fruit of Caribbean fruit fly (Sharp, 1992), or with 50°C air for 133 or more minutes until seed surface temperatures are 48°C to disinfest the fruit of West Indian fruit fly (Mangan and Ingle, 1992).

Japan requires mangoes from the Philippines or Thailand to be quarantine treated for fruit flies using vapour heat until seed surface temperatures are 46.0°C (Philippines) or 46.5°C (Thailand) and held at the respective temperature for 10 min (Japan Ministry of Agriculture, Forestry and Fisheries, 1975, 1987).

New Zealand requires EDB fumigation of mangoes from Australia, Cook Islands and the Philippines using 33, 29, or 22 g m^{-3} EDB at 10–15, 15.5–19.5, or 20°C or above, respectively, at NAP to disinfest fruit flies from the fruit before entry (New Zealand Ministry of Agriculture and Fisheries, 1992f). Mangoes from Thailand must be vapour heat treated for fruit flies by heating the fruit until seed surface temperatures are 46.5°C, then held at that temperature for 10 min (New Zealand Ministry of Agriculture and Fisheries, 1992i), and mangoes from Mexico must receive a hot-water quarantine treatment for fruit flies using the same treatment schedule as the US (Animal and Plant Health Inspection Service, 1992l).

Mangoes imported into Australia from countries where fruit flies occur must be treated using 16–35 g m^{-3} EDB fumigation for 2 h and

21–26°C or above depending on the country of origin and how the fruit is packed (Australian Quarantine and Inspection Service, 1985, 1988).

Mangosteen, **Garcinia mangostana** *L.*

Mangosteen, like guava, is a relatively new entry to export markets. Presently, it is exported from Thailand as a frozen product (Chamlong Chettanachitara, personal communication). However, Burikam *et al.* (1992) recently developed a cold treatment at 6°C for 13 days to disinfest mangosteen of oriental fruit fly.

Melons, **Cucumis melo** *L. and* **Citrullus lanatus** *(Thunb.) Matsum. and Nakai cultivars*

Cantaloupe melon, honeydew melon and rockmelon imported into Australia must be fumigated with 8 g m^{-3} EDB for 2 h at 21°C or above, and watermelon must be fumigated with 32 g m^{-3} for 4 h at 21°C or above (Australian Quarantine and Inspection Service, 1985, 1988).

Watermelon entering New Zealand from the Kingdom of Tonga must be fumigated with 32 g m^{-3} MB for 4 h at 21–26°C (New Zealand Ministry of Agriculture and Fisheries, 1992c), and watermelon from Australia must be fumigated with the same treatment schedule used for mangoes and papaya from Australia (New Zealand Ministry of Agriculture and Fisheries, 1992f). Both treatments are used to disinfest the fruit of tephritid fruit flies.

Melons entering the US must be disinfested of surface feeders, such as noctuid moths, thrips and *Copitarsia* species. Honeydew melon, muskmelon (or watermelon) must be fumigated with 48 or 64 g m^{-3} MB at 50–59 or 40–49°C, respectively, for 2 h (Animal and Plant Health Inspection Service, 1992i). Cantaloupes must be fumigated with MB using the same treatment schedule for disinfesting surface feeders from bananas (Animal and Plant Health Inspection Service, 1992c).

Papaya, **Carica papaya** *L.*

Papaya exported from Hawaii to the US mainland must be treated to disinfest the fruit of Mediterranean fruit fly, melon fly, and oriental fruit fly. Two quarantine treatments used for this purpose are forced hot-air and vapour heat (Animal and Plant Health Inspection Service, 1992o,p). Both treatments consist of heating papayas until the fruit centre temperatures are 47.2°C. The major difference between the two quarantine treatments is that the fruit surfaces remain dry during forced hot-air treatment whereas they are wet from condensation during vapour heat treatment (Armstrong *et al.*, 1989).

Japan requires the same vapour heat quarantine treatment as the US as a condition of entry for papaya from Hawaii (Japan Ministry of Agriculture, Forestry and Fisheries 1972; Animal and Plant Health Inspection Service, 1992p).

New Zealand requires that papayas from Australia, Cook Islands, and Fiji receive EDB fumigation to disinfest the fruit of fruit flies using the same treatment schedules as required for litchis from Cook Islands and guava from Fiji (New Zealand Ministry of Agriculture and Fisheries, 1992f).

Passionfruit, **Passiflora** *species and cultivars*

Passionfruit entering New Zealand from Fiji must be EDB fumigated for fruit flies using the same treatment schedule as used for litchis from Cook Islands and guava from Fiji (New Zealand Ministry of Agriculture and Fisheries, 1992f).

Passionfruit entering the US from South Africa must receive the same cold treatment as used for grapes and stone fruits to disinfest the fruit of Mediterranean fruit fly (Animal and Plant Health Inspection Service, 1992q).

Persimmon, **Diospyros kaki** *L.*

Persimmon entering the US from Italy or Jordan must receive the same cold treatment schedule as used for citrus and other fruits from countries where Mediterranean fruit fly occurs (Animal and Plant Health Inspection Service, 1992q).

Pineapple, **Ananas comosus** *L.*

Pineapple entering Australia must be disinfested using EDB fumigation at 8 g m^{-3} for 2 h at 21°C or above (Australian Quarantine and Inspection Service, 1985, 1988).

Pineapple entering the US infested with internally feeding insects must be fumigated with 32 g m^{-3} MB for 6 h at 21.7°C or above (Animal and Plant Health Inspection Service, 1992j). Pineapples with externally feeding insects must be fumigated with 24–64 g m^{-3} MB with the MB concentration dependent on fruit temperatures at the time of fumigation (Animal and Plant Health Inspection Service, 1992j). A vapour heat treatment is required for pineapple from Hawaii (other than the 'Smooth Cayenne' cultivar) to disinfest the fruit of Mediterranean fruit fly, melon fly, and oriental fruit fly. The treatment consists of heating the fruit pulp to 44.6°C in air saturated with water vapour, then holding the fruit at 44.6°C for 8.75 h (Animal and Plant Health Inspection Service, 1992p).

Pomegranate,* Punica granatum *L.

Pomegranate imported into the US from Israel must receive a cold treatment to disinfest the fruit of Mediterranean fruit fly (Animal and Plant Health Inspection Service, 1992q), and pomegranate from Argentina or Haiti must receive a slightly more severe cold treatment schedule to disinfest the fruit of *Anastrepha* fruit fly species (Animal and Plant Health Inspection Service, 1992q).

Conclusion

An overview of available quarantine disinfestation treatments for 19 tropical and subtropical fruits or fruit groups has been given. The removal of the fumigation treatment schedules that use ethylene dibromide or methyl bromide would leave only heat or cold treatments, which could severely limit international trade for many fruits. It is obvious that a major research effort to develop alternative quarantine treatments to replace fumigants by using heat, cold or other treatment technologies is of immediate international concern to maintain the trade in fresh horticultural commodities through quarantine barriers.

References

Animal and Plant Health Inspection Service (1992a) T101(c^1): Methyl bromide fumigation of avocado from Hawaii. In: *Plant Protection and Quarantine Treatment Manual.* US Department of Agriculture, Hyattsville, Maryland, p. 5.4.

Animal and Plant Health Inspection Service (1992b) T101(d^1): Methyl bromide fumigation of banana for surface feeders such as *Noctuidae* spp., *Thrips* spp., *Copitarsia* spp. In: *Plant Protection and Quarantine Treatment Manual.* US Department of Agriculture, Hyattsville, Maryland, p. 5.4.

Animal and Plant Health Inspection Service (1992c) T101(k^1): Methyl bromide fumigation of cantaloupes for surface and external feeders. In: *Plant Protection and Quarantine Treatment Manual.* US Department of Agriculture, Hyattsville, Maryland, p. 5.9.

Animal and Plant Health Inspection Service (1992d) T101(q^1): Methyl bromide fumigation of cherimoya for *Brevipalpus chilensis.* In: *Plant Protection and Quarantine Treatment Manual.* US Department of Agriculture, Hyattsville, Maryland, p. 5.12.

Animal and Plant Health Inspection Service (1992e) T101(h^2): Methyl bromide fumigation of grapes for *Lobesia botrana* (vine moth). In: *Plant Protection and Quarantine Treatment Manual.* US Department of Agriculture, Hyattsville, Maryland, p. 5.23.

Animal and Plant Health Inspection Service (1992f) T101(i^2): Methyl bromide fumigation of grapes for other insects. In: *Plant Protection and Quarantine Treatment Manual*. US Department of Agriculture, Hyattsville, Maryland, p. 5.23.

Animal and Plant Health Inspection Service (1992g) T101(j^2): Methyl bromide fumigation of grapefruit for *Aleurocanthus woghumi* (citrus blackfly). In: *Plant Protection and Quarantine Treatment Manual*. US Department of Agriculture, Hyattsville, Maryland, p. 5.24.

Animal and Plant Health Inspection Service (1992h) T101(m^2): Methyl bromide fumigation of kiwi for external feeders and surface feeders. In: *Plant Protection and Quarantine Treatment Manual*. US Department of Agriculture, Hyattsville, Maryland, p. 5.26.

Animal and Plant Health Inspection Service, (1992i) T101(o^2): Methyl bromide fumigation of honeydew melon, muskmelon, and watermelon for surface feeders such as noctuid species, thrips species, and *Copitarsia* species. In: *Plant Protection and Quarantine Treatment Manual*. US Department of Agriculture, Hyattsville, Maryland, p. 5.29.

Animal and Plant Health Inspection Service (1992j) T101(r^2) and (s^2): Methyl bromide fumigation of pineapples for internal and external feeders. In: *Plant Protection and Quarantine Treatment Manual*. US Department of Agriculture, Hyattsville, Maryland, pp. 5.30–5.31.

Animal and Plant Health Inspection Service (1992k) T101(t^2): Methyl bromide fumigation of plantains for surface feeders such as noctuid species, thrips species, and *Copitarsia* species. In: *Plant Protection and Quarantine Treatment Manual*. US Department of Agriculture, Hyattsville, Maryland, p. 5.31.

Animal and Plant Health Inspection Service (1992l) T102(a): Hot water treatment for mangoes. In: *Plant Protection and Quarantine Treatment Manual*. US Department of Agriculture, Hyattsville, Maryland, pp. 5.42–5.43.

Animal and Plant Health Inspection Service (1992m) T102(b): Soapy water and wax treatment to remove *Brevipalpus chilensis* from cherimoya. In: *Plant Protection and Quarantine Treatment Manual*. US Department of Agriculture, Hyattsville, Maryland, p. 5.45.

Animal and Plant Health Inspection Service (1992n) T103(a): High-temperature forced-air treatment of grapefruit for *Anastrepha ludens*. In: *Plant Protection and Quarantine Treatment Manual*. US Department of Agriculture, Hyattsville, Maryland, p. 5.46.

Animal and Plant Health Inspection Service (1992o) T103(b): High-temperature forced-air treatment of papaya for *Ceratitis capitata, Bactrocera dorsalis, B. cucurbitae*. In: *Plant Protection and Quarantine Treatment Manual*. US Department of Agriculture, Hyattsville, Maryland, p. 5.47.

Animal and Plant Health Inspection Service (1992p) T106(a), (b), and (c): Vapour heat treatment of grapefruit, orange, tangerine, and mango from Mexico for *Anastrepha ludens*; bell pepper, eggplant, pineapple (other than smooth Cayenne), squash, tomato, and zucchini for *Ceratitis capitata, Bactrocera dorsalis*, and *B. cucurbitae*; and mangoes from Taiwan for *Bactrocera dorsalis*. In: *Plant Protection and Quarantine Manual*. US Department of Agriculture, Hyattsville, Maryland, pp. 5.52–5.53.

Animal and Plant Health Inspection Service (1992q) T107(a), (b), (c), and (d): Cold treatment for *Ceratitis capitata* (Mediterranean fruit fly), *Anastrepha ludens* (Mexican fruit fly), species of *Anastrepha* other than *A. ludens*, and *Bactrocera tryoni* (Queensland fruit fly). In: *Plant Protection and Quarantine Treatment Manual.* US Department of Agriculture, Hyattsville, Maryland, pp. 5.55–5.59.

Animal and Plant Health Inspection Service (1992r) T108(a)(1) and (2): Fumigation plus refrigeration. In: *Plant Protection and Quarantine Treatment Manual.* US Department of Agriculture, Hyattsville, Maryland, pp. 5.63–5.64.

Anonymous (1985) Ninety Seventh Session of the National Health and Medical Research Council, Sydney, Australia, June 1984. Australian Government Printing Office, Canberra.

Anonymous (1986) Irradiation in the production, processing, and handling of food. Final Rule. *Congressional Federal Register* 51, 13379. US Government Printing Office, Washington, DC.

Anonymous (1992) Ethylene di-bromide residue limit reduction. *Sentinel* 24, 8.

Armstrong, J.W. (1982) Development of a hot-water immersion quarantine treatment for Hawaiian-grown 'Brazilian' bananas. *Journal of Economic Entomology* 75, 787–790.

Armstrong, J.W. (1983) Infestation biology of three fruit fly species on 'Brazilian', 'Valery', and 'William's' cultivars of banana in Hawaii. *Journal of Economic Entomology* 76, 539–543.

Armstrong, J.W., Hansen, J.D., Hu, B.K.S. and Brown, S.A. (1989) High-temperature forced-air quarantine treatment for papayas infested with tephritid fruit flies. *Journal of Economic Entomology* 82, 1667–1674.

Australian Quarantine and Inspection Service (1985) Treatments for fresh fruit and vegetables. In: *Plant Quarantine Treatment Schedule.* Australian Quarantine and Inspection Service, Canberra, pp. 21–23.

Australian Quarantine and Inspection Service (1988) Conditions for importation of fruit and vegetables. In: *Plant Quarantine Manual,* section E. Australian Quarantine and Inspections Service, Canberra, pp. 1–59.

Back, E.A. and Pemberton, C.E. (1916) Banana as a host fruit of the Mediterranean fruit fly. *Journal of Agricultural Research* 5, 793–804.

Baker, A.C. (1952) The vapor-heat process. In: *US Department of Agriculture Yearbook.* US Government Printing Office, Washington, DC, pp. 401–404

Balock, J.W., Burditt, A.K., Seo, S.T. and Akamine, E.K. (1966) Gamma radiation as a quarantine treatment for Hawaiian fruit flies. *Journal of Economic Entomology* 59, 202–204.

Burikam, I., Sarnthoy, O., Charernsom, K., Kanno, T. and Homma, H. (1992) Cold temperature treatment for mangosteens infested with the oriental fruit fly. *Journal of Economic Entomology* 85, 2298–2301.

Couey, H.M. and Hayes, C.F. (1986) Quarantine procedure for Hawaiian papaya using fruit selection and a two-stage hot-water immersion. *Journal of Economic Entomology* 79, 1307–1314.

Drew, R.A.I., Hooper, G.H.S. and Bateman, M.A. (1978) *Economic Fruit Flies of the South Pacific Region.* Watson, Ferguson and Co., Brisbane, Australia, 137pp.

Gilmore, J. E. (1983) Introduction: The medfly in California: The threat, defense strategies, and public policy. *HortScience* 18, 39.

Gould, W.P. and Sharp, J.L. (1990) Cold-storage quarantine treatment for carambolas infested with the Caribbean fruit fly (Diptera: Tephritidae). *Journal of Economic Entomology* 83, 458–460.

Gould, W.P. and Sharp, J.L. (1992) Hot-water immersion quarantine treatment for guavas infested with Caribbean fruit fly. *Journal of Economic Entomology* 85, 1235–1239.

Hallman, G.J. (1990) Vapor-heat treatment of carambolas infested with Caribbean fruit fly. *Journal of Economic Entomology* 83, 2340–2342.

Hallman, G.J. and Sharp, J.L. (1990) Hot-water immersion quarantine treatment for carambolas infested with Caribbean fruit fly. *Journal of Economic Entomology* 83, 1471–1474.

Hawkins, L.A. (1932) Sterilization of citrus fruit by heat. *Citriculture* 9, 7–8, 21–22.

Hooper, T. (1907) Cold storage and fruit fly. *Journal of the Department of Agriculture, Western Australia* 15, 252–253.

Japan Ministry of Agriculture, Forestry and Fisheries (1972) Standards established by the Minister of Agriculture, Forestry and Fisheries on fresh Solo type papayas from Hawaiian Islands. Notification No. 798 In: *Collection of Ministry Notifications of Plant Protection Laws and Regulations*. Plant Protection Division, Japan Plant Quarantine Association, Tokyo, pp. 153–154.

Japan Ministry of Agriculture, Forestry and Fisheries (1973) Standards established by the Minister of Agriculture, Forestry and Fisheries on fresh sweet oranges, lemons and grapefruit from Republic of South Africa. Notification No. 1045. In: *Collection of Ministry Notifications of Plant Protection Laws and Regulations*. Plant Protection Division, Japan Plant Quarantine Association, Tokyo, pp. 159–160.

Japan Ministry of Agriculture, Forestry and Fisheries (1975) Standards established by the Minister of Agriculture, Forestry and Fisheries on fresh Manila Super variety of mangoes from Philippines. Notification No. 693. In: *Collection of Ministry Notifications of Plant Protection Laws and Regulations*. Plant Protection Division, Japan Plant Quarantine Association, Tokyo, pp. 162–163.

Japan Ministry of Agriculture, Forestry and Fisheries (1980) Standards established by the Minister of Agriculture, Forestry and Fisheries on fresh Ponkan, Tankan, Liutin variety of sweet oranges, and litchi from Taiwan. Notification No. 437. In: *Collection of Ministry Notifications of Plant Protection Laws and Regulations*. Plant Protection Division, Japan Plant Quarantine Association, Tokyo, pp. 155–156.

Japan Ministry of Agriculture, Forestry and Fisheries (1987) Standards established by the Minister of Agriculture, Forestry and Fisheries on fresh Nang Klarngwun variety of mangoes from Thailand. Notification No. 187. In: *Collection of Ministry Notifications of Plant Protection Laws and Regulations*. Plant Protection Division, Japan Plant Quarantine Association, Tokyo, pp. 177–178.

Liquido, N.J. (1990) Survival of oriental fruit fly and melon fly eggs oviposited in morphologically defective blossom end of papaya following two-stage hot-water immersion treatment. *Journal of Economic Entomology* 83, 2327–2330.

Mangan, R.L. and Ingle, S.J. (1992) Forced hot-air quarantine treatment for mangoes infested with West Indian fruit fly. *Journal of Economic Entomology* 85, 1859–1864.

Mangan, R.L. and Ingle, S.J. (1993) Forced hot-air quarantine treatment for grapefruit infested with Mexican fruit fly. *Journal of Economic Entomology* (in press).

New Zealand Ministry of Agriculture and Fisheries (1992a) Treatment requirements, NASS:152.02, Border protection specifications: Clearance of fresh produce, Section 2.0, T2. Wellington, NZ.

New Zealand Ministry of Agriculture and Fisheries (1992b) Treatment requirements, NASS:152.02, Border protection specifications: Clearance of fresh produce, Section 2.0, T3. Wellington, NZ.

New Zealand Ministry of Agriculture and Fisheries (1992c) Treatment requirements, NASS:152.02, Border protection specifications: Clearance of fresh produce, Section 2.0, T5. Wellington, NZ.

New Zealand Ministry of Agriculture and Fisheries (1992d) Treatment requirements, NASS:152.02, Border protection specifications: Clearance of fresh produce, Section 2.0, T6. Wellington, NZ.

New Zealand Ministry of Agriculture and Fisheries (1992e) Treatment requirements, NASS:152.02, Border protection specifications: Clearance of fresh produce, Section 2.0, T7. Wellington, NZ.

New Zealand Ministry of Agriculture and Fisheries (1992f) Treatment requirements, NASS:152.02, Border protection specifications: Clearance of fresh produce, Section 2.0, T8. Wellington, NZ.

New Zealand Ministry of Agriculture and Fisheries (1992g) Treatment requirements, NASS:152.02, Border protection specifications: Clearance of fresh produce, Section 2.0, T9. Wellington, NZ.

New Zealand Ministry of Agriculture and Fisheries (1992h) Treatment requirements, NASS:152.02, Border protection specifications: Clearance of fresh produce, Section 2.0, T11. Wellington, NZ.

Ruckelshaus, W.D. (1984) Ethylene dibromide, amendment of notice of intent to cancel registration of pesticide products containing ethylene dibromide. *Congressional Federal Register* 49, 14182–14185. US Government Printing Office, Washington, DC.

Saunders, G.W. and Elder, R.J. (1966) Sterilization of banana fruit infested with banana fruit fly. *Queensland Journal of Agricultural Science* 23, 81–85.

Sharp, J.L. (1992) Hot-air quarantine treatment for mango infested with Caribbean fruit fly. *Journal of Economic Entomology* 85, 2302–2304.

Sharp, J.L. and Hallman, G.J. (1992) Hot-air quarantine treatment for carambolas infested with Caribbean fruit fly. *Journal of Economic Entomology* 85, 168–171.

Shellie, K.C., Firko, M.J. and Mangan, R.L. (1993) Phytotoxic response of 'Dancy' tangerine to high-temperature moist forced-air treatment for fruit fly infestation. *Journal of the American Society for Horticultural Science* (in press).

Umeya, K. and Yamamoto, H. (1971) Studies on the possible attack of the Mediterranean fruit fly (*Ceratitis capitata* (Wiedemann)) on green bananas. *Research Bulletin of Plant Protection* 9, 6–17.

United Nations Environment Programme (1992) Methyl bromide: Its atmospheric science, technology and economics. Montreal Protocol Assessment Supplement, United Nations Headquarters, Ozone Secretariat, PO Box 30552, Nairobi, Kenya, 41pp.

US Department of Agriculture (1940) Administrative instructions amending authorization of the shipment of fruits and vegetables from Hawaii to the mainland subject to fumigation with methyl bromide under supervision. USDA, Bureau of Entomology and Plant Quarantine Memorandum No. 510. Washington, DC.

US Department of Agriculture (1951) Ethylene dibromide fumigation approved as a condition for certification of certain fruits and vegetables under Hawaiian fruit and vegetable quarantine No. 13. USDA, Bureau of Entomology and Plant Quarantine Memorandum No. 592. Washington, DC.

Voss, H.J. (1990) Permit QC 222, master permit for the shipment of citrus fruit from the state of Florida to California, amended 20 Sept., 1990. California Department of Food and Agriculture, Sacramento, 14pp.

Williamson, D.L., Summy, K.R., Hart, W.G., Sanchez, R., Wolfenbarger, D.A. and Burton, B.D. (1986) Efficacy and phytotoxicity of methyl bromide as a fumigant for the Mexican fruit fly in grapefruit. *Journal of Economic Entomology* 79, 172–175.

Zee, F.T., Nishina, M.S., Chan, H.T. and Nishijima, K.A. (1989) Blossom end defects and fruit fly infestation in papayas following hot water quarantine treatment. *HortScience* 24, 323-325.

Temperate Fruits 15

MICHAEL MEHERIUK and ALAN P. GAUNCE
Agriculture Canada, Research Station, Summerland, BC, Canada, V0H 1Z0.

Introduction

One of the earliest reports on chemical fumigation of temperate fruits was that of Quaintance (1909). Several apple cultivars were fumigated with HCN to eliminate San Jose scale. This was also the first report on injury to fruit as one lot of 'Rhode Island Greening' showed browning of the skin especially at the calyx end. Carbon disulphide was tried as a fumigant but was ineffective against the insect. The earliest report of methyl bromide on temperate tree fruits is that of Vayssiere (1934) in which pears showed no adverse effects from the fumigation treatment. However, Francolini (1935) noted accelerated softening in pears treated with $60\,g\,m^{-3}$ of methyl bromide. Injury to apples (skin browning) from methyl bromide fumigation was first reported by Monro (1938).

An early attempt to use cold temperature as a means of disinfestation was tried by Lounsbury (1907) on peaches infested with the Mediterranean fruit fly. All of the larvae were dead after 3 weeks at 4–5°C. A subsequent study by Back and Pemberton (1916) indicated a period of 10 days at 0°C was needed to kill the pupae. Temperature and chemical fumigation continue to be the main focus for insect disinfestation in temperate fruits. Excellent reviews by Couey (1983, 1989) provide valuable information on current quarantine procedures. Vapour heat was shown by Claypool and Vines (1956) to be quite harmful to temperate fruits and little work has been done with this procedure in that regard.

Although γ-irradiation is an effective disinfestation procedure (Burditt, 1982, 1985), it has had limited application because of consumer concerns. At present, there are no approved commercial irradiation

From: Paull, R.E. and Armstrong, J.W. (eds) (1994) *Insect Pests and Fresh Horticultural Products: Treatments and Responses.* CAB INTERNATIONAL, Wallingford, UK.

treatments for insect disinfestation of temperate fruit. An excellent review of effects of irradiation on quality of pome fruits, stone fruits, and berries is that of Thomas (1986).

Current procedures or studies on insect disinfestation of temperate fruits are discussed under the headings of atmosphere, chemical, gamma irradiation, temperature, and others. The scientific and common names of insects pertinent to this chapter are listed in Table 15.1.

Table 15.1. Common and scientific names for insects and mites mentioned in this chapter.

Apple maggot:	[Diptera: Tephritidae] *Rhagoletis pomonella* (Walsh)
Brown-headed leafroller:	[Lepidoptera: Tortricidae] *Ctenopseustis obliquana* (Walker)
Caribbean fruit fly:	[Diptera: Tephritidae] *Anastrepha suspensa* (Loew)
Cherry fly:	[Diptera: Tephritidae] *Rhagoletis cerasi* L.
Cherry fruit fly:	[Diptera: Tephritidae] *Rhagoletis cingulata* (Loew)
Codling moth:	[Lepidoptera: Tortricidae] *Cydia pomonella* (L.)
European red mite:	[Acarina: Tetranychidae] *Panonchus ulmi* (Koch)
Green-headed leafroller:	[Lepidoptera: Tortricidae] *Planotortrix excessana* (Walker)
Light-brown apple moth:	[Lepidoptera: Tortricidae] *Epiphyas postvittana* (Walker)
McDaniel spider mite:	[Acarina: Tetranychidae] *Tetranychus mcdanieli* McGregor
Mediterranean fruit fly:	[Diptera: Tephritidae] *Ceratitis capitata* (Wiedemann)
Melon fly:	[Diptera: Tephritidae] *Dacus cucurbitae* Coquillett
Oriental fruit fly:	[Diptera: Tephritidae] *Dacus dorsalis* Hendel
Oriental fruit moth:	[Lepidoptera: Tortricidae] *Graphalita molesta* (Busck)
Plum curculio:	[Coleoptera: Curculionidae] *Corotrachelus nenuphar* (Herbst)
Queensland fruit fly:	[Diptera: Tephritidae] *Dacus tryoni* (Froggatt)
San Jose scale:	[Homoptera: Diaspididae] *Aspidiotus perniciosus* (Comst.)
Western cherry fruit fly:	[Diptera: Tephritiobe] *Rhagoletis indifferens* Curran
Western flower thrips:	[Thysanoptera: Thripidae] *Frankiniela occidentalis* Pergande

Apples

Atmosphere

An atmosphere of >90% CO_2 (<1% O_2) was lethal to San Jose scale if infested fruit were exposed for 1 or 2 days at 12 or 22°C but some survival was noted when a temperature of 2°C was used (Morgan and Gaunce, 1975). Low oxygen atmospheres (100% N_2 or CO_2) killed European red mite eggs and diapausing larvae in 2 days at 21–24°C (Gaunce *et al.*, 1982) but 3 days was needed with 90% CO_2 at 18°C (Herne and Bond, 1981). Complete mortality was also obtained after 210 days in 1% O_2 + 1–1.5% CO_2 at 5°C (Lidster *et al.*, 1981, 1984). Some survival was reported for

McDaniel spider mites after 7 days in 100% CO_2 at 21–24°C (Gaunce *et al.*, 1982). Codling moth eggs and larvae succumbed within 2.5 and 10 days, respectively, in an atmosphere of 60% CO_2 at 25°C (Soderstrom *et al.*, 1990). Apple maggot eggs and larvae required 8 days in 100% N_2 at 20°C or 24 days at 0°C (AliNiazee *et al.*, 1989). Controlled atmosphere (CA) storage has been more effective than regular cold storage for mortality of leafroller (Batchelor *et al.*, 1985) and San Jose scale (Morgan, 1967). Whiting *et al.* (1991) observed high mortality of the light brown apple moth in an atmosphere of 0.4% O_2 + 5% CO_2 at 20°C for 22.5 h.

Both internal and external injuries due to high CO_2 (Meheriuk, 1977) or low O_2 (Gaunce *et al.*,1982) have been reported for apples. Unacceptable off-flavours in apples stored in low oxygen (0–0.25%) atmospheres at 0 or 5°C for 14 days or 10°C for 7 days were attributed to the accumulation of ethanol (Ke *et al.*, 1991a).

Chemical

Codling moth larvae underwent complete mortality with a methyl bromide dose of 32 g m^{-3} for 2 h at 17 or 26°C followed by 30 or 60 days of storage near 0°C (Moffitt, 1971; Morgan *et al.*, 1974) but it is likely that fumigation alone may have been sufficient for complete larval mortality. Eggs, which are less susceptible to the fumigant (Gaunce *et al.*, 1980; Tebbets *et al.*, 1986), were controlled by the same dosage at 23°C without subsequent cold storage. Diapausing larvae (5th instar) required a dose of 56 g m^{-3} for 2 h at 6°C (Moffitt *et al.*, 1983) but this larval stage is unlikely to be present in fruit at harvest. Methyl bromide at 32 g m^{-3} for 2 h at 10 or 17°C combined with a cold storage period was effective against San Jose scale (Morgan *et al.*, 1975; Angerilli *et al.*, 1986) and the Queensland and Mediterranean fruit flies (Rippon *et al.*, 1982). Apple maggot larvae and eggs were killed by 16 g m^{-3} for 2 h at 30°C and 20 g m^{-3} at 21°C (Chapman, 1940; Richardson, 1955). Oriental fruit moth larvae and eggs respectively required 16 and 25 g m^{-3} for 2 h at 21°C (Johnson *et al.*, 1947). Larvae of the plum curculio needed a dose of 32 g m^{-3} for 2 h at 23°C or higher doses when lower temperatures were utilized (Richardson and Roth, 1966).

Fruit injury from methyl bromide fumigation includes skin browning (Phillips *et al.*, 1939; Chapman, 1940; Meheriuk *et al.*, 1990) and internal browning (Phillips *et al.*, 1939; Galletti and Berger, 1987; Drake *et al.*, 1990; Meheriuk *et al.*, 1990). Methyl bromide fumigation increased the incidence of scald and exacerbated softening in apples (Galletti and Berger, 1987). Cultivar response differences were demonstrated by Meheriuk *et al.* (1990) who found a high, moderate, and low tolerance of 'Golden Delicious', 'Red Delicious' and 'Spartan' apples, respectively, to methyl bromide fumigation treatments. The authors also found that high doses and longer fumigation periods increased the incidence and severity of internal and external injury in treated apples. Fumigation followed immediately by CA storage was extremely injurious to the fruit (Meheriuk, unpublished data). Skin injury in 'Fuji' apples was avoided if adsorbents for methyl bromide and carbon dioxide were placed in the fruit carton after fumigation (Kawakami and Soma, 1991). Apples are less prone to injury if dipped in a diphenylamine solution prior to fumigation (Sproul *et al.*, 1976; Meheriuk *et al.*, 1990).

Ethylene dibromide gave complete mortality of apple maggot larvae at a rate of 4 g m^{-3} for 2 h at 21–25°C (Richardson, 1955) or 6 g m^{-3} at 10–12°C (Sanford, 1962). Diapausing European red mite eggs were killed by 6–8 g m^{-3} for 4–4.5 h at 25°C (Sanford *et al.*, 1976; Herne and Bond, 1981), Queensland fruit flies by 16 g m^{-3} for 2 h (Anon., 1955), and plum curculio by 32 g m^{-3} of a 1 : 3 mixture of ethylene dibromide and methyl bromide for 2 h at 23°C (Richardson and Roth, 1966).

High mortality of apple maggot larvae was obtained with ethylene chlorobromide (Sanford, 1962). Carroll *et al.* (1980) observed complete mortality of apple maggot with 4.2 g m^{-3} of 2-propenyl isothiocyanate for 3 h at 25°C but the treated 'York' apples were severely injured. One report of hydrogen cyanide fumigation of apples infested with San Jose scale was noted in the literature (Quaintance, 1909).

Gamma irradiation

The FAO/IAEC (1986) recommends dosages of 150 Gy for fruit flies (Tephritidae) and 250 Gy for codling moth as acceptable quarantine procedures. Doses that prevented development of viable adults were 153 Gy for codling moth (Burditt and Hungate, 1989) and 500 Gy for Mediterranean fruit fly (Kaneshiro *et al.*, 1985). No fruit injury was reported with dosages up to 225 Gy on three varieties of apple (Rigney and Wills, 1985a) but Massey *et al.* (1964) found some loss of varietal flavour in 'Cortland' and skin injury in 'McIntosh' with dosages of 500 Gy.

Temperature

Storage temperatures near 0°C are lethal to many insects. Complete mortality of codling moth eggs (Moffitt and Burditt, 1989) and oriental fruit moth larvae (Dustan, 1963) were obtained after storage periods of 55 and 56 days, respectively, at or near 0°C. Lethal cold storage treatments for other insects include 14–16 days for larvae and eggs of the Mediterranean fruit fly (Sproul, 1976), 33 days for apple maggot and plum curculio larvae (Glass *et al.*, 1961), 90 days for larvae of three leaf roller species (Batchelor *et al.*, 1985), and 90 days for codling moth larvae (Toba and Moffitt, 1991). All stages of San Jose scale were killed by a cold storage period of 54–154 days (Morgan, 1967; Morgan and Angle, 1967; Ferrari,

1975; Dickler, 1976). Apples dipped in diphenylamine, ethoxyquin or alcohol required a shorter cold storage period than undipped fruit to achieve complete mortality of San Jose scale (Morgan and Angle, 1967).

Apricots

Chemical

Mediterranean fruit fly eggs and larvae were killed by methyl bromide dosages ranging from 24 g m^{-3} for 2.5 h to 48 g m^{-3} for 1 h at $\geqslant$30°C (Armstrong and Couey, 1984). Oriental fruit moth larvae were killed by 16 g m^{-3} for 2 h at 15–18°C (Johnson *et al.*, 1947). No fruit injury was reported in either study.

Cherries

Chemical

Codling moth larvae and eggs suffer complete mortality when exposed to methyl bromide at 32 g m^{-3} for 2 h at 20–24°C (Anthon *et al.*, 1975; Gaunce *et al.*, 1981). Conditions needed for the Western cherry fruit fly were 40, 46, and 64 g m^{-3} at 16, 10 and 6°C, respectively (Moffitt *et al.*, 1983). The Mediterranean fruit fly was controlled by 48 g m^{-3} for 2 h or by 32 g m^{-3} for 4 h, both at 21°C (Spitler and Couey, 1983). No fruit injury was found with comparable treatments (Harvey and Harris, 1982). Others have reported an increase in stem abscission (Jessup, 1988) and a higher incidence of fruit pitting and stem browning (Drake *et al.*, 1991) in cherries treated with methyl bromide than in non-treated fruit.

Ethylene dibromide (8 g m^{-3} for 2 h at 18–25°C) controlled cherry fruit fly and had a minor effect on taste (Jones, 1955) but severe phytotoxicity was reported by Wade and Rigney (1979) when 32 g m^{-3} of the fumigant were used.

Gamma irradiation

Disinfestation of fruit with Queensland fruit fly larvae and eggs was obtained with a dosage of 75 Gy (Rigney and Wills, 1985b) but cherry fruit fly larvae required 100 Gy (Jona and Arzone, 1979). Massey *et al.* (1965) reported the threshold for fruit softening to be approximately 500 Gy.

Temperature and atmosphere

Insecticidal atmospheres of 0.02 and 0.25% O_2 at 0 and 5°C have been investigated by Ke and Kader (1992) and fruit were tolerant of such conditions for up to 40 days. Such conditions appear to have potential for insect disinfestation.

Grapes

Grapes were disinfested of Queensland fruit fly with an ethylene dibromide treatment combined with cold temperatures (Anon., 1955). No injury was reported in grapes irradiated with a dose up to 225 Gy (Rigney and Wills, 1985a), a dosage sufficient to render eggs and larvae of most insects non-viable. Fumigation with methyl bromide (48 g m^{-3} for 3 h at 16°C) caused off-flavours in treated grapes during subsequent storage unless an adsorbent was used to remove the fumigant after treatment (Soma *et al., 1991).*

Kiwifruit

Jessup and Baheer (1990) reported no survival of Queensland fruit fly larvae or eggs in kiwifruit after cold storage at 1°C for 12 days.

Nectarines

Atmosphere

Insecticidal atmospheres of 0.5% O_2 for 3–9 days at 5–15°C were investigated by Smilanick and Fouse (1989) but development of high ethanol and acetaldehyde content and other internal problems precluded the treatment for insect control. However, Ke and Kader (1992) noted that nectarines tolerated a 0.25% atmosphere for 20 days or more if the temperature was 0°C. Higher temperatures resulted in significant flesh browning of the fruit.

Chemical

Larvae and eggs of the Mediterranean and oriental fruit flies are controlled by 24, 32 or 48 g m^{-3} of methyl bromide, respectively, for 2.5, 1.5, and 1 h at 30°C (Armstrong and Couey, 1984; Armstrong *et al.*, 1988).

At 21°C the Mediterranean fruit fly required 32 or 48 g m^{-3} for 4 or 2 h respectively (Spitler and Couey, 1983). Dosages effective against the Mediterranean fruit fly also controlled the Caribbean fruit fly (Benschoter, 1988). Nectarines infested with oriental fruit moth required 32 g m^{-3} for 2 h at 16–26°C (Yokoyama *et al.*, 1987a). No codling moth eggs hatched when subjected to 48 g m^{-3} of methyl bromide for 2 h at 21°C (Yokoyama *et al.*, 1987b, 1990a,b).

Methyl bromide did not cause fruit injury at doses of 48 g m^{-3} for 2 h at 21°C but skin browning and pitting were evident with 64 g m^{-3} (Harvey *et al.*, 1989). Internal breakdown and mealiness were reported by Harman *et al.* 1990). Phytotoxicity to the fumigant was cultivar dependent in both studies.

Ethylene dibromide fumigation (16 g m^{-3} for 2 h) of nectarines was effective against Queensland fruit fly (Anon., 1955).

Gamma irradiation

Doses of 300–500 Gy were sufficient for fruit flies (Moy *et al.*, 1983; FAO/IAEC, 1986) but 500 Gy affected aroma in treated fruit (Moy and Nagai, 1985).

Temperature

Cold storage at 0°C for 12 days was effective against Queensland fruit-fly-infested nectarines (Anon., 1955). Hot water treatments were very injurious to the fruit and not recommended as a quarantine procedure (Sharp, 1990).

Peaches

Atmosphere

Peaches have been successfully stored for 40 days at 0°C in an atmosphere of 0.25% O_2 or 14 days at 5°C in 0.02 or 0.25% O_2 (Ke *et al.*, 1991b). Longer storage periods led to an accumulation of ethanol and acetaldehyde and development of off-flavours. Both atmospheres are likely to be lethal to many insects.

Chemical

Disinfestation of peaches infested with Mediterranean, Caribbean and oriental fruit flies was obtained with the same methyl bromide treatments that were used for nectarines (Armstrong and Couey, 1984; Armstrong *et al.*, 1988; Benschoter, 1988; Spitler and Couey,1983). Dosages effective against the oriental fruit moth caused no injury to the fruit (Johnson *et al.*,

1947). Methyl bromide fumigation enhanced red colour development but adversely affected flavour when used at 32 g m^{-3} for 1 h (Claypool, 1941). Some peach cultivars are particularly sensitive to methyl bromide but others showed no injury to 48 g m^{-3} for 2 h at 20°C (Harvey and Harris, 1982).

Ethylene dibromide at 16 g m^{-3} for 2 h was effective against the Queensland fruit fly in peaches (Anon., 1955).

Gamma irradiation

Doses of 300–500 Gy are lethal to fruit fly eggs and larvae in peaches (Moy *et al.*, 1983; FAO/IAEC, 1986) but colour, texture and aroma can be adversely affected by the upper limit (Moy *et al.*, 1983; Moy and Nagai, 1985).

Temperature

The time required to achieve larval mortality near or at 0°C was 21 days for the Mediterranean fruit fly (Lounsbury, 1907; Pettey and Griffiths, 1931), 12 days for the Queensland fruit fly (Anon., 1955), and 8 days for Caribbean fruit fly (Benschoter, 1988). Hot water treatments are not promising because of pronounced external and internal injury (Kerbel *et al.*, 1985; Sharp, 1990).

Pears

Atmosphere

'Bartlett' pears tolerated atmospheres of 1%, 0.5% and 0.25% O_2 for up to 10 days at temperatures between 0 and 10°C but off-flavours and injury were observed after 4 to 6 days in high CO_2 environments (Ke *et al.*, 1990). Asian pears tolerated atmospheres of 0.25 and 0.02% O_2 at 0°C for 14 and 7 days, respectively (Ke *et al.*, 1991a) but these atmospheres were tolerated for only 7 days at 5°C. These findings suggest a potential for insect control without incurring fruit injury.

Chemical

Methyl bromide doses of 32 or 48 g m^{-3} for 4 or 2 h, respectively, at 20–21°C are recommended against the Mediterranean fruit fly (Spitler and Couey, 1983). Pears have been successfully fumigated to control codling moth (Mackie and Carter, 1940; Palladini and Martelli, 1956) and plum curculio (Richardson and Roth, 1966). A dosage of 48 g m^{-3} for 2 h controlled the Caribbean fruit fly (Benschoter, 1988). Fumigation to control the oriental fruit moth caused extensive internal injury to 'Bartlett' pears (Johnson *et al.*, 1947). A dose of 48 g m^{-3} for 2 h at 6°C increased scald in 'Anjou' and internal discoloration in 'Bosc' (Drake and Moffitt, 1992). Harvey and Harris (1982) reported no phytotoxicity in 'Bartlett' fumigated at 21°C with 48 or 32 g m^{-3} for 3 or 4 h.

Gamma radiation

Pears were not injured by a dose of 600 Gy (Kaneshiro *et al.*, 1985), a dose lethal to the eggs and larvae of insects.

Temperature

Low temperatures (near 0°C) controlled the Queensland fruit fly in pears after 12 days (Anon., 1955) and the Caribbean fruit fly after 8 days (Benschoter, 1988).

Other

Postharvest spraying of fruit with summer oil was reported to prevent codling moth eggs from hatching (Newcomer, 1930). This treatment also caused a delay in ripening.

Plums

Atmosphere

Insecticidal low-oxygen (0.25%) atmospheres were assessed for one variety of plum (Ke and Kader, 1992) and periods of up to 40 days were tolerated before any detrimental effects became obvious.

Chemical

Quarantine treatments effective for peaches and nectarines are also applicable to plums (Spitler and Couey, 1983; Armstrong and Couey, 1984; Benschoter, 1988) except that a longer treatment period is needed for the oriental fruit fly in plums (Armstrong *et al.*, 1988). Fumigation to control the oriental fruit moth produced no injury in plums (Johnson *et al.*, 1947). However, some cultivars are more sensitive to methyl bromide than others (Harvey and Harris, 1982).

Gamma irradiation

Doses of 300–500 Gy are tolerated by plums (Moy *et al.*, 1983; FAO/IAEC, 1986) but skin colour changes were noted at 500 Gy (Moy and Nagai, 1985).

Temperature

Hot water treatments (49–51°C) were injurious to the fruit (Sharp, 1990) and therefore not suitable for quarantine purposes.

Strawberries

Mediterranean fruit fly in strawberries was controlled with 48 g m^{-3} of methyl bromide for 3 h at 15°C (Armstrong *et al.*, 1984). Fumigation of strawberries with acetaldehyde or ethyl formate has been reported to control the western flower thrips (Aharoni *et al.*, 1979; Aharoni and Stewart, 1980). The ethyl formate treatment had no effect on the fruit but the acetaldehyde treatment resulted in taste detection of the fumigant up to 7 days after treatment. High CO_2 (about 90%) atmospheres combined with low oygen (<2%) were necessary to obtain total mortality of the thrips but off-flavours from the treatment persisted in the fruit (Aharoni *et al.*, 1981). Strawberries tolerated up to 10 days in low O_2 atmospheres (0.25%) at 0 or 5°C without any apparent effect on flavour (Ke *et al.*, 1991c).

Summary

Because ethylene dibromide is banned almost universally, there is little point in commenting upon its use. Methyl bromide is the predominant fumigant in quarantine procedures and doses of 24–48 g for 2 h at 5–30°C appear to be tolerated by most deciduous fruits. Higher concentrations may have applicability to some commodities. Since injury can occur with methyl bromide fumigation, each cultivar should be tested individually and a protocol developed for it. Cold storage is effective for many insects and periods may vary from 12 to 150 days, the treatment being dependent upon temperature, insect species, and fruit commodity. Methyl bromide fumigation often accompanies cold storage treatments because some stages of the insect tolerate cold conditions. Vapour heat has not been recommended for temperate fruit because of severe injury. Gamma irradiation is a promising procedure for disinfestation but public reluctance to accept the procedure still hampers its implementation. Electron beam and X-ray irradiation are currently being investigated and either may be more acceptable publically than gamma-ray irradiation. Low O_2 or high CO_2 atmospheres may be another option in some cases. Sterile insect release, pheromone trapping, and insect mating disruption are additional approaches that would aid in meeting quarantine requirements.

References

Aharoni,Y. and Stewart, J.K. (1980) Thrips mortality and strawberry quality after vacuum fumigation with acetaldehyde or ethyl formate. *Journal of the American Society for Horticultural Science* 105, 926–929.

Aharoni,Y., Hartsell, P.L., Stewart, J.K. and Young, D.K. (1979) Control of western flower thrips on harvested strawberries with acetaldehyde in air, 50% carbon dioxide or 1% oxygen. *Journal of Economic Entomology* 72, 820–822.

Aharoni, Y., Stewart, J.K. and Guardagni, D.G. (1981) Modified atmospheres to control western flower thrips on harvested strawberries. *Journal of Economic Entomology* 74, 338–340.

AliNiazee, M.T., Richardson, D.G., Kossittrakun, M. and Mohammad, A.B. (1989) Non-insecticidal quarantine treatments for apple maggot control in harvested fruit. *Proceedings of the Fifth International Controlled Atmosphere Conference*, Wenatchee, Washington, pp. 193–205.

Angerilli, N.P.D., Gaunce, A.P. and Logan, D.M. (1986) Some effects of post-harvest fumigation, controlled-atmosphere storage, and cold storage on San Jose scale (Homoptera: Diaspididae) survival on two varieties of apples. *Canadian Entomologist* 118, 493–497.

Anon. (1955) Treatment of fruit-fly infested fruit. Rural Research CSIRO No. 13, p.13.

Anthon, E.W., Moffitt, H.R., Couey, H.M. and Smith, L.O. (1975) Control of codling moth in harvested sweet cherries with methyl bromide and effects upon quality and taste of treated fruit. *Journal of Economic Entomology* 68, 524–526.

Armstrong, J.W. and Couey, H.M. (1984) Methyl bromide fumigation treatments at 30°C for California stone fruits infested with the Mediterranean fruit fly (Diptera: Tephritidae). *Journal of Economic Entomology* 77, 1229–1232.

Armstrong, J.W., Schneider, E.L., Garcia, D.L. and Couey, H.M. (1984) Methyl bromide quarantine fumigation for strawberries infested with Mediterranean fruit fly (Diptera: Tephritidae). *Journal of Economic Entomology* 77, 680–682.

Armstrong, J.W., Harvey, J.M., Garcia, D.L., Menezes, T.D. and Brown, S.A. (1988) Methyl bromide fumigation for control of oriental fruit fly (Diptera: Tephritidae) in California stone fruits. *Journal of Economic Entomology* 81, 1120–1123.

Back, E.A. and Pemberton, C.E. (1916) Effect of cold storage temperatures upon the pupae of the Mediterranean fruit fly. *Journal of Agricultural Research* 6, 251–260.

Batchelor, T.A., O'Donnell, R.L. and Roby, J.J. (1985) The efficacy of controlled atmosphere cool storage in controlling leafroller species. *Proceedings of the 38th New Zealand Weed and Pest Control Conference* 38, 53–55.

Benschoter, C.A. (1988) Methyl bromide fumigation and cold storage as treatments for California stone fruits and pears infested with the Caribbean fruit fly (Diptera: Tephritidae). *Journal of Economic Entomology* 81, 1665–1667.

Burditt, A.K., Jr (1982) Food irradiation as a quarantine treatment of fruits. *Food Technology* 36, 51–62.

Burditt, A.K., Jr (1985) Present status of research on efficacy of quarantine treatment for horticultural crops in the U.S.A. with special reference to the use of irradiation as a potential quarantine treatment. International Atomic Energy Agency Document No. 326, pp.17–42.

Burditt, A.K., Jr and Hungate, F.P. (1989) Gamma irradiation as a quarantine treatment for apples infested by codling moth (Lepidoptera: Tortricidae). *Journal of Economic Entomology* 82, 1386–1390.

Carroll, J.F., Morgan, N.O. and Weber, J.D. (1980) Bioassay of three isothiocyanates as fumigants against larvae of a Caribbean fruit fly and the apple maggot. *Journal of Economic Entomology* 73, 321–323.

Chapman, P.J. (1940) Effect of methyl bromide on apple maggot in apples. *Journal of Economic Entomology* 33, 817.

Claypool, L.L. (1941) Red color increase in fruits after harvest following treatment with methyl bromide. *Proceedings of the American Society for Horticultural Science* 38, 289–290.

Claypool, L.L. and Vines, H.M. (1956) Commodity tolerance studies of deciduous fruits to moist heat and fumigants. *Hilgardia* 24, 297–355.

Couey, H.M. (1983) Development of quarantine systems for host fruits of the medfly. *HortScience* 18, 45–47.

Couey, H.M. (1989) Heat treatment for control of postharvest diseases and insect pests of fruits. *HortScience* 24, 198–202.

Dickler, E. von. (1976) Influence of standard cold storage and controlled atmosphere storage of apples from Italy on the mortality and the fecundity of the San Jose scale (Quadraspidiotus: perniciosus Comst.). *Nachrichtenblatt des Deutschen Pflanzenschutzdienstes* 28, 67–73.

Drake, S.R. and Moffitt, H.R. (1992) Winter pear (Anjou and Bosc) response to methyl bromide fumigation. *HortScience* 27, 813–816.

Drake, S.R., Moffitt, H.R. and Mattheis, J.P. (1990) Methyl bromide, time, and temperature of exposure on apple quality. *Journal of Food Processing and Preservation* 14, 85–92.

Drake, S.R., Moffitt, H.R. and Kupferman, E.M. (1991) Quality characteristics of 'Bing' and 'Rainier' sweet cherries treated with gibberellic acid, following fumigation with methyl bromide. *Journal of Food Quality* 14, 119–125.

Dustan, G.G. (1963) The effect of standard cold storage and controlled atmosphere storage on survival of larvae of the oriental fruit moth, *Grapholitha molesta*. *Journal of Economic Entomology* 56, 167–169.

FAO/IAEC (1986) Excerpt of the task force meeting on irradiation as a quarantine treatment. *Food Irradiation Newsletter* 10, 5–10.

Ferrari, R. (1975) Tests on the survival and fecundity of the apple infestant *Q. perniciosus* after controlled atmosphere storage. *Redia (Giornale di Zoologia)* 56, 391–400.

Francolini, J. (1935) Action sur les produits vegetaux du bromine de methyle en fumigation sous vide partiel. *Revue de Pathologie Végétale et d'Entomologie Agricole de France* 22, 9–12.

Galetti, G.L. and Berger, S.H. (1987) Effects of methyl bromide on apples intended for export. *Simiente* 57, 201–206.

Gaunce, A.P., Madsen, H.F., McMullen, R.D. and Hall, J.W. (1980) Dosage response of the stages of codling moth, *Laspeyresia pomonella* (Lepidoptera: Olethreutidae), to fumigation with methyl bromide. *Canadian Entomologist* 112, 1033–1038.

Gaunce, A.P., Madsen, H.F. and McMullen, R.D. (1981) Fumigation with methyl bromide to kill larvae and eggs of the codling moth in Lambert cherries. *Journal of Economic Entomology* 74, 154–157.

Gaunce, A.P., Morgan, C.V.G. and Meheriuk, M. (1982) Control of tree fruit insects with modified atmospheres. *Proceedings of the 3rd National Controlled Atmosphere Conference*, Corvallis, Oregon, pp. 383–390.

Glass, E.H., Chapman, P.J. and Smock, R.M. (1961) Fate of apple maggot and plum curculio larvae in apple fruits held in controlled atmosphere storage. *Journal of Economic Entomology* 54, 915–918.

Harman, J.E., Lay-Yee, M., Billing, D.P., Yearsley, C.W. and Jackson, P.J. (1990) Effects of methyl bromide fumigation, delayed cooling, and controlled atmosphere storage on the quality of Redgold and Fantasia nectarine fruit. *New Zealand Journal of Crop and Horticultural Science* 18, 197–203.

Harvey, J.M. and Harris, C.M. (1982) Phytotoxic responses of cherries, nectarines, peaches, pears, and plums fumigated with methyl bromide for control of Mediterranean fruit fly. *Journal of the American Society for Horticultural Science* 107, 993–996.

Harvey, J.M., Harris, C.M. and Hartsell, P.L. (1989) Tolerances of California nectarine cultivars to methyl bromide quarantine treatments. *Journal of the American Society for Horticultural Science* 114, 626–629.

Herne, D.C. and Bond, E.J. (1981) Control by fumigants of winter eggs of European red mite, *Panonychus ulmi* (Acarina: Tetranychidae), on harvested apples. *Canadian Entomologist* 113, 121–126.

Jessup, A.J. (1988) Response of 'Lambert' and 'Ron's Seedling' sweet cherries to fumigation with methyl bromide plus cold storage. *Australian Journal of Experimental Agriculture* 28, 431–434.

Jessup, A.J. and Baheer, A. (1990) Low-temperature storage as a quarantine treatment for kiwifruit infested with *Dacus tryoni* (Diptera: Tephritidae). *Journal of Economic Entomology* 83(6), 2317–2319.

Johnson, A.C., Bowen, C.H. and Phillips, G.L. (1947) Report on cooperative experiments conducted by the Bureau of Entomology and Plant Quarantine and the California State Department of Agriculture relative to the fumigation of fruit and nurserystock hosts of the oriental fruit moth. United States Department of Agriculture Publication. E-711, 44pp.

Jona, R. and Arzone, A. (1979) Control of *Rhagoletis cerasi* in cherries by γ-irradiation. *Journal of Horticultural Research* 54, 167–170.

Jones, S.C. (1955) Fumigation tests with ethylene dibromide for the control of cherry fruit fly eggs, larvae and puparia. *Journal of Economic Entomology* 48, 618.

Kaneshiro, K.Y., Ohta, A.T., Kurihara, J.S., Kanegawa, K.M. and Nagamine, L.R. (1985) Gamma-radiation treatment for disinfestation of the medfly in thirty-five varieties of California-grown fruits. In: Moy, J. (ed.), *Radiation Disinfestation of Food and Agricultural Products*. Hawaii Institute for Tropical Agriculture and Human Resources, pp. 98–110.

Kawakami, R. and Soma, Y. (1991) Some factors causing chemical injury of apples fumigated with methyl bromide and injury protection by gas-adsorbents. *Research Bulletin of the Plant Protection Service, Japan* 27, 41–46.

Ke, D. and Kader, A.A. (1992) Factors that determine fruit tolerance to low-oxygen atmospheres. *Journal of the American Society for Horticultural Science* 117, 913–918.

Ke, D., van Gorsel, H. and Kader, A.A. (1990) Physiological and quality responses of 'Bartlett' pears to reduced O_2 and enhanced CO_2 levels and storage temperature. *Journal of the American Society for Horticultural Science* 115, 435–439.

Ke, D., Rodriguez-Sinobas, L. and Kader, A.A. (1991a) Physiology and prediction of fruit tolerance to low-oxygen atmospheres. *Journal of the American Society for Horticultural Science* 116, 253–260.

Ke, D., Rodriguez-Sinobas, L. and Kader, A.A. (1991b) Physiological responses and quality attributes of peaches kept in low oxygen atmospheres. *Scientia Horticulturae* 47, 295–303.

Ke, D., Goldstein, L., O'Mahony, M. and Kader, A.A. (1991c) Effects of short-term exposure to low O_2 and high CO_2 atmospheres on quality attributes of strawberries. *Journal of Food Science* 56 (1), 50–54.

Kerbel, E.L., Mitchell, F.G. and Mayer, G. (1985) Effect of postharvest heat treatments for insect control on the quality and market life of peaches. *HortScience* 20, 725–727.

Lidster, P.D., Sanford, K.H. and McRae, K.B. (1981) Effects of modified atmosphere storage on overwintering populations of the apple rust mite and European red mite eggs. *HortScience* 16, 328–329.

Lidster, P.D., Sanford, K.H. and McRae, K.B. (1984) Effects of temperature and controlled atmosphere on the survival of overwintering populations of European red mite eggs on stored 'McIntosh' apples. *HortScience* 19, 257–258.

Lounsbury, C.P. (1907) The fruit fly (*Ceratitis capitato*). *Agricultural Journal of the Cape of Good Hope* 31, 186–187.

Mackie, D.B and Carter, W.B. (1940) The fumigation of fresh fruit with methyl bromide under industrial conditions. *California Department of Agriculture Bulletin* 29, 78–86.

Massey, L.M., Jr, Parsons, G.F. and Smock, R.M. (1964) Some effects of gamma radiation on the keeping quality of apples. *Agricultural and Food Chemistry* 12, 268–274.

Massey, L.M., Jr, Robinson, W.B., Spaind, J.F., Splittstoesser, D.F., van Buren, J.P. and Kertesz, Z.I. (1965) Effect of gamma radiation upon cherries. *Journal of Food Science* 30, 759–765.

Meheriuk, M. (1977) Treatment of Golden Delicious apples with CO_2 prior to CA storage. *Canadian Journal of Plant Science* 57, 467–471.

Meheriuk, M., Gaunce, A.P. and Dyck, V.A. (1990) Response of apple cultivars to fumigation with methyl bromide. *HortScience* 25, 538–540.

Moffitt, H.R. (1971) Methyl bromide fumigation combined with storage for control of codling moth in apples. *Journal of Economic Entomology* 64, 1258–1260.

Moffitt, H. R. and Burditt, A.K., Jr (1989) Effects of low temperatures on three embryonic stages of the codling moth (Lepidoptera: Tortricidae). *Journal of Economic Entomology* 82, 1379–1381.

Moffitt, H.R., Burditt, A.K., Jr and Sell, C.R. (1983) Quarantine barriers and export markets for deciduous fruits. *Proceedings of the Oregon State Horticultural Society* 74, 38–45.

Monro, H.A.U. (1938) Some observations on the use of methyl bromide as a fumigant. *Annual Report of the Entomological Society of Ontario* 68, 72.

Morgan, C.V.G. (1967) Fate of the San Jose scale and the European fruit scale (Homoptera: Diaspididae) on apple and prunes held in standard cold storage and controlled atmosphere storage. *Canadian Entomologist* 99, 650–659.

Morgan, C.V.G. and Angle, B.J. (1967) Mortality of the San Jose scale (Homoptera: Diaspididae) on stored apples of different varieties and harvest dates. *Canadian Entomologist* 99, 971–974.

Morgan, C.V.G. and Gaunce, A.P. (1975) Carbon dioxide as a fumigant against the San Jose scale (Homoptera: Diaspididae) on harvested apples. *Canadian Entomologist* 107, 935–936.

Morgan, C.V.G., Gaunce, A.P. and Jong, C. (1974) Control of codling moth larvae in harvested apples by methyl bromide fumigation and cold storage. *Canadian Entomologist* 106, 917–920.

Morgan, C.V.G., Gaunce, A.P. and Madsen, B.J. (1975) Control of the San Jose scale on harvested apples by methyl bromide fumigation and cold storage. *Canadian Entomologist* 107, 367–368.

Moy, J.H. and Nagai, N.Y. (1985) Quality of fresh fruits irradiated at disinfestation doses. In: Moy, J.H. (ed.), *Radiation Disinfestation of Food and Agricultural Products*. Hawaii Institute of Tropical Agriculture and Human Resources, pp. 135–147.

Moy, J.H., Kaneshiro, K.Y., Ohta, A.T. and Nagai, N. (1983) Radiation disinfestation of California stone fruits infested by medfly – effectiveness and fruit quality. *Journal of Food Science* 48, 928–931.

Newcomer, E.J. (1930) Experiments in killing eggs of the codling moth on harvested fruit. *Journal of Economic Entomology* 23, 798–802.

Palladini, G. and Martelli, G.P. (1956) First test of vacuum disinfestation made in Abruzzo with a mobile apparatus. *Annelidella Sperimentazione Agraria (N.S.)* 9, XCV–CV.

Pettey, F.W. and Griffiths, E.A. (1931) Effective control of fruit fly by refrigeration. South African Department of Agricultural and Technical Services. *Science Bulletin* 99, 1–9.

Phillips, W.R., Monro, H.A.U. and Allen, C.E. (1939) Some observations on the fumigation of apples with methyl bromide. *Scientific Agriculture* 19, 7–20.

Quaintance, A.L. (1909) Fumigation of apples for the San Jose scale. *United States Department of Agriculture Bulletin* No. 84, 43 pp.

Richardson, H.H. (1955) Ethylene dibromide and methyl bromide fumigation of apples infested with apple maggots. *Journal of Economic Entomology* 48, 483–484.

Richardson, H.H. and Roth, H. (1966) Methyl bromide, ethylene dibromide, and other fumigants for control of plum curculio in fruit. *Journal of Economic Entomology* 98, 1149–1152.

Rigney, C.J. and Wills, P.A. (1985a) Suitability of low-dose gamma irradiation for disinfestation of several fruits. In: Moy, J. (ed.), *Radiation Disinfestation of Food and Agricultural Products*. Hawaii Institute of Tropical Agriculture and Human Resources, pp. 129–134.

Rigney, C.J. and Wills, P.A. (1985b) Efficacy of gamma irradiation as a quarantine treatment against Queensland fruit fly. In: Moy, J. (ed.), *Radiation Disinfestation of Food and Agricultural Products*. Hawaii Institute of Tropical Agriculture and Human Resources, pp. 116–120.

Rippon, L.E., Singh, G., Sproul, A.N. and Gilbert, W.S. (1982) Methyl bromide fumigation and cold storage for disinfestation of Granny Smith apples against Queensland and Mediterranean fruit flies. *Australian Journal of Experimental Agriculture and Animal Husbandry* 22, 116–123.

Sanford, K.H. (1962) Fumigation of apples to control the apple maggot *Rhagoletis pommonella*. *Journal of Economic Entomology* 55, 724–727.

Sanford, K.H., Stewart, D.K.R. and Poapst, P.A. (1976) Fumigation of apples under plastic tarpaulins with ethylene dibromide to control diapausing European red mite (Acarina: Tetranychidae) eggs. *Canadian Entomologist* 108, 217–223.

Sharp, J.L. (1990) Immersion in heated water as a quarantine treatment for California stone fruits infested with the Caribbean fruit fly (Diptera: Tephritidae). *Journal of Economic Entomology* 83, 1468–1470.

Smilanick, J.L. and Fouse, D.C. (1989) Quality of nectarines stored in insecticidal low-O_2 atmospheres at 5 and 15°C. *Journal of the American Society for Horticultural Science* 114, 431–436.

Soma, Y., Sunagawa, K., Kurokawa, K., Nakamura, M., Misumi, T. and Kawakami, F. (1991) Chemical injury of Kyoho grapes fumigated with methyl bromide. *Japanese Plant Protection Service Research Bulletin* 27, 83–86.

Soderstrom, E.L., Brandl, D.G. and Mackey, B. (1990) Response of the codling moth (Lepidoptera: Tortricidae) life stages to high carbon dioxide or low oxygen atmospheres. *Journal of Economic Entomology* 83, 472–475.

Spitler, G.H. and Couey, H.M. (1983) Methyl bromide fumigation treatments of fruits infested by the Mediterranean fruit fly (Diptera: Tephritidae). *Journal of Economic Entomology* 76, 547–550.

Sproul, A.N. (1976) Disinfestation of western Australian Granny Smith apples by cold treatment against the egg and larval stages of the Mediterranean fruit fly (*Ceratitis capitata* (Wied.)). *Australian Journal of Experimental Agriculture and Animal Husbandry* 16, 280–285.

Sproul, A.N., O'Loughlin, J.B., Terauds, A. and Hardisty, S. (1976) The use of diphenylamine to protect apples from methyl bromide injury. *Australian Journal of Agricultural Research* 27, 541–545.

Tebbets, J.S., Vail, P.V., Hartsell, P.L. and Nelson, H.D. (1986) Dose/response of codling moth (Lepidoptera: Tortricidae) eggs and nondiapausing and diapausing larvae to fumigation with methyl bromide. *Journal of Economic Entomology* 79, 1039–1043.

Thomas, P. (1986) Radiation preservation of foods of plant origin. Part V.

Temperate fruits: pome fruits, stone fruits, and berries. *CRC Critical Reviews in Food Science and Nutrition* 24, 357–400.

Toba, H.H. and Moffitt, H.R. (1991) Controlled-atmosphere cold storage as a quarantine treatment for nondiapausing codling moth (Lepidoptera: Tortricidae) larvae in apples. *Journal of Economic Entomology* 84, 1316–1319.

Vayssiere, P. (1934) Les stations de désinfection des végétaux sous vide partiel. *Review of Applied Entomology* 23, 194–195.

Wade, N.L. and Rigney, C.J. (1979) Phytotoxicity of ethylene dibromide to cherry and banana fruit. *Journal of the American Society for Horticultural Science* 104, 900–903.

Whiting, D.C., Foster, S.P. and Maindonald, J.H. (1991) Effects of oxygen, carbon dioxide, and temperature on the mortality responses of *Epiphyas postvittana* (Lepidoptera: Tortricidae). *Journal of Economic Entomology* 84, 1544–1549.

Yokayama, V.Y., Miller, G.T. and Hartsell, P.L. (1987a) Methyl bromide fumigation to control the Oriental fruit moth (Lepidoptera: Tortricidae) in nectarines. *Journal of Economic Entomology* 80, 1226–1228.

Yokoyama, V.Y., Miller, G.T. and Hartsell, P.L. (1987b) Methyl bromide fumigation for quarantine control of codling moth (Lepidoptera: Tortricidae) on nectarines. *Journal of Economic Entomology* 80, 840–842.

Yokayama, V.Y., Miller, G.T. and Hartsell, P.L. (1990a) Evaluation of methyl bromide quarantine treatment to control codling moth (Lepidoptera: Tortricidae) on nectarine cultivars proposed for export to Japan. *Journal of Economic Entomology* 83, 466–471.

Yokoyama, V.Y., Miller, G.T. and Hartsell, P.L. (1990b) A methyl bromide quarantine treatment to control codling moth (Lepidoptera: Tortricidae) on nectarines packed in shipping containers for export to Japan and effect on fruit attributes. *Journal of Economic Entomology* 83, 2335–2339.

Vegetables 16

JEFFREY K. BRECHT
Horticultural Sciences Department, University of Florida, Gainesville, FL 32611-0690, USA.

Introduction

Research and development of postharvest controls for insect pests of horticultural crops has been predominately focused on the tephritid fruit flies and their primary host crops, tropical and subtropical fruits (Robinson and Hooper, 1989). The postharvest insect pests of vegetable crops are fewer, or at any rate of less economic importance, than those of fruits, although some vegetable crops, notably tomato, pepper, squash, cucumber, and melons, are considered hosts for one or more fruit fly species (Bateman, 1972). Thus, less effort has been directed towards development of quarantine disinfestation treatments for vegetables. In this chapter, treatments that have been developed for fruit crops will of necessity be mentioned in terms of a treatment's potential application to vegetables.

The decision in the USA to cancel the registration of ethylene dibromide (EDB) in 1984 (US Government, 1984) has spurred the search for alternative quarantine treatments and approaches to insect control. Since that time, methyl bromide (MB) has become the fumigant of choice for quarantine treatment of a wide range of crops (USDA-APHIS, 1985, 1992). The adoption of other alternatives to EDB has been slower.

Brown (1985) summarized the quarantine situation in Florida soon after the loss of EDB. Among vegetables, fumigation with MB could be used for cucumber, squash, sweet potato, and tomato, although damage to sweet potatoes by MB precluded its use, and delayed ripening (Brecht *et al.*, 1986) and marginal resistance of tomatoes to the phytotoxic effects of MB make its use problematic. Vapour heat treatment was available for

From: Paull, R.E. and Armstrong, J.W. (eds) (1994) *Insect Pests and Fresh Horticultural Products: Treatments and Responses.* CAB INTERNATIONAL, Wallingford, UK.

pepper, eggplant, squash and tomato, and required for those crops when shipped from Hawaii. Certification that Florida peppers and tomatoes were commercially grown and packed was accepted by other states and required by California for cucumbers. Organic growers of those crops are required to set out traps for Caribbean fruit fly (caribfly) to demonstrate that the production area is a fly-free zone (P.L. Hornby, personal communication). There has been little in the way of significant change to that picture in the intervening years, although a number of promising new approaches have been investigated (Mitchell and Saul, 1990), primarily for fruits.

The focus of this chapter will be on those commodity treatments for vegetables that have come into use or been investigated since the loss of EDB, with particular reference to the situation in the USA. Commodity treatments involving application of chemicals, including fumigation, as well as dips and sprays, will be covered first. Treatments involving manipulation of the environment, including temperature, atmospheric gases, and ionizing radiation, will be discussed next, followed by consideration of the potential for development of commodity treatments for vegetables that combine chemical and environmental factors, with the intention of improving the effectiveness of both against the insects, while hopefully reducing the severity of the stress experienced by the commodity.

Chemical Treatments

Fumigants

Methyl bromide

Fumigation with MB has been approved in the United States as a quarantine treatment for a large number of vegetable crops and related insect pests (Table 16.1). The severe phytotoxic responses shown by some commodities following exposure to MB levels required for insect mortality (Knight *et al.*, 1980; Delate and Brecht, 1989) are a major disadvantage of MB. Other fruits and fruit-type vegetables like melons are marginally tolerant of MB (Lipton and Tebbets, 1984; Cowley *et al.*, 1991), or, like tomato, may exhibit significantly attenuated rates of ripening following MB exposure (Spalding *et al.*, 1978; Lipton *et al.*, 1982; Brecht *et al.*, 1986). The ripening delay in tomato following MB exposure is due to inhibition of ethylene production (Brecht *et al.*, 1986) and reduced sensitivity of the tomatoes to ethylene (Brecht, 1988), and is most obvious in green fruit. Cucumbers showed small areas of slight yellowing after MB fumigation at 32 or 48 g m^{-3} at 21°C for 4 or 2 h, respectively, but the blemishes were occasionally observed on control cucumbers also, and were considered

Table 16.1. Vegetables for which MB has been approved as a quarantine treatment in the USA and the targeted insects.

Commodities	Targeted insects
Asparagus	Red-legged earth mite[1], *Noctuidae* spp., *Thrips* spp., *Copitarsia* spp.[2]
Beans (snap, string)	*Maruca testulalis, Epinotia aporema, Cydia fabivora*
Cabbage[2]	Surface feeders
Cantaloupe[2]	Surface feeders, external feeders
Carrot[2]	External feeders
Cassava[2]	External feeders
Celery[2]	Surface feeders
Chayote[2]	Surface feeders
Chicory[2]	Not specified
Cipollini	*Exosoma lusitanica*
Corn[1]	Not specified
Cucumber[2]	Not specified
Dasheen (malanga)[2]	External and internal feeders
Endive[2]	Surface feeders
Faba bean	Bruchidae[2]
Garlic	*Brachycerus* spp. and *Dyspessa ulula*
Ginger[2]	External and internal feeders
Horseradish	Imported crucifer weevil
Leafy vegetables	Surface feeders[2]
Melons (honeydew, muskmelon, watermelon)	Surface feeders such as *Noctuidae* spp., melon, *Thrips* spp., *Copitarsia* spp.
Okra	Pink bollworm
Onion	Internal feeders[2]
Potato[1]	*Graphognathus* spp., *Ostrinia nubilalis, Phthorimaea operculela*
Pumpkin	Surface feeders[2]
Squash (summer, winter, zucchini)[2]	Surface feeders[2]
Sweet potato	Surface feeders[2]
Tomato	Fruit flies
Yam	Internal feeders[2]

Source: USDA-APHIS PPQ *Treatment Manuals*, 1985, 1992.
[1] Mentioned in 1985 PPQ *Treatment Manual* only.
[2] Mentioned in 1992 PPQ *Treatment Manual* only.

too minor to affect marketability (Armstrong and Garcia, 1985). Peppers were severely damaged when exposed to MB doses over 40 g m^{-3} for longer than 1.5 h, showing surface pitting and decay (Lipton *et al.*, 1982).

Asparagus spears were damaged by fumigation with MB at concentrations as low as 48 g m^{-3} for 2 h, but the damage, which was expressed as increased decay, was not evident until after 5 or 6 days at 20°C (Beever *et al.*, 1985). However, asparagus fumigated with 60 g m^{-3} also developed off-flavour. It seems likely that asparagus stored at lower temperatures following MB fumigation would have sufficient shelf life for normal marketing.

Stechmann *et al.* (1988) and Cowley *et al.* (1991) treated watermelons with 32 g m^{-3} MB for 4 h at 21 to 26°C, without fruit damage, after fruit from Tonga was found to contain *Bactrocera xanthodes* maggots upon arrival in New Zealand. Although it was found that fruit flies cannot normally lay eggs into watermelons unless rind injuries or blossom end imperfections are present, and maggots normally cannot leave the fruit for pupation and subsequently drown inside the fruits, opening of infested fruits for consumption could allow survival of the insect.

Phosphine

Fumigation with phosphine (PH_3) is a much longer process than MB fumigation. Whereas MB treatment times are of the order of a few hours, PH_3 must be applied for 2 or more days in most cases to be effective. Because of the time required for treatment, there has been little commercial use of PH_3. Nevertheless, PH_3 has been shown to be an effective fumigant against fruit flies (von Windeguth *et al.*, 1977; Seo *et al.*, 1979a) and cabbage looper (Leesch, 1984). Phytotoxity was not a problem when lettuce was treated with PH_3 at 0.28, 0.56, or 0.83 g l^{-1} at 4.4 or 24°C, except when the air was circulated at 24°C (Leesch, 1984). Tomato and bell pepper were uninjured after PH_3 fumigation at 5500 to 20 000 ppm h^{-1} and eggplant tolerated 5500 ppm h^{-1} at ambient temperatures of 18.5 to 25.1°C (Seo *et al.*, 1979a), although phytotoxity was observed in tomato exposed to 1700 ppm PH_3 for 4 days at 24°C compared to 250 or 350 ppm for 4 days at 14°C (Spalding *et al.*, 1978). Phosphine has been considered as a quarantine treatment for sweet potato weevil on sweet potatoes in Louisiana (P.L. Hornby, personal communication). The most likely use of PH_3 may be as a treatment applied during transport of vegetables in containers constructed for controlled atmosphere applications.

Hydrogen cyanide

Hydrogen cyanide (HCN) is not well suited for use as a fumigant due to its solubility in water and relatively high boiling point (26°C). Fumigation of vegetables with HCN seems to have been limited in recent years to disinfestation of products exported to Japan that have been found to harbour live insects (Stewart and Aharoni, 1983; Beever *et al.*, 1985). The treatment for asparagus in Japan was reported to be 1.8 g m^{-3} for 30 min (Beever *et al.*, 1985), while asparagus was reported to be damaged by exposure for 1 h at 20°C to 6 g m^{-3} but not 3 g m^{-3}. Lettuce was damaged by 2, 4, and 8 mg l^{-1} concentrations of HCN when applied for 0.5 or 1 h at 24°C (Leesch, 1984).

Acetaldehyde

Acetaldehyde vapour has been shown to be an effective fumigant for green peach aphid (Aharoni *et al.*, 1979) and cabbage looper (Leesch, 1984) on lettuce. There was no phytotoxicity on lettuce from acetaldehyde at 1.7, 6.7, or 13.4 mg l^{-1} at either 4 or 24°C as long as the atmosphere surrounding the product was circulated (Leesch, 1984). Vacuum fumigation has been shown to improve the penetration of acetaldehyde into wrapped, packaged heads of lettuce (Stewart *et al.*, 1980). Acetaldehyde also seems to possess antimicrobial properties. Deterioration of lettuce was reduced in acetaldehyde-treated heads compared to untreated heads (Leesch, 1984).

Ethyl formate

Ethyl formate (EF) has been tested for control of green peach aphid on lettuce (Stewart and Aharoni, 1983; Stewart and Mon, 1984) without much success. Concentrations of 0.5 to 2.0% EF resulted in equal aphid mortality, but the highest mortality was only 98% and concentrations greater than 0.5% were severely phytotoxic to the lettuce (Stewart and Aharoni, 1983). However, the work of Stewart and Mon (1984) was notable for demonstrating the feasibility of commercial-scale vacuum fumigation for non-gaseous or low volatility chemicals.

Other chemicals

Dimethoate and fenthion

The insecticides dimethoate and fenthion have been tested in Australia as postharvest dips or sprays for control of Queensland fruit fly in tomato (Swaine *et al.*, 1984; Heather *et al.*, 1987), and cucumber fly in zucchini and rockmelon (Heather et al., 1992). A concentration of 400 mg l^{-1} of either chemical gave essentially 100% mortality of eggs and larvae of both insects. There were no adverse effects on taste and residue levels were less than 2 mg kg^{-1}.

Heather *et al.* (1992) pointed out that there can be advantages to dip and spray treatments carried out in the packinghouse compared to fumigation, including the greater likelihood that packinghouse treatments will be carried out immediately after harvest. Delays before availability of fumigation facilities usually require commodities to be rewarmed, which can reduce storage life.

Environmental Treatments

Temperature

Vapour heat

Vapour heat (VH) treatments for disinfestation were developed for fruits over 60 years ago, but were subsequently overshadowed by fumigant treatments, especially those using EDB (Couey, 1989). Regulatory restrictions and the cost of developing new chemical fumigants are combining to generate renewed interest in VH and heat treatments in general (Couey, 1989). There is developing evidence that short-term high temperature treatments can be beneficial to fruits and fruit type vegetables by inhibiting ripening and decreasing decay (Paull, 1990). Standard VH treatments generally consist of a gradual warming to the treatment temperature over a number of hours, followed by holding at that temperature for the time necessary to kill the insect pest, then rapid cooling of the commodity back to ambient temperature (Couey, 1989; USDA-APHIS, 1992). The efficiency of heat transfer is increased over dry air by the latent heat of condensation of water vapour on the commodity surface. An alternative approach consists of more rapid heating to a specified temperature, followed by gradual warming to the treatment temperature. This has been called the 'quick run up' treatment (Couey, 1989).

Vapour heat treatments for vegetables have been developed for bitter momordica (bitter melon), eggplant, pepper, squash (including zucchini), and tomato (Sugimoto *et al.*, 1983; Furusawa *et al.*, 1984; Sungawa *et al.*, 1988; USDA-APHIS, 1992; Corcoran *et al.*, 1993). These treatments are all directed against fruit flies and use similar treatment temperatures of 43–45°C. Commodity injury is affected by heating rate, treatment temperature, and exposure time, with more gradual heating rates tending to reduce injury, perhaps due to a conditioning effect (Couey, 1989).

High-temperature forced-air

High-temperature, forced-air (HTFA) treatments have been recently developed for papaya (Armstrong *et al.*, 1989) and mango (Mangan and Ingle, 1992; Sharp, 1992) infested with fruit flies. This approach is inherently less efficient than VH in warming commodities, but may be more easily adapted to commercial conditions and existing equipment in those cases where temperatures lethal to insects can be achieved without damaging vegetables.

Hot water

Hot water (HW) is the most efficient heat transfer medium compared to VH and HTFA. Insect quarantine treatments using HW were developed for banana, mango, and papaya (Couey, 1989). Some fruit flies may survive in HW-treated papaya when blossom-end defects allow eggs to be laid within the fruit cavity, since the treatment is intended to kill eggs and larvae only within 5 mm of the fruit surface (Zee *et al.*, 1989). Therefore, papayas were screened for ripeness stage to limit potential fruit fly development, and for blossom end defects to avoid resistant insects in the fruit cavity. This situation is similar to that described in Tongan watermelons above (Stechmann *et al.*, 1988; Cowley *et al.*, 1991), and similar precautions will have to be taken when developing HW treatments for cucurbits and other vegetables. The registration of the hot water treatment for papaya was rescinded in July 1992 due to system failures (United States Animal and Plant Health Inspection Service, 1991).

Heat–mortality relationships for immersion in HW have been established for several fruit fly species (Jang, 1986; Sharp and Chew, 1987; Heard *et al.*, 1991). Hot water treatments of fruits and vegetables developed for decay control indicate that most commodities are able to tolerate exposure to 45–55°C water for 10–20 min (Couey, 1989). Peppers were able to tolerate 51°C for only 2.5 min, however, when treated with HW for control of *Erwinia* soft rot (Johnson, 1964). A 30–45 min dip in 48°C water was used to control imported crucifer weevil in horseradish (Foster *et al.*, 1979).

Inaba and Crandall (1988) determined the time–temperature tolerance of tomatoes to HW treatment by measuring electrolyte leakage as an indication of tissue damage. They found that the critical exposure times before injury occurred were 166 min at 45°C, 105 min at 50°C, and 34 min at 55°C, which is well beyond the tolerance of fruit flies (Jang, 1986; Sharp and Chew, 1987; Heard *et al.*, 1991). Cucumbers immersed in HW at temperatures, and for times, sufficient to kill fruit flies, however, are severely injured and show surface pitting (Chan and Linse, 1989). Chan and Linse (1989) also observed enhanced yellowing of cucumbers following HW treatment at 45 and 46°C for 30 min, but yellowing decreased as immersion time was increased to 60 min. This behaviour is probably related to stimulation of ethylene production by short HW treatments, with inhibition occurring after longer exposures. Inhibition of ethylene-forming enzyme (EFE) has been observed in HW-treated cucumber (Chan, 1986b), and also in papaya (Chan, 1986a). Conditioning cucumbers at 32.5°C for 24 h increased their tolerance of HW treatment from 30 to 60 min at 45°C and from less than 30 to 50 min at 46°C (Chan and Linse, 1989).

Cold treatment

Low temperatures are lethal to insects, such as fruit flies, that are of tropical origin (Benschoter, 1983; Benschoter and Witherell, 1984). Quarantine cold treatments (CT) are available (USDA-APHIS, 1992) or have been reported (Burditt and McAlister, 1982; Burditt and Balock, 1985) for treatment of fruits infested with fruit flies and false codling moth. These range from 10 days to as long as 22 days at temperatures ranging from slightly less than 0°C to about 2°C. Unfortunately, most of the vegetables of quarantine significance are chilling sensitive, and would be expected to be severely injured by such treatments. The length of exposure to cold required for quarantine security makes these treatments impractical even for temperate crops, except for export situations that involve extended transit periods, or for crops that are normally stored for some time before marketing.

An approach to reducing chilling injury that involves exposing commodities to temperatures near their chilling threshold for some period of time prior to exposure to chilling temperatures ('conditioning') has been successfully used to allow CT of grapefruit (Hatton and Cubbedge, 1982, 1983). Early-season Florida grapefruit are normally stored and transported at 16°C (the chilling threshold temperature) or above, but were able to tolerate up to 21 days at 1°C if first exposed to 16°C for 7 days, allowing CT during shipment from Florida to Japan. A large number of chilling-sensitive vegetables have been reported to respond favourably to conditioning, including cucumber, eggplant, pepper, tomato, and watermelon (Hatton, 1990). Thus, CT may be a viable option for quarantine treatment of some vegetables. The main drawback to the use of conditioning is that shelf life may be unacceptably shortened by storage at the relatively high conditioning temperatures prior to the extended storage at the low temperatures required for CT.

Atmospheric composition

Reduced oxygen and elevated carbon dioxide

The potential for the use of controlled atmospheres (CA) for disinfestation of horticultural crops has been pointed out by Davis and Jay (1977) and Klag (1985). Unfortunately, there is little information available on the tolerance of either vegetables or their insect pests to CA treatments that might be used for this purpose. The use of CA for insect control in grain storage, on the other hand, has been extensively studied (Jay, 1971; Bailey and Banks, 1975, 1980; Banks, 1984). Atmospheres containing less than about 2% O_2 or more than 60% CO_2 are lethal to stored product pests if exposure time and temperature are of the order of 4 days and 27°C, respectively (Jay, 1971). Treatment of fresh products at such high temperatures for extended periods is probably not practical, however. Studies with grain storage pests also indicate that the toxicity of both low O_2 and high CO_2 is increased in the presence of modified levels of the other gas. For example, Harein and Press (1968) found that atmospheres containing as much as 15% O_2 were lethal to larvae of two storage pests when about 36% CO_2 was present, and Calderon and Navarro (1979) found that the exposure times necessary for 100% mortality of red flour beetle adults in 2–5% O_2 atmospheres at 26 and 30°C were reduced from over 120 h to as little as 48 h when from 5 to 35% CO_2 was added to the atmosphere. Increasing the CO_2 level from 10 to 39% also increased the toxicity of low O_2 (0.5–2%) at 15°C on adults of five species of beetles in stored grain (Khrishnamurthy *et al.*, 1986).

The beneficial effects of CA and the levels of O_2 and CO_2 tolerated by most vegetables during long-term storage are well known (Saltveit, 1989). However, such tolerances have little relevance to potential insect disinfestation treatments, which would preferably be of limited duration. Recent work with sweet potato (Delate and Brecht, 1989) and sweet potato weevil (Delate *et al.*, 1990) illustrates the difference in tolerance to CA between long-term storage and short-term disinfestation treatments. The limits of tolerance of sweet potato to low O_2 and high CO_2 during storage have been reported as 7% and 10%, respectively, due to development of off-flavours and odours (Mattus and Hassan, 1968). Chang and Kays (1981) reported that sweet potato tolerated 5% O_2 for 1 week at 25°C and showed evidence of anaerobiosis at 2.5% O_2. In developing a disinfestation treatment for sweet potato weevil, Delate and Brecht (1989) found that sweet potato is able to tolerate 2 or 4% O_2 plus 40% CO_2, or 4% O_2 plus 40 or 60% CO_2 for 1 week at 25°C without injury. Exposure of lettuce to the equivalent of 0.26 or 0.40% O_2 for 24–52 h at 2°C in a low pressure system resulted in 100% mortality of green peach aphid with no detrimental effects to the lettuce (Aharoni *et al.*, 1986). Similar responses are likely with other vegetables.

Ke *et al.* (1991) have recently developed a model for predicting fruit tolerance to low O_2 atmospheres based on ethanol formation and off-flavour development. Similar, systematic approaches are needed to develop information on O_2 and CO_2 tolerances of vegetables, as well as the insects.

Semipermeable barriers

MODIFIED ATMOSPHERE PACKAGING

Modified atmosphere packaging (MAP) refers to the procedure of surrounding a commodity with a barrier to diffusion (usually a semipermeable polymeric film) in order to establish modified levels of O_2, CO_2, and

RH that are beneficial to the storage life of the commodity (Kader et al., 1989). Shetty *et al.* (1989) recently proposed that MAP may have potential for insect disinfestation. The levels of O_2 and CO_2 that could be used in MAP systems would be similar to those recommended for long-term CA storage. Thus, the atmospheres would in most cases not be optimal for insect mortality. Despite this, there are two possible advantages of MAP over CA for insect disinfestation that might tend to negate the drawback of suboptimal atmospheres: (i) when used for individual items or single-serving packages, the MAP would allow maintenance of the MA conditions through the entire postharvest system, up to the point of consumption, ensuring longer treatment times; and (ii) the MAP would provide a barrier to escape by emerging insects.

Shetty *et al.* (1989) and Jang (1990) applied a polyolefin shrink wrap (Cryovac D-955) to papayas infested with fruit fly eggs and larvae, and held the fruit for up to 6 days at about 24°C. No oriental fruit flies survived after 5 days in MAP (Shetty *et al.*, 1989). Although some medfly and melon fly larvae remained after 6 days, Jang (1990) noted that over 90% of the infested fruit had holes in the wrap or were loosely wrapped. He also observed that larvae moved out of the fruit within 30–60 min after the fruit were wrapped, and suggested that the larvae were reacting to a build-up of gases or volatile components within the fruit.

The limited information so far available indicates that MAP can be a useful commodity treatment. With the large number of films available, with wide ranges of O_2 and CO_2 permeabilities (Kader *et al.*, 1989), it should be possible to obtain insecticidal atmospheres applicable to most vegetables. In some cases, it may be desirable to expose vegetables in MAP to high temperatures or modified external atmospheres for some period initially in order to create a temporary insecticidal atmosphere that will ensure 100% insect mortality. Gould and Sharp (1990) estimated that mangoes would need to be held in shrink wrap over 16 days to achieve probit 9 mortality of caribfly, which could be too long for successful marketing of some more perishable vegetables.

Waxes and Other Coatings

Hallman *et al.* (1991) recently reported that the use of coatings dramatically reduced the length of the cold treatment necessary to kill caribfly in grapefruit. A mortality level of 100% was reached in 3 days when grapefruit were coated with FMC 360 wax and 7 days when coated with a polysaccharide-based edible film, compared to 10 days for uncoated fruit.

The presence or absence of coatings is a factor that has been typically ignored in commodity treatment research. Since most fruit-type vegetables are typically coated with any of several different waxes or oils in commercial operations, this factor needs to receive increased attention in future research. Edible coatings are available that possess differential permeability to O_2 and CO_2 (Nisperos-Carriedo *et al.*, 1991) and may be used to create an internal modified atmosphere.

Relative humidity

Insects are sensitive to dehydration and, thus, low relative humidity (RH) is damaging to insects. The effectiveness of CA in causing insect mortality is generally increased as RH is lowered (Davis and Jay, 1977). In fact, it has been suggested that the mode of action of high CO_2 against insects involves interference with spiracle function resulting in dehydration of the insects (Bell, 1984). However, the potential effects of low RH on insects infesting fresh products are probably limited to emerged stages due to the high water content and water-saturated intercellular spaces of plant tissues. Also, since vegetables tend to be relatively sensitive to water loss compared to fruits, there is probably little reason to consider low RH as a component of vegetable disinfestation treatment systems.

Ionizing radiation

The possible use of ionizing radiation (IR) for insect disinfestation has been proposed as one of its most promising applications (Burditt, 1982; Kader, 1986). Fresh fruits and vegetables may be treated with up to 100 krad of ionizing radiation (IR) in the USA (US Government, 1986). Much lower doses of IR (*ca.* 10–20 krad) are effective in sterilizing many insects such as fruit flies (Balock *et al.*, 1963, 1966) and pickleworm (Elsey and Brower, 1984) or their progeny; a dose of 40 krad was necessary to sterilize European corn borer (Anwar, 1968). This is the basis of the sterile insect release method that has been used in recent fruit fly eradication programmes (Mitchell and Saul, 1990).

Ionizing radiation in the range 15–50 krad will kill fruit flies (Seo *et al.*, 1973; Mitchell and Saul, 1990). Halfhill (1988) reported that three- to sixfold greater IR dose or time after treatment were required when IR was applied at 2.2°C compared to 20°C to asparagus infested with *Brachycorynella asparagi* aphids. Seo *et al.* (1973) found that subsequent storage of pepper and eggplant at 7.2°C after an IR treatment increased the mortality of the fruit flies compared to storage at 16–31°C. Some vegetables may be injured by levels of IR that kill insects, but most vegetables are able to tolerate at least 100 krad, which would result in complete mortality of insects (Kader, 1986). Tomatoes tolerated 25–100 krad with no impairment of quality, while cucumbers tolerated up to 50 krad without damage, but had a slight loss of crispness and flavour at 100 krad (Balock *et al.*, 1966).

Acceptance of sterilizing doses of IR as a quarantine treatment would require a change in the criterion for acceptability of a treatment from probit 9 mortality to a more reasonable 'inability to reproduce'. The use of IR for quarantine treatments has been questioned on several grounds: economic considerations, logistical and operational problems, phytotoxicity in some commodities, and consumer acceptance (Sommer and Mitchell, 1986). The need for specialized facilities to conduct IR treatments makes the likelihood of widespread adoption of IR as a commercial commodity treatment unlikely in locations other than those that are permanently affected by quarantine restrictions on exported produce.

Combined Treatments

Fumigation plus heat or cold

Combining fumigation with heat or cold treatments has been found to reduce the level of fumigant required for effective insect control. A HW treatment at 49°C for 20 min prior to fumigation of papayas with EDB at ambient temperature reduced the effective concentration of EDB from 16 to 8 g m^{-3} (Seo *et al.*, 1979b; Couey *et al.*, 1985). Fumigating at 33°C further reduced the effective EDB concentration to 4 g m^{-3}. A similar result was found for MB fumigation of stone fruits (Armstrong and Couey, 1984). Combining MB fumigation with heat treatments may be a viable approach for developing treatments for vegetables such as tomato that are only marginally tolerant of MB but which can withstand relatively severe HW treatments by allowing lower concentrations of MB to be used. Alternatively, a commodity such as pepper that is sensitive to both MB and HW treatments may be effectively treated with less extreme levels of fumigant and temperature in a combination treatment than would be necessary in separate treatments.

While the CT for fruit flies requires 10 to 16 days' exposure to 0–2.2°C, the CT required following fumigation with MB is only 4–10 days at 0.6–13.3°C (USDA-APHIS, 1985). Treatments combining MB and cold treatment also require less MB or shorter exposure time than treatments in which MB is used alone (Couey, 1983). Seo *et al.* (1971) found that the treatment (pre- or post-fumigation CT or MB) that is least effective should be given first and the most effective treatment given later to obtain the required insect mortality in the least amount of time. Chilling-sensitive vegetables might be more likely to tolerate the less extreme cold treatments required when MB is used. Shortening the length of the CT by using a combination treatment could make CT more commercially practicable for vegetables with relatively short storage life or which will be marketed after short transit periods.

Fumigation plus modified atmospheres

Bond (1984) suggested that low O_2 or high CO_2 may greatly increase the effectiveness of fumigants in grain storages. The increased effectiveness of low O_2 and high CO_2 when combined has been mentioned. Mixtures of air, CO_2, and either MB or PH_3 have been shown with grain storage insects to allow shorter treatment times than high CO_2 alone and, at the same time, lower concentrations of the fumigant to be used. This approach may be especially useful for PH_3 treatment of vegetables, where the main limitation to use of PH_3 seems to be the long treatment time.

Heat plus cold treatments

Couey *et al.* (1984) used a standard HW treatment of 49°C for 20 min that is used for decay control in papaya to reduce the CT necessary to kill fruit flies to 10 days at 8 or 9°C compared to 10–16 days at 0–2.2°C. The HW treatment alone reduced survival of fruit fly eggs in papayas to 3–4% of the control. A cold treatment alone of 5 or 7 days at 10°C reduced survival to 1.5 and 0.27%, respectively. Larvae were more resistant than eggs, so one-quarter or less ripe fruit was specified for the treatment to avoid fruit with insects at the larval stage. This approach would require little or no modification of normal handling procedures for many vegetables. The HW treatment serves a dual purpose of disinfestation and decay control, and allows CT at temperatures tolerated by chilling-sensitive crops. Some manipulation of HW temperature may allow the period of CT to be tailored to different transit times.

Heat and cold plus modified atmospheres

The effectiveness of CA is increased as the exposure temperature is increased (Davis and Jay, 1977). Fruit flies and other storage insects are killed at low temperatures below about 10°C (Couey *et al.*, 1984). These two observations led to the possibility that various combinations of heat, cold, and modified atmospheres, applied simultaneously or sequentially, may have merit as quarantine treatments. Benschoter (1987) exposed caribfly eggs and larvae to combinations of 2, 10, or 20% O_2 and 20, 50, or 80% CO_2 at 10 or 15.6°C for 3, 5, 7, or 10 days. Mortality was greater at the higher temperature, but was not much affected by the O_2 level. Carbon dioxide concentration and exposure time had the greatest effects on mortality. These results suggest that high-temperature CA, perhaps followed by CT, might be the most likely combination for maximum effectiveness of these types of treatment. This approach could be compatible with MAP systems on vegetables.

Conclusion

With the exception of insecticides that specifically target the insect pest, the effectiveness of the chemical and environmental treatments used for disinfestation of vegetables depends on the condition that the relative tolerance of the insect must be less than that of the commodity. Some stress to the commodity is generally unavoidable when using such methods, however. The deterioration in consumer tolerance for chemical residues on fresh products strongly favours the development of residue-free environmental quarantine treatments for vegetables. Advantages of heat, cold, or atmosphere treatments over fumigation and other chemical treatments also include increased worker safety, and little or no requirement for aeration following treatment. The likelihood that insects could develop resistance to environmental treatments is very small due to the many biochemical alterations that would be necessary for such an occurrence.

Combination treatments that debilitate and then kill the insect pests or eliminate the possibility of subsequent reproduction are probably most likely to result in avoidance of damage to the commodity. High temperature and modified atmosphere treatments that can potentially have beneficial effects on the commodity beyond disinfestation are especially worthy of further investigation.

References

Aharoni, Y., Stewart, J.K., Hartsell, P.O. and Young, D.K. (1979) Acetaldehyde – a potential fumigant for control of the green peach aphid on harvested head lettuce. *Journal of Economic Entomology* 72, 493–495.

Aharoni, Y., Apelbaum, A. and Copel, A. (1986) Use of reduced atmospheric pressure for control of the green peach aphid on harvested head lettuce. *HortScience* 21, 469–470.

Anwar, M. (1968) Sterilization of the European corn borer (*Ostrinia nubilalis* Hb.) by gamma irradiation. In: *Isotopes and Radiation in Entomology*. STI/PUB/166, IAEA, Vienna, Austria.

Armstrong, J.W. and Couey, H.M. (1984) Methyl bromide treatments at 30°C for California stone fruits infested with the Mediterranean fruit fly. *Journal of Economic Entomology* 77, 1229–1232.

Armstrong, J.W. and Garcia, D.L. (1985) Methyl bromide quarantine fumigations for Hawaii-grown cucumbers infested with melon fly and oriental fruit fly (Diptera: Tephritidae). *Journal of Economic Entomology* 78, 1308–1310.

Armstrong, J.W., Hansen, J.D., Hu, B.K.S. and Brown, S.A. (1989) High-temperature, forced-air quarantine treatment for papayas infested with tephritid fruit flies (Diptera:Tephritidae). *Journal of Economic Entomology* 82, 1667–1674.

Bailey, S.W. and Banks, H.J. (1975) The use of controlled atmospheres for the storage of grain. *Proceedings of the First International Working Conference on Stored-Product Entomology*, Savannah, Georgia, pp. 362–374.

Bailey, S.W. and Banks, H.J. (1980) A review of recent studies of the effects of controlled atmospheres on stored products pests. In: Shejbal, J. (ed.), *Controlled Atmosphere Storage of Grains*. Elsevier, Amsterdam, pp. 101–118.

Balock, J.W., Burditt, A.K., Jr and Christenson, L.D. (1963) Effects of gamma radiation on various stages of three fruitfly species. *Journal of Economic Entomology* 56, 42–46.

Balock, J.W., Burditt, A.K., Jr, Seo, S.T. and Akamine, E.K. (1966) Gamma radiation as a quarantine treatment for Hawaiian fruit flies. *Journal of Economic Entomology* 59, 202–204.

Banks, H.J. (1984) Current methods and potential systems for production of controlled atmospheres for grain storage. In: Ripp, B.E. (ed.), *Controlled Atmosphere and Fumigation in Grain Storage*. Elsevier, Amsterdam, pp. 523–528.

Bateman, M.A. (1972) The ecology of fruit flies. *Annual Review of Entomology* 17, 493–518.

Beever, D.J., Yearsley, C.W. and Hogg, M.G. (1985) Effect of post-harvest fumigation on quality of asparagus spears. *New Zealand Journal of Agricultural Research* 28, 537–543.

Bell, C.H. (1984) Effects of oxygen on the toxicity of carbon dioxide to storage insects. In: Ripp, B.E. (ed.), *Controlled Atmosphere and Fumigation in Grain Storages*. Elsevier, Amsterdam, pp. 67–73.

Benschoter, C.A. (1983) Lethal effects of cold storage temperatures on Caribbean fruit fly in grapefruit. *Proceedings of the Florida State Horticultural Society* 96, 318–319.

Benschoter, C.A. (1987) Effects of modified atmospheres and refrigeration temperatures on survival of eggs and larvae of Caribbean fruit fly (Diptera: Tephritidae) in laboratory diet. *Journal of Economic Entomology* 80, 1223–1225.

Benschoter, C.A. and Witherell, P.C. (1984) Lethal effects of suboptimal temperatures on immature stages of *Anastrepha suspensa*. *Florida Entomologist* 67, 189–193.

Bond, E.J. (1984) Fumigation as part of an integrated pest management program. In: Ripp, B.E. (ed.), *Controlled Atmosphere and Fumigation in Grain Storages*. Elsevier, Amsterdam, pp. 723–735.

Brecht, J.K. (1988) Attenuation of methyl-bromide-induced ripening inhibition in tomato fruit by propylene. *Plant Physiology* 86, S-44 (Abstract).

Brecht, J.K., Huber, D.J., Sherman, M. and Lee, J. (1986) Methyl bromide inhibits ripening and ethylene production in tomato (*Lycopersicon esculentum* Mill.) fruit. *Journal of Plant Growth Regulation* 5, 29–35.

Brown, R.E. (1985) Approved quarantine treatments: the Florida situation. *Citrograph* 70, 64–65.

Burditt, A.K., Jr (1982) Food irradiation as a quarantine treatment of fruits. *Food Technology* 36, 51–62.

Burditt, A.K., Jr and Balock, J.W. (1985) Refrigeration as a quarantine treatment for fruits and vegetables infested with eggs and larvae of *Dacus dorsalis* and *Dacus cucurbitae* (Diptera: Tephritidae). *Journal of Economic Entomology* 78, 885–887.

Burditt, A.K., Jr and McAlister, L.C., Jr (1982) Refrigeration as a quarantine treatment for fruit infested with eggs and larvae of *Anastrepha* species. *Proceedings of the Florida State Horticultural Society* 95, 224–226.

Calderon, M. and Navarro, S. (1979) Increased toxicity of low-oxygen atmospheres supplemented with carbon dioxide on *Tribolium castaneum* adults. *Entomologia Experimentalis et Applicata* 25, 39–44.

Chan, H. T., Jr (1986a) Effects of heat treatments on the ethylene forming enzyme system in papayas. *Journal of Food Science* 51, 581–583.
Chan, H. T., Jr (1986b) Heat inactivation of the ethylene forming enzyme system in cucumbers. *Journal of Food Science* 51, 1491–1493.
Chan, H.T., Jr and Linse, E. (1989) Conditioning cucumbers for quarantine heat treatments. *HortScience* 24, 985–989.
Chang, L.A. and Kays, S.J. (1981) Effect of low oxygen storage on sweet potato roots. *Journal of the American Society for Horticultural Science* 106, 481–483.
Corcoran, R.J., Heather, N.W. and Heard, T.A. (1993) Vapor heat treatment for zucchini infested with *Bactrocera cucumis* (Diptera: Tephritidae). *Journal of Economic Entomology* 86, 66–69.
Couey, H. M. (1983) Development of quarantine systems for host fruits of the medfly. *HortScience* 18, 45–47.
Couey, H.M. (1989) Heat treatment for control of postharvest diseases and insect pests of fruits. *HortScience* 24, 198–202.
Couey, H.M., Linse, E.S. and Nakamura, A.N. (1984) Quarantine procedure for Hawaiian papayas using heat and cold treatments. *Journal of Economic Entomology* 77, 984–988.
Couey, H.M., Armstrong, J.W., Hylin, J.W., Thornburg, W., Nakamura, A.N., Linse, E.S., Ogata, J. and Vetro, R. (1985) Quarantine procedure for Hawaii papaya, using a hot-water treatment and high-temperature, low-dose ethylene dibromide fumigation. *Journal of Economic Entomology* 78, 879–884.
Cowley, J.M., Baker, R.T., Englberger, K.G. and Langi, T.G. (1991) Methyl bromide fumigation of Tongan watermelons against *Bactrocera xanthodes* (Diptera: Tephritidae) and analysis of quarantine security. *Journal of Economic Entomology* 84, 1763–1767.
Davis, R. and Jay, E. G. (1977) The current status of controlled atmospheres as a method of insect control. In: Dewey, D.H. (ed.), *Controlled Atmospheres for the Storage and Transport of Perishable Agricultural Commodities.* Horticultural Report 28. Michigan State University, East Lansing, pp. 207–211.
Delate, K.M. and Brecht, J.K. (1989) Quality of tropical sweet potatoes exposed to controlled-atmosphere treatments for postharvest insect control. *Journal of the American Society for Horticultural Science* 114, 963–968.
Delate, K.M., Brecht, J.K. and Coffelt, J.A. (1990) Controlled atmosphere treatments for control of sweet potato weevil (Coleoptera: Curculionidae) in stored tropical sweet potatoes. *Journal of Economic Entomology* 83, 461–465.
Elsey, K.D. and Brower, J. (1984) Sterilization of the pickleworm (Lepidoptera: Pyralidae) by ionizing radiation or heat. *Journal of Economic Entomology* 77, 1236–1239.
Foster, R.E., Sherrod, D.E., Eastman, C.E. and Randell, R. (1979) Treatment of horseradish propagative stocks for control of the imported crucifer weevil. *Journal of Economic Entomology* 72, 555–556.
Furusawa, K., Sugimoto, T. and Gaja, T. (1984) The effectiveness of vapor heat treatment against the melon fly, *Dacus cucurbitae* Coquillett in eggplant and fruit tolerance to the treatment. *Research Bulletin Plant Protection Service Japan* 20, 17–24.
Gould, W.P. and Sharp, J.L. (1990) Caribbean fruit fly (Diptera: Tephritidae) mortality induced by shrink-wrapping infested mangoes. *Journal of Economic Entomology* 83, 2324–2326.

Halfhill, J. E. (1988) Irradiation disinfestation of asparagus spears contaminated with *Brachycorynella asparagi* (Mordvilko) (Homoptera: Aphididae). *Journal of Economic Entomology* 81, 873–876.

Hallman, G.J., Nisperos-Carriedo, M. and Baldwin, E.A. (1991) Use of edible vegetable oil coatings in postharvest insect quarantine treatments of fruits. *Florida State Horticultural Society Program and Abstracts* 104, 47 (Abstract).

Harein, P.K. and Press, A.F. (1968) Mortality of stored-peanut insects exposed to mixtures of atmospheric gases at various temperatures. *Journal of Stored Product Research* 4, 77–82.

Hatton, T.T. (1990) Reduction of chilling injury with temperature manipulation. In: Wang, C.Y. (ed.), *Chilling Injury of Horticultural Crops*. CRC Press, Boca Raton, Florida, pp. 269–280.

Hatton, T.T. and Cubbedge, R.H. (1982) Conditioning Florida grapefruit to reduce chilling injury during low-temperature storage. *Journal of the American Society for Horticultural Science* 107, 57–60.

Hatton, T.T. and Cubbedge, R.H. (1983) Preferred temperature for prestorage conditioning of 'Marsh' grapefruit to prevent chilling injury at low temperatures. *HortScience* 18, 721–722.

Heard, T.A., Heather, N.W. and Corcoran, R.J. (1991) Dose–mortality relationships for eggs and larvae of *Bactrocera tryoni* (Diptera: Tephritidae) immersed in hot water. *Journal of Economic Entomology* 84, 1768–1770.

Heather, N.W., Hargreaves, P.A., Corcoran, R.J. and Melksham, K.J. (1987) Dimethoate and fenthion as packing line treatments for tomatoes against *Dacus tryoni* (Froggatt). *Australian Journal of Experimental Agriculture* 27, 465–469.

Heather, N.W., Walpole, D.E., Corcoran, R.J., Hargreaves, P.A. and Jordan, R.A. (1992) Postharvest quarantine disinfestation of zucchinis and rockmelons against *Bactrocera cucumis* (French) using insecticide dips of fenthion or dimethoate. *Australian Journal of Experimental Agriculture* 32, 241–244.

Inaba, M. and Crandall, P. G. (1988) Electrolyte leakage as an indicator of high-temperature injury to harvested mature green tomatoes. *Journal of the American Society for Horticultural Sciences* 113, 96–99.

Jang, E.B. (1986) Kinetics of thermal death in eggs and first instars of three species of fruit flies. *Journal of Economic Entomology* 79, 700–705.

Jang, E.B. (1990) Fruit fly disinfestation of tropical fruits using semipermeable shrinkwrap film. *Acta Horticulturae* 269, 453–458.

Jay, E.G. (1971) Suggested conditions and procedures for using carbon dioxide to control insects in grain storage facilities. USDA, ARS Miscellaneous Publication 51–46.

Johnson, H.B. (1964) Effect of hot water treatments and hydrocooling on postharvest bacterial soft rot in bell peppers. *USDA Marketing Research Report 517*.

Kader, A.A. (1986) Potential applications of ionizing radiation in postharvest handling of fresh fruits and vegetables. *Food Technology* 40, 117–121.

Kader, A.A., Zagory, D. and Kerbel, E.L. (1989) Modified atmosphere packaging of fruits and vegetables. *Critical Reviews in Food Science and Nutrition* 28, 1–30.

Ke, D., Rodriguez-Sinobas, L. and Kader, A.A. (1991) Physiology and prediction of fruit tolerance to low oxygen atmospheres. *Journal of the American Society for Horticultural Science* 116, 253–260.

Khrishnamurthy, T.S., Spratt, E.C. and Bell, C.H. (1986) The toxicity of carbon dioxide to adult beetles in low oxygen atmospheres. *Journal of Stored Products Research* 22, 145–152.

Klag, N. (1985) Use of CA for quarantine control of insects on fresh fruits and vegetables. In: Blankenship, S.M. (ed.), *Controlled Atmospheres for Storage and Transport of Perishable Agricultural Commodities.* Horticultural Report 126. North Carolina State University, Raleigh, pp. 199–206.

Knight, R.J., Spalding, D.W., King, J.R., von Windeguth, D.L., Benschoter, C.A., Burditt, A.J., Jr and Fons, J. (1980) Results of fumigation of fruits and vegetables of southern Mexico to control the Mediterranean fruit fly. *Proceedings of the American Society for Horticultural Science Tropical Region* 24, 117–126.

Leesch, J.G. (1984) Fumigation of lettuce: efficacy and phytotoxicity. *Journal of Economic Entomology* 77, 142–150.

Lipton, W.J. and Tebbets, J.S. (1984) Methyl bromide and ethylene dibromide as potential quarantine treatments of cantaloupe and honey dew muskmelons against Tephritid fruit flies. *Acta Horticulturae* 157, 161–186.

Lipton, W.J., Tebbets, J.S., Spitler, G.H. and Hartsell, P.L. (1982) Commodity treatments: responses of tomatoes and green peppers to fumigation with methyl bromide or ethylene dibromide. *USDA Marketing Research Report 1125.*

Mangan, R.L. and Ingle, S.J. (1992) Forced hot-air quarantine treatment for mangoes infested with West Indian fruit fly (Diptera: Tephritidae). *Journal of Economic Entomology* 85, 1859–1864.

Mattus, G.E. and Hassan, F.M. (1968) Controlled atmosphere storage studies with sweet potatoes. *HortScience* 3, 90 (Abstract).

Mitchell, W.C. and Saul, S.H. (1990) Current control methods for the Mediterranean fruit fly, *Ceratitis capitata*, and their application in the USA. *Review of Agricultural Entomology* 78, 923–940.

Nisperos-Carriedo, M.O., Baldwin, E.A. and Shaw, P.E. (1991) Development of an edible coating for extending postharvest life of selected fruits and vegetables. *Proceedings of the Florida State Horticultural Society* 104, 122–125.

Paull, R. E. (1990) Postharvest heat treatments and fruit ripening. *Postharvest News and Information* 1, 355–363.

Robinson, A.F. and Hooper, G.H.S. (eds) (1989) *Fruit Flies: Their Biology, Natural Enemies and Control*. Elsevier, Amsterdam.

Saltveit, M.E. (1989) A summary of requirements and recommendations for the controlled atmosphere storage of harvested vegetables. In: Fellman, J.K. (ed.), *Proceedings Fifth International Controlled Atmosphere Research Conference.* Washington State University, Wenatchee, pp. 329–352.

Seo, S.T., Kobayashi, R.T., Steiner, L.F., Balock, J.W., Komura, M. and Lee, C.Y.L. (1971) Fumigation with methyl bromide plus refrigeration to control infestation of fruit flies in agricultural commodities. *Journal of Economic Entomology* 64, 1270–1274.

Seo, S.T., Kobayashi, R.M., Chambers, D.L., Dollar, A.M. and Hanaoka, M. (1973) Hawaiian fruit flies in papaya, bellpepper, and eggplant: quarantine treatment with gamma irradiation. *Journal of Economic Entomology* 66, 937–939.

Seo, S.T., Akamine, E.K., Goo, T.T.S., Harris, E.J. and Lee, C.Y.L. (1979a) Oriental and Mediterranean fruit flies: fumigation of papaya, avocado, tomato, bell pepper, eggplant, and banana with phosphine. *Journal of Economic Entomology* 72, 354–359.

Seo, S.T., Chambers, D.L., Akamine, E.K., Harris, E.J., Lee, C.Y.L. and Komura, M. (1979b) Oriental fruit fly: factors affecting ethylene dibromide fumigation of infested papayas. *Journal of Economic Entomology* 72, 350–353.

Sharp, J.L. (1992) Hot-air quarantine treatment for mango infested with Caribbean fruit fly (Diptera: Tephritidae). *Journal of Economic Entomology* 85, 2302–2304.

Sharp, J.L. and Chew, V. (1987) Time/mortality relationships for *Anastrepha suspensa* (Diptera: Tephritidae) eggs and larvae submerged in hot water. *Journal of Economic Entomology* 80, 646–649.

Shetty, K. ., Klowden, M.J., Jang, E.B. and Kochan, W.J. (1989) Individual shrink wrapping: a technique for fruit fly disinfestation in tropical fruits. *HortScience* 24, 317–319.

Sommer, N.F. and Mitchell, F.G. (1986) Gamma irradiation – a quarantine treatment for fresh fruits and vegetables? *HortScience* 21, 356–360.

Spalding, D.H., King, J.R., Benschoter, C.A., von Windeguth, D.L., Reeder, W.F. and Burditt, A.K., Jr (1978) Ethylene dibromide, methyl bromide and phosphine fumigation of tomatoes. *Proceedings of the Florida State Horticultural Society* 91, 156–158.

Stechmann, D.H., Englberger, K. and Langi, T.F. (1988) Estimation of mortality of *Dacus xanthodes* (Broun) maggots in fumigated and non-fumigated watermelons, a fruitfly (Dipt.: Tephritidae) of plant quarantine importance in the Pacific region. *Anzeiger für Schadlingskunde, Pflanzenschutz, Umweltschutz* 61, 125–129.

Stewart, J.K. and Aharoni, Y. (1983) Vacuum fumigation with ethyl formate to control the green peach aphid in packaged head lettuce. *Journal of the American Society for Horticultural Science* 108, 295–298.

Stewart, J.K. and Mon, T.R. (1984) Commercial-scale vacuum fumigation with ethyl formate for postharvest control of the green peach aphid (Homoptera: Aphididae) on film-wrapped lettuce. *Journal of Economic Entomology* 77, 569–573.

Stewart, J.K., Aharoni, Y., Hartsell, P.L. and Young, D.K. (1980) Acetaldehyde fumigation at reduced pressures to control the green peach aphid on wrapped and packed head lettuce. *Journal of Economic Entomology* 73, 149–152.

Sugimoto, T., Furusawa, K. and Mizobuchi, M. (1983) The effectiveness of vapor heat treatment against the oriental fruit fly, *Dacus dorsalis* Hendel, in green pepper and fruit tolerance to the treatment. *Research Bulletin Plant Protection Service Japan* 19, 81–88.

Sungawa, K., Kume, K. and Ishikawa, R. (1988) Efficacy of vapor heat treatment for bitter momordica fruit infested with melon fly *Dacus cucurbitae* (Coquillet), (Diptera: Tephritidae). *Research Bulletin Plant Protection Service Japan* 24, 1–5.

Swaine, G., Hargreaves, P.A., Jackson, D.E. and Corcoran, R.J. (1984) Dimethoate dipping of tomatoes against Queensland fruit fly, *Dacus tryoni* (Froggatt). *Australian Journal of Experimental Agriculture and Animal Husbandry* 24, 447–449.

USDA-APHIS (United States Department of Agriculture. Animal and Plant Health Inspection Service) (1985) Plant protection and quarantine treatment manual. USDA. APHIS. Frederick, Maryland.

USDA-APHIS (United States Department of Agriculture, Animal and Plant Health Inspection Service) (1991) Papayas from Hawaii, Final Rule Docket No. 91-139 U.S. Dept Agric. *Federal Register* 56(227), 59205.

USDA-APHIS (United States Department of Agriculture. Animal and Plant Health Inspection Service) (1992) Plant protection and quarantine treatment manual. USDA. APHIS. Frederick, Maryland.

United States Government (1984) Ethylene dibromide; amendment of notice to cancel registrations of pesticide products containing ethylene dibromide. *Federal Register* 49, 14182–14185.

United States Government (1986) Irradiation in the production, processing, and handling of foods; final rule. *Federal Register* 51, 13376–13399.

von Windeguth, D.L., Arner, A., Burditt, A.K., Jr and Spalding, D.H. (1977) Phosphine as a fumigant for grapefruit infested by Caribbean fruit fly larvae. *Proceedings of the Florida State Horticultural Society* 90, 144–147.

Zee, F.T., Nishina, M.S., Chan, H.T., Jr and Nishijima, K.A. (1989) Blossom end defects and fruit fly infestation in papayas following hot water quarantine treatment. *HortScience* 24, 323–325.

Ornamentals and Flowers 17

ARNOLD H. HARA
Department of Entomology, University of Hawaii, Beaumont Agricultural Research Center, 461 W Lanikaula Street, Hilo, HI 96720, USA.

The enormous increase in per capita consumption of floriculture products during the past four decades is expected to continue throughout the world (Tayama, 1991). This worldwide growth in exports and imports of floriculture products has in turn caused increases in shipment rejections of flowers and foliage contaminated with insect pests of quarantine significance. In addition, propagative cuttings of flowers such as chrysanthemum and poinsettia are shipped regularly worldwide, and insect-infested cuttings have been implicated as a source of spread of serious pests. For example, Spencer (1981) proposed that the agromyzid leafminer (*Liriomyza trifolii* (Burgess)), originally a major pest of chrysanthemums in Florida, was introduced into numerous nurseries in England, France, and The Netherlands in 1977 from a propagating nursery in Kenya that was regularly supplied with infested plants from Florida. Furthermore, the original population in Florida had probably already developed resistance to commonly used insecticides (Nakahara, 1982). In many importing countries of chrysanthemums and gypsophila, the agromyzid leafminer is now a quarantine pest, and these countries ban the entry of infested flowers (Yathom *et al.*, 1990). To prevent spread of serious pests, certain countries such as Japan and the US have extremely strict quarantine regulations. Of the 162 million stalks of cut flowers imported into Japan from throughout the world in 1986, pests and diseases were found in 32.5% of the total inspected flowers and foliage (Agricultural Production Bureau, 1988). Infested flowers are either fumigated with hydrogen cyanide or methyl bromide, or they are destroyed.

Certified quarantine treatments and procedures are essentially focused on quarantine pests of edible fruits, nuts, and vegetables (Animal

From: Paull, R.E. and Armstrong, J.W. (eds) (1994) *Insect Pests and Fresh Horticultural Products: Treatments and Responses.* CAB INTERNATIONAL, Wallingford, UK.

and Plant Health Inspection Service, 1978, 1985) with quarantine research on disinfestation of floriculture crops primarily limited to commodities suffering shipment rejections. Current quarantine treatments approved for cut flowers by the Animal and Plant Health Inspection Service (APHIS) (1978, 1986) are hand removal, chemical dips, and methyl bromide fumigation. This review attempts to summarize the research status of quarantine procedures and postharvest disinfestation treatments for floriculture crops and their responses to these treatments.

Aerosol/Fog

Aerosols produce droplets below 10 μm in diameter (Ware, 1989) and are not as effective as fumigants in penetrating flowers. Their effectiveness depends on the location of insects in flowers, the complexity of flower structure, and the arrangement of flower bunches during spraying (Seaton and Joyce, 1988). Unsatisfactory results obtained with permethrin or pyrethrin applied as aerosols can be overcome through the use of forced draft application and/or vaporized formulation of the insecticides (Seaton and Joyce, 1989). The use of a thermal fogging device with forced draft (created by a blower circulating the fog within the chamber) was tested on dendrobium orchids infested with western flower thrips (*Frankliniella occidentalis* (Pergande)) (Hara, 1992, unpublished data). Effective control is obtained with abamectin 0.5 EC and chlorpyrifos 4.0 EC with 94% and 99% thrips reduction, respectively, after a 3 h fumigation period (Hara, 1992 unpublished data). No significant phytotoxicity or vase life reduction is observed with either insecticide. Wang and Lin (1984b) tested organophosphates and synthetic pyrethroids delivered as a fog (smoke) for control of aphids and thrips on chrysanthemums and concluded that the synthetic pyrethroids, namely, deltamethrin, fenvalerate, and permethrin give the most effective control.

Controlled Atmospheres

High carbon dioxide or nitrogen with low oxygen and carbon dioxide at low temperatures were investigated for flowers. High carbon dioxide or nitrogen is unsatisfactory for controlling earwigs and garden weevil on protea (Maughan, 1986). However, based on the test insects, confused flour beetle (*Tribolium confusum* Jacquelin du Val) and the Mediterranean fruit fly (*Ceratitis capitata* (Wiedemann)), a promising controlled atmosphere treatment for Australian cut flowers is 30–45% CO_2 at 0–1°C for 1 week; though some damage occurs on cut flowers. Confused flour beetle and Mediterranean fruit fly were chosen as the test insects because they are easily reared and do not naturally infest cut flowers and foliage, thus avoiding the situation where international trade of Australian products can be restricted on the basis of the results (Seaton and Joyce, 1989). This treatment is suited to sea freighted flowers, since treatment time is 1 week or longer (Seaton and Joyce, 1989). Some tropicals, however, are subject to cold injury at storage temperatures below 10°C (Halevy *et al.*, 1978).

Fumigation

Aluminium or magnesium phosphide (phosphine)

Aluminium or magnesium phosphide reacts with atmospheric moisture to release hydrogen phosphide (PH_3, phosphine), an effective insect-killing gas. Phosphine is extensively used in insecticidal fumigation of animal feed, bulk grain, cottonseed, peanuts, and processed food and in space fumigation of flour mills, railcars, and warehouses (*Farm Chemicals Handbook*, 1992). Researchers have investigated phosphine in disinfestation of fresh fruits, but fruit quality is adversely affected (Seo *et al.*, 1979; Soderstrom *et al.*, 1991). Also, long exposures are needed to achieve sufficient control compared to other fumigants (APHIS, 1980; Wit and van de Vrie, 1985b). Wang and Lin (1984b) in Taiwan found that the combination of xylene (8.5 ml m^{-3}) and magnesium phosphide (16.6 g m^{-3}) for 4 h kills 100%, 96.9%, and 98% of thrips, aphids, and spider mites, respectively, with no phytotoxicity to chrysanthemum cut flowers; higher concentrations are phytotoxic, and magnesium phosphide without xylene is not as effective. Phytotoxicity will probably be a limiting factor on other flowers.

Hydrogen cyanide

Hydrogen cyanide (HCN) is a highly toxic fumigant that has been used to control insects in dormant nursery stock, ornamentals, and glasshouse plants (Monro, 1969). HCN fumigation is a treatment that is currently approved for specific insect pests of cotton, grain, and tobacco (Animal and Plant Health Inspection Service, 1986); cut flowers and foliage are not included. In Japan, during the warm growing season, infested imported living plants and fresh fruits are fumigated with 2500 ppm HCN for 30 min; the rate is doubled during the colder winter months (Japanese Fumigation Engineering Society, 1981). Although HCN is very toxic to humans and mammals, it is readily detectable by a characteristic odour, and safety procedures for its use have been established (Monro, 1969; Animal and Plant Health Inspection Service, 1980; Japanese Fumigation Engineering Society, 1981). HCN fumigation of *Cacoeciomorpha pronubana* (Hub.) larvae on carnations results in total mortality of larvae at 2 g m^{-3} but induces damage in the pale-coloured variety 'Anita' when water droplets are present on the flowers (Wit and van de Vrie, 1985b). Hansen *et al.* (1991a) measured HCN phytotoxicity towards various Hawaiian cut flowers and foliage; all tested tropical foliage and most heliconia are undamaged at fumigation levels of 5500 ppm HCN; most protea and 'Midori' anthuriums (*Anthurium andraeanum* Andre) are uninjured at 4600 ppm HCN; pink and red ginger (*Alpinia purpurata* (Vieil.) K. Schum) are uninjured at 3700 ppm HCN; some moist tropical flowers and all pincushion protea (*Leucospermum* sp.) show phytotoxic reaction to HCN. Efficacy of 85–100% mortality is obtained at 3700 ppm HCN for 30 min against banana aphid (*Pentalonia nigronervosa* (Coquerel)), Cockerell scale or magnolia white scale (*Pseudaulacaspis cockerelli* (Cooley)), green scale (*Coccus viridis* (Green)), coconut mealybug (*Nipaecoccus nipae* (Maskell)), citrus mealybug (*Planococcus citri* (Risso)), obscure mealybug (*Pseudococcus affinis* (Maskell)), and longtailed mealybug (*Pseudococcus longispinus* (Targioni-Tozzetti)) (Hansen *et al.*, 1991b). In the US, HCN fumigation for agricultural use is no longer available (Degesch America, Inc., personal communication) due to EPA requirements for additional toxicological data; the cost of providing these data exceeds the company's financial commitment to HCN.

Methyl bromide

Methyl bromide fumigation without chloropicrin is widely used by many countries as a standard quarantine treatment for various infested commodities including flowers and foliage (Animal and Health Inspection Service, 1988). Pure methyl bromide, circulated thoroughly, is used by Thailand on orchids exported to Europe and America (Sanguthai, 1991). Fumigation at or below the federally approved dosage of 32 g m^{-3} methyl bromide for 2 h causes damage or reduces vase life of certain floral products (Animal and Plant Health Inspection Service, 1978; Wit and van de Vrie, 1985b; Maughan, 1986; Seaton and Joyce, 1988, 1989). Yathom *et al.* (1990) concluded that the use of methyl bromide on cut flowers, for example gypsophila, poses some problems because the margin between the effective and phytotoxic doses is narrower than for other plant material. Wit and van de Vrie (1985b) demonstrated that, as the treatment temperature increases from 17 to 23°C at 30 g m^{-3} methyl bromide for 1.5 h, phytotoxicity significantly increases in cut flower species such as alstroemeria, chrysanthemums, and carnations. Cultivar differences in sensitivity to methyl bromide are observed in roses, alstroemeria, chrysanthemum, and carnation (Wit and van de Vrie, 1985b). Therefore, pure methyl bromide is an effective fumigant that can be safely used only on certain cultivars of flowers and foliage at specific temperature and dosage. The continued availability of methyl bromide in the US and elsewhere is questionable due to its ozone depletion potential (ODP). The US Environmental Protection Agency (EPA) has stated in the Clean Air

Act that methyl bromide has an ODP of between 0.44 and 0.69, and the Clean Air Act will require all Class I ozone depletors (any substance with an ODP of 0.2 or greater) phased out of production no later than the year 2000 (*Pesticide and Toxic Chemical News*, 1992).

Other fumigants

Sulfur dioxide fumigation of pink mink protea (*Protea neriifolia* R.Br.) provides 73% mortality of earwigs after a 1.5 h exposure with some bleaching of flowers and foliage; the treatment shows promise but needs further testing (Maughan, 1986). Dichlorvos (DDVP), delivered in impregnated polyvinyl chloride (PVC) resin strips, has been tested on cut flowers and potted foliage plants (Hamlen and Henley, 1979; Seaton and Joyce, 1988) with 94–99% reduction of a solanum mealybug (*Phenacoccus solani* Ferris), the green peach aphid (*Myzus persicae* Sulzer), and the two-spotted spider mite (*Tetranychus urticae* Koch). The most effective method of dichlorvos exposure is multiple fumigations at 7 day intervals in sealed polyethylene bags, however total elimination of infestations is never achieved (Hamlen and Henley, 1979). Dichlorvos is phytotoxic to certain tropical foliage such as *Aphelandra squarrosa* Nees., *Chamaedorea elegans* Mart., *Nephrolepis exaltata* (L.) Schott, and *Peperomia obtusifolia* (L.) A. Dietr. (Hamlen and Henley, 1979). Dichlorvos propelled by CO_2 and applied as an aerosol (Insectigas, Commonwealth Industrial Gases, Ltd (CIG), Perth, Australia) is not 100% effective against a bioassay insect, confused flour beetle (*T. confusum*), probably due to poor fumigant penetration into the tightly packed flower bunches (Seaton and Joyce, 1989). Coetzee and Wright (1992) reported that dichlorvos fumigation of protea flowers for 16 h provides 91.6% mortality of insects. When protea flowers are treated in an enclosed chamber for 2 h with a combination of pyrethrin (Pestigas, CIG, Perth, Australia) and dichlorvos (Insectigas, CIG, Perth, Australia), both propelled by carbon dioxide, the combination is more efficacious than the gases used alone (Wood and Wood, 1991).

Polyethylene bags impregnated with 1% chlorpyrifos (Polyethylene-D, DowElanco, Midland, MI), developed for insect control on bananas in the field (Hirae *et al.*, 1987), were used to wrap protea flowers (*Banksia menziesii* R.Br.) in shipping boxes to control Argentine ants (*Iridomyrmex humilis* (Mayr)). After 24 h, 95–100% mortality of Argentine ants was achieved with no phytotoxicity (Hara, 1992, unpublished data). Tjosvold and Ali (1989) tested chlorpyrifos-impregnated polyethylene bags and showed 50% reduction in western flower thrips (*F. occidentalis*) with no phytotoxicity to carnation flowers. These impregnated bags can provide a safe and effective postharvest treatment for certain insect pests on flowers, but the bags are not currently registered for use in the US, and the probability of registration for use on minor crops such as flowers and bananas is low.

Insecticidal Dips

Insecticidal dipping is one of the most common postharvest disinfestation treatments for cut flowers, especially tropicals, practised by growers and shippers in tropical areas. For plant material not tolerant to fumigation, the Animal and Plant Health Inspection Service (1978) recommends hand removal of pests plus a malathion–cabaryl chemical dip for 30 s with agitation which eliminates air pockets and aids in complete coverage of plants. Malathion alone as a dip does not provide satisfactory control of thrips on dendrobium in Thailand and insects on protea (Paull, 1992 personal communication). Many effective insecticides registered in the US for field use on floriculture crops may also be used as postharvest insecticidal dips as long as: (i) the pesticide label directions do not prevent use as a dip; (ii) the dip is properly disposed of in registered sites; (iii) proper protective clothing and equipment are worn. Insecticides effective on specific pest species in the field should also be effective when used as dips. The synthetic pyrethroid, fluvalinate (Mavrik Aquaflow, Sandoz Crop Protection Corp., Des Plaines, IL), is one of the few insecticides specifically registered in the US as a dip for flower and foliage cuttings. Osborne (1986) dipped cuttings of tropical ornamentals in fluvalinate for 1 min to control two-spotted spider mites (*T. urticae*), solanum mealybug (*P. solani*), green peach aphid (*M. persicae*), and greenhouse whitefly (*Trialeurodes vaporariorum* (Westwood)) and achieved at least 70% reduction of all pest species; fluvalinate is non-phytotoxic on 11 (approximately 92%) of the tropical foliage plants tested and is proposed as a regulatory treatment for infested plant material coming into the US from other countries. Dipping red ginger infested with banana aphids (*P. nigronervosa*) into fluvalinate for 1 min causes 92.1 and 84.7% mortality of nymphs and adults, respectively, while dipping for 15 min causes 100% mortality of nymphs and adults (Tenbrink *et al.*, 1990). Fluvalinate in combination with insecticidal soap (potassium salts of fatty acids) for 5 min provides the most efficacious dip treatment against mealybugs in red ginger (95–100% mortality) over fluvalinate (20–100% mortality) or soap alone (11–20% mortality) (Tenbrink *et al.*, 1991a); the combination is safe for most tropical foliages and flowers except for sago palm (*Cycas revoluta* Thumb.), dendrobium orchid, (*Dendrobium* 'Jaequelyn Thomas'), and anthuriums (Tenbrink *et al.*, 1991b). Hern *et al.* (1988) demonstrated the potential use of the repellent activity or irritant action of fluvalinate as a protection against two-spotted spider mite (*T. urticae*) on kiwifruits; the mite is a cause of quarantine interceptions on fruits exported from New Zealand to Japan.

Wang and Lin (1984a) used polyethylene perforated bags to cover chrysanthemum flower buds and a postharvest insecticidal dip to eliminate insects on leaves; this combination provides insect-free chrysanthemum cut flowers and reduces the amount of insecticides used in the field. Recent research demonstrates that a double dipping with a 2 h waiting period between dips significantly increases efficacy over a single dip against western flower thrips (*F. occidentalis*) and melon thrips (*Thrips palmi* Karny) on dendrobium orchids (Hata *et al.*, 1993). The first insecticidal dip, a pyrethroid or organophosphate insecticide, probably causes nerve poisoning symptoms of excitability and restlessness that draw insects out of cracks and crevices to be exposed to the second insecticidal dip for final kill. Another possible explanation is a repelling and killing effect of the first insecticide that decreases the load, or number of thrips per flower, for the second dip. Hata *et al.* (1992) showed that a postharvest insecticidal dip of fluvalinate and insecticidal soap is 100% effective against various pests of red ginger as long as field populations are reduced with preharvest insecticide applications.

Insecticides used as postharvest dips should be different from those used in the field and the dipping procedure should be conducted in the packinghouse away from the field. These safeguards are necessary so that survivors of the dip treatment will be minimal and the probability of these individuals returning to the field population will be low, thus avoiding development of resistance. Insecticidal dipping appears promising as a postharvest disinfestation treatment but is not without disadvantages: (i) flowers must be dried after dipping to avoid fungal problems (Seaton and Joyce, 1988); (ii) efficacy depends on the susceptibility of the insect pest species to the particular insecticide and contact with the insecticide, which is especially difficult with certain insects, for example mealybugs occurring in tight cracks and crevices of the flower; (iii) dipping is labour intensive and poses a potential health risk to the operator if proper protective clothing and equipment are not worn at all times (Seaton and Joyce, 1988); (iv) similar to other postharvest disinfestation treatments, phytotoxicity is a limiting factor for certain species or cultivars of flowers and foliage that are sensitive to certain insecticides.

Irradiation

The use of ionizing radiation as a postharvest treatment of cut flowers and agricultural products has been reviewed (Moy, 1985; Wit and van de Vrie, 1985a). The effects of ionizing radiation on insects can be categorized as: (i) induction of mortality; or (ii) induction of sterility (Nelson, 1967). Efficacy tests on pests of floriculture crops, for example beet armyworm (*Spodoptera exigua* (Hubner)), green peach aphid (*M. persicae*), and a thrips (*Frankliniella pallida* (Uzel)) have demonstrated that dosages of 10–20 krad are needed to arrest development or reproduction (Wit and van de Vrie, 1985a). Emergence of agromyzid leafminer (*L. trifolii*) adults is prevented at doses of 7.5, 7.5, and 75 krad for eggs, larvae, and pupae, respectively (Yathom *et al.*, 1990). Higher doses are necessary for acute mortality. For example, 100 krad are needed for acute mortality of larvae and pupae of *L. trifolii*, and 150 krad are needed to kill the melon thrips inside dendrobium orchid flowers, but significant effects on orchid vase life are observed at 50–100 krad (Piriyathamrong *et al.*, 1985; Yathom *et al.*, 1990; Mersino and Moy, 1992, unpublished data). Pincushion protea (*Leucospermum cordifolium* (Salisb. ex Knight) Fourcade), exhibits reduced vase life as doses increase from 13.8 to 40 krad, and leaf damage occurs at 13.8 krad with increasing damage to leaves as the dose increases; blue spotting of the spathe occurs on 'Ozaki' anthurium at 7 krad, although vase life is not measurably reduced (Paull, 1992, unpublished data). Doses that kill insects cause unacceptable damage to native Australian flowers (Seaton and Joyce, 1988). Reactions of other cut flowers to ionizing radiation are variable; some species tolerate high levels of radiation, while other species show damage at levels as low as 5 krad. In a number of cut flower species, seasonal variation in phytotoxicity of radiation occurs, with deleterious effects found during the period between August and April but absent between May and July (Wit and van de Vrie, 1985a; Maughan, 1986).

Radiation disinfestation has many characteristics of a good quarantine treatment (Brower and Tilton, 1985) including: (i) a rapid processing rate that does not unduly delay shipment of the commodity; (ii) complete effectiveness as long as proper doses are selected and applied; (iii) effectiveness in treating sealed packages and containers. Major disadvantages include: (i) intolerance of some cut flower species at dosages necessary to adversely affect development, reproduction, or mortality of insect pests; (ii) low consumer acceptance of irradiation as safe to humans and the environment; (iii) high cost of irradiation facilities; (iv) lack of approval by quarantine officials of a treatment that causes sterility and arrested development but not acute mortality at the probit 9 level.

Systems Approach to Quarantine Security

A systems approach of preharvest crop pest management, postharvest culling, and final product inspection is successful against several quarantine pests on various fruit crops (Moffitt, 1989). This approach to quarantine security unifies pest management practices and treatments both before and after harvest into a unified system. Treatments or management procedures that are not effective alone can, therefore, meet quarantine security. Wood and Wood (1991) emphasized the necessity to combine an effective fumigation procedure with a regular and effective spray programme in protea plantations, since any sign of scales or borers, dead or alive, does not pass Japanese quarantine inspectors. Coetzee and Wright (1992) emphasized that no postharvest fumigation treatment tested on protea flowers gives 100% mortality; however, if used in conjunction with preharvest spraying, fewer insects will be present, and acceptable results can be achieved. Hata *et al.* (1992) demonstrated the systems approach using a tropical floriculture crop, red ginger. The systems approach to quarantine security is based on the fact that control measures before harvest, for example applying chemical insecticides, can reduce pests to a level at which an insecticidal dip after harvest is 100% effective (Fig. 17.1). This management strategy requires that pests be identified, population levels assessed, and the determination made if the levels are above quarantine thresholds. If a pest population is at or above quarantine thresholds, the pest population should be made the target of a specific, effective control measure, for example application of an effective chemical insecticide. Thereafter, the harvested flowers should be treated with an effective postharvest treatment followed by a final visual inspection to ensure quarantine security.

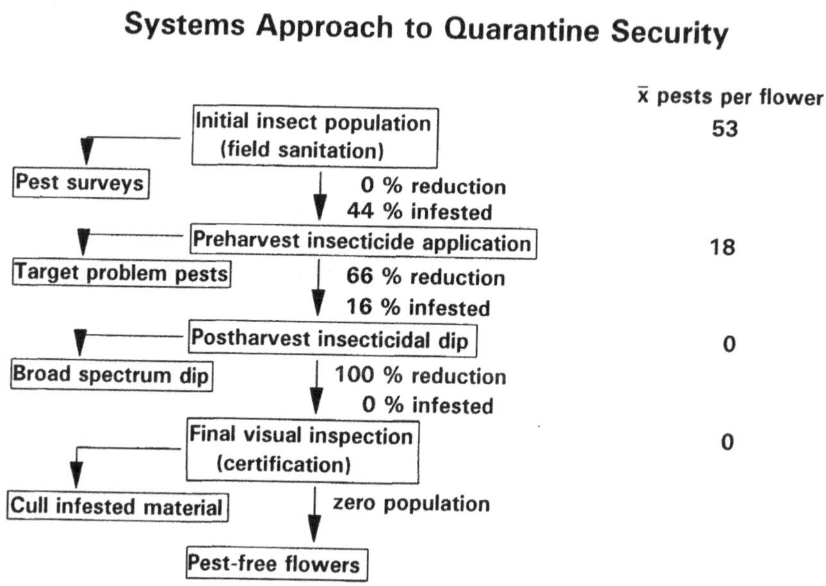

Fig. 17.1 Sequence of pre- and postharvest procedures used in a systems approach to quarantine on red ginger (*Alpinia purpurata*). Mean number of all pests of red ginger per flower including aphids, ants, mealybugs, thrips, and earwigs. Preharvest insecticidal application consisting of chlorpyrifos DF (Pageant, DowElanco, Midland, MI) applied to the foliage at 2 week intervals. A 5 min postharvest insecticidal dip consisting of fluvalinate (Mavrik Aquaflow, Sandoz Crop Protection Corp., Des Plaines, IL) and insecticidal soap (potassium salts of fatty acids) with agitation kills any remaining survivors (from Hata *et al.*, 1992).

Temperature

Cold

Use of cold is limited to floriculture crops that can tolerate low temperatures, for example temperate flowers. Low temperature (0–1°C) may not be a satisfactory method of disinfestation for flowers because at least 8–13 days are required for substantial insect kill, and some flowers are damaged (Seaton and Joyce, 1988, 1989). Storing bird of paradise below 5°C results in necrotic lesions and failure of bud opening (Halevy *et al.*, 1978), and red ginger flowers should be stored above 12°C (Broschat and Donselman, 1988).

Heat

The use of vapour heat and hot water is well documented for treatment of fruit flies on various fruits (Sinclair and Lindgren, 1955; Seo *et al.*, 1974; Armstrong, 1982; Sharp and Spalding, 1984; Sharp, 1986; Couey and Hayes, 1986; Hayes *et al.*, 1987; Sharp *et al.*, 1988; Chan and Linse, 1989; Heard *et al.*, 1991; Nascimento *et al.*, 1992) but not as well studied on floriculture crops. One of the early uses of heat for control of insect pests on floriculture crops is vapour heat for the control of bulb flies (*Eumerus* sp.) on narcissus bulbs at 43.3°C for 3 h (Spruijt and Blanton, 1933) and control of the gladiolus thrips (*Thrips simplex* (Morison) = *Taeniothrips gladioli* Moulton and Steinweden) on gladiolus corms at 43.3°C for 20–30 min (Doucette, 1933). Hansen *et al.* (1992) investigated vapour heat treatments to disinfest tropical cut flowers and foliage using a commercial facility designed for disinfesting papayas of fruit flies. At 46.6°C, 90–98% relative humidity for 1 h, 90–100% of nymphs and adults of banana aphid (*P. nigronervosa*), green scale (*C. viridis*), Cockerell scale (*P. cockerelli*), cardamom thrips (*Sciothrips cardamomi* (Ramakrishna)), and mealybugs (*N. nipae*, *P. citri*, *P. affinis*, and *P. longispinus*) are killed. Most heliconias, red ginger, bird of paradise (*Strelitzia reginae* Ait.) flowers and leaves, and most tropical foliage are not damaged; anthurium, pincushion protea and dendrobium orchid flowers are very sensitive to vapour heat. Hansen *et al.* (1992) concluded that vapour heat is a promising postharvest treatment to disinfest tropical cut flowers and foliage, with research needed on decreasing heat stress in temperature-sensitive floral commodities.

APHIS (1978) recommends hot water at 44°C for 20 min for plant material not tolerant to fumigation but warns that this treatment may not be effective against all insects and some plants may not be tolerant. The use of hot water on banksia protea is documented by Seaton and Joyce (1988) who showed that banksia quality deteriorates following a 20 min dip at 46°C or a 10 min dip at 56°C; the hot water treatment causes many florets to open and pollen is washed from the banksia flowers. Dipping Geraldton wax (*Chamelaucium unicinatum* Schau.) for 20 min at 46°C reduces vase life by 26% (Seaton and Joyce, 1988). Hot water treatment of Cockerell scale (*P. cockerelli*) on bird of paradise is a potential quarantine treatment that can be adjusted to any specific risk level including zero tolerance (Hara *et al.*, 1993). Exposure to 49°C for 5, 5, and 6 min results in 100% mortality of adult females, nymph, and crawler life stages, respectively (Hara *et al.*, 1993). Most mealybugs on red ginger are effectively killed in water at 49°C for 12 min without significant effects on vase life (Hara, unpublished data). Tolerance of other tropical flowers and foliage to hot water appears similar to tolerance to vapour heat (Hara, unpublished data).

Microwave heating at 600 W for 10 s causes extensive damage to Geraldton wax and 81% loss in vase life; it is not effective in controlling insects (Seaton and Joyce, 1988).

Conclusions

Each postharvest treatment has its advantages and disadvantages (Table 17.1) with no single effective treatment for the various pest species and floral crops. A procedure or treatment for quarantine security will have to be tailored to a particular pest species and host species. In addition, a single postharvest treatment may not provide the level of security demanded by certain countries and states (zero tolerance or probit 9 mortality for insects); this will require both before harvest pest management measures and after harvest treatments. Presently, most quarantine protocols require evidence that a probit 9 level of security can be achieved by a treatment or procedure (Landolt *et al.*, 1984; Jang, 1991). However, as suggested by several US Department of Agriculture researchers (Chew and Ouye, 1985; Ouye and Gilmore, 1985; Jang, 1991), quarantine treatments based on the probability of an actual introduction in a non-infested geographical area will more realistically relate to a given treatment for quarantine security. Consequently, some of the discussed postharvest disinfestation treatments or procedures may not meet the stringent probit 9 level of security but meet a realistic 'negligible pest risk', or a low probability of introducing a pest into a non-infested area, based on infestation level of the commodity and on survival, feeding, dispersal, mate-finding, and host-finding of the pest (Chew and Ouye, 1985). Quarantine officials should recognize the difference in probability of introducing a pest on propagative material versus cut flowers or foliage into a non-infested area. The probability of introducing a pest on propagative materials for use by commercial nurseries or farms is much higher than

Table 17.1. Advantages and disadvantages of disinfestation treatments and procedures for quarantine security of floriculture crops.

Treatment	Advantage	Disadvantage
Aerosol/fog	1. Efficacious with the combined use of pyrethrin and dichlorvos propelled by CO_2 2. Use of abamectin or chlorpyrifos delivered by thermal fogging device with forced draft is effective in reducing thrips by 94–99%	1. Not as effective as fumigants in penetrating flowers 2. Long exposure time ranging from 3 to 16 h for efficacious results 3. Permethrin or pyrethrin applied as an aerosol produces unsatisfactory results
Controlled atmospheres	1. CO_2 at 0–1°C for 1 week is promising for protea flowers	1. CO_2 or nitrogen alone gives unsatisfactory efficacy results 2. Long treatment exposure suited for sea freighted flowers only 3. Tropicals are subject to chilling injury
Fumigation		
Aluminium phosphide (phosphine)	1. Extensively used in insecticidal fumigation of animal feed, bulk grain, flour mills, and processed foods	1. Long exposures needed to achieve sufficient control 2. Phytotoxicity will probably be major limiting factor on flowers and foliage
Chlorpyrifos (impregnated bags)	1. Polyethylene bags impregnated with chlorpyrifos reduce ants and thrips by 95–100% and 50%, respectively 2. No phytotoxicity observed	1. Bags are currently not registered for use in the US 2. Probability of obtaining registration from EPA for minor use is low
Dichlorvos (DDVP in PVC)	1. When delivered in polyvinyl chloride (PVC) resin strips 94–99% reduction in aphids, mealybugs, and spider mites	1. Only partial insect control and infestations are never eliminated 2. Phytotoxic on certain tropical foliage plants
Hydrogen cyanide	1. Effective against aphids, scales, mealybugs 2. Used in Japan on infested imported living plants and fresh fruits	1. No longer available in the US for agricultural use due to EPA requirements for additional toxicological data

Table 17.1. *contd*

Treatment	Advantage	Disadvantage
	3. Readily detectable by a characteristic odour, and safety procedures for its use have been established	2. Phytotoxic on certain cut flowers such as pincushion protea and some moist flowers such as carnations
Methyl bromide	1. Approved quarantine treatment for many countries	1. Margin between effective and phytotoxic dose is narrow
		2. Continued availability questionable due to its ozone depletion potential (ODP)
Sulfur dioxide	1. Significant mortality can be achieved after 1.5 h	1. Slight bleaching of protea flowers and foliage
		2. Further testing needed to confirm efficacy and phytotoxicity
Insecticidal dips	1. Effective and readily available for use by growers and shippers	1. Flowers must be dried after dipping to avoid fungal problems
	2. Insecticides effective on specific pest species in the field should also be effective when used as a dip	2. Efficacy depends on contact with insect which is difficult with certain insects which occur in tight cracks and crevices of the flower
	3. If an insecticide is not used to control field pest populations, and used only as a dip, pesticide resistance can be minimized	3. Dipping is labour intensive and poses a potential health risk to the operator if proper protective clothing and equipment are not worn at all times
	4. Synthetic pyrethroids, which have repellent activity, act as a protectant and are especially effective in combination with insecticidal soap	4. Phytotoxicity is a limiting factor for use on certain cultivars of flowers and foliage
Irradiation	1. Rapid process that does not delay shipment	1. Not all cut flower species are tolerant to dosages necessary to affect adversely development, reproduction or mortality of insects
	2. Completely effective by arresting development of reproduction of pests as long as proper doses are selected	2. Consumer acceptance of irradiation as safe to humans and the environment is low
	3. Sealed packages and containers can be treated effectively	3. High cost of irradiation facilities
		4. Approval needed by quarantine officials of a treatment that causes

Table 17.1. *contd*

Treatment	Advantage	Disadvantage
		sterility and arrested development rather than acute mortality at the probit 9 level
Systems approach	1. Demonstrated approach on a floriculture crop, red ginger, that provides complete quarantine security 2. Field pest management reduces pest population to a level at which postharvest treatments are 100% effective	1. Requires development of field pest management practices 2. Preharvest and postharvest procedures must be implemented by growers and snippers for quarantine security
Temperature		
Cold	1. Non-toxic to humans and the environment	1. Limited to crops that can tolerate low temperature 2. At least 8 to 10 days needed for substantial kill and some flowers are damaged
Hot water	1. Low cost of equipment as compared with vapour heat, radiation, radiation, and fumigation 2. Effective against armoured scales, mealybugs, and other insects with similar activity and phytotoxic reaction as vapour heat 3. A relatively short treatment of 49°C for 10 min exceeds the probit 9 mortality level for armoured scales	1. Certain flowers and foliage are sensitive to heat 2. Research needed on decreasing heat stress in temperature-sensitive commodities
Microwave	1. None	1. Extensive damage to flowers with 81% loss in vase life 2. Not effective in controlling insects
Vapour heat	1. Widely accepted as a quarantine treatment for fruit flies on various fruits 2. Effectiveness 90–100% against scales, thrips, mealybugs at 46.6°C, 90–98% RH, for 1 h 3. Non-phytotoxic to most heliconia, red ginger, bird of paradise and most tropical foliage	1. High cost of vapour heat equipment 2. Phytotoxic to anthurium, dendrobium orchids, pincushion protea, and bamboo orchid foliage 3. Relatively long exposure time of 1 h or longer 1 h or longer

the probability of introducing a pest on cut flowers for retail consumption by business establishments or home residences. For example, Beardsley and Gonzalez (1975) stated that the probability of armoured scale crawlers establishing a new infestation by non-propagative plant material is unlikely because crawlers must be in close proximity to suitable growing hosts. These factors should be considered in developing quarantine procedures for floriculture crops, rather than basing procedures solely on the customary probit 9 security level.

Acknowledgements

We thank Ms Victoria L. Tenbrink, Mr Trent Y. Hata, and Mr Benjamin K.S. Hu for their critical review, comments and generous assistance on this manuscript; Dr Robert E. Paull for encouragement, critical review and comments, and providing literature from Australia. This is the Hawaii Institute of Tropical Agriculture and Human Resources Journal Series No. 3863.

References

Agricultural Production Bureau (1988) Guide to import plant quarantine in Japan (cut flowers and fresh vegetables). Agricultural Production Bureau, Ministry of Agriculture, Forestry and Fisheries, Japan.

Animal and Plant Health Inspection Service (APHIS) (1978) Section VI, T200–201 – Schedules for propagative plant material. *PPQ Treatment Manual.* US Department of Agriculture, Washington, DC.

Animal and Plant Health Inspection Service (APHIS) (1980) Section I – Fumigants. *PPQ Treatment Manual.* US Department of Agriculture, Washington, DC.

Animal and Plant Health Inspection Service (APHIS) (1985) Code of Federal Regulations (CFR), Title 7 – Agriculture, Part 318 – Hawaiian and territorial quarantine notices. US Department of Agriculture, Washington, DC.

Animal and Plant Health Inspection Service (APHIS) (1986) Section VI, T300 – Schedules for miscellaneous plant products. *PPQ Treatment Manual.* US Department of Agriculture, Washington, DC.

Animal and Plant Health Inspection Service (APHIS) (1988) *Export Certification Manual,* Vols 1–6. PPQ, US Department of Agriculture, Washington, DC.

Armstrong, J.W. (1982) Development of a hot-water immersion quarantine treatment for Hawaiian-grown 'Brazilian' bananas. *Journal of Economic Entomology* 75, 787–790.

Beardsley, J.W., Jr and Gonzalez, T.H. (1975) The biology and ecology of armored scales. *Annual Review of Entomology* 20, 47–73.

Broschat T.K. and Donselman, H. (1988) Production and postharvest culture of red ginger in South Florida. *Proceedings of the Florida State Horticultural Society* 101, 326–327.

Brower, J.H. and Tilton, E.W. (1985) The potential of irradiation as a quarantine treatment for insects infesting stored-food commodities. In: Moy, J.H. (ed.), *Radiation Disinfestation of Food and Agricultural Products. Proceedings of an International Conference.* 14–18 November 1983, Hawaii Institute of Tropical Agriculture and Human Resources, University of Hawaii at Manoa, Honolulu, Hawaii, pp. 75–86.

Chan, H.T. and Linse, E. (1989) Conditioning cucumbers to increase heat resistance in the EFE system. *Journal of Food Science* 54, 1375–1376.

Chew, V. and Ouye, M.T. (1985) Statistical basis for quarantine treatment schedule and security. In: Moy, J.H. (ed.), *Radiation Disinfestation of Food and Agricultural Products. Proceedings of an International Conference.* 14–18 November 1983, Hawaii Institute of Tropical Agriculture and Human Resources, University of Hawaii at Manoa, Honolulu, Hawaii, pp. 70–74.

Coetzee, J.H. and Wright, M.G. (1992) Post-harvest treatment and disinfestation of protea cut flowers. *Protea News* 12, 19–20, Pretoria, South Africa.

Couey, H.M. and Hayes, C.F. (1986) Quarantine procedure for Hawaiian papaya using fruit selection and a two-stage hot-water immersion. *Journal of Economic Entomology* 79, 1307–1314.

Doucette, C.F. (1933) The gladiolus thrips. *Proceedings of the Washington State Horticultural Association* 28, 290–293.

Farm Chemicals Handbook (1992) *Pesticide Dictionary.* Meister Publishing Co., Willoughby, Ohio.

Halevy, A., Kofranek, A.M. and Besemer, S.T. (1978) Postharvest handling methods for bird-of-paradise flowers (*Strelitzia reginae* Ait.). *Journal of the American Society for Horticultural Science* 103, 165–169.

Hamlen, R.A. and Henley, R.W. (1979) Pest control and phytotoxicity on indoor ornamental plants with dichlorvos fumigation. *Journal of the American Society for Horticultural Science* 104, 136–138.

Hansen, J.D., Chan, H.T., Jr, Hara, A.H. and Tenbrink, V.L. (1991a) Phytotoxic reaction of Hawaiian cut flowers and foliage to hydrogen cyanide fumigation. *HortScience* 26, 53–56.

Hansen, J.D., Hara, A.H., Chan, H.T., Jr and Tenbrink, V.L. (1991b) Efficacy of hydrogen cyanide fumigation as a treatment for pests of Hawaiian cut flowers and foliage after harvest. *Journal of Economic Entomology* 84, 532–536.

Hansen, J.D., Hara, A.H. and Tenbrink, V.L. (1992). Vapor heat: a potential treatment to disinfest tropical cut flowers and foliage. *HortScience* 27, 139–143.

Hara, A.H., Hata, T.Y., Hu, B.K.S. and Tenbrink, V.L. (1993) Hot water immersion as a potential quarantine treatment against *Pseudaulacaspis cockerelli* (Homoptera: Diaspididae). *Journal of Economic Entomology* 86, 1167–1170.

Hata, T.Y., Hara, A.H. Jang, E.B., Imaino, L.S., Hu, B.K.S. and Tenbrink, V.L. (1992) Pest management before harvest and insecticidal dip after harvest as a systems approach to quarantine security for red ginger. *Journal of Economic Entomology* 85, 2310–2316.

Hata, T.Y., Hara, A.H., Hu, B.K.S., Kaneko, R.T. and Tenbrink, V.L. (1993) Field sprays and insecticidal dips after harvest for pest management of *Frankliniella occidentalis* (Pergande), and *Thrips palmi* Karny (Thysanoptera: Thripidae) on orchids. *Journal of Economic Entomology* 86, 1483–1489.

Hayes, C.F., Chingon, H.T.G, Nitta, F.A. and Leung, A.M.T. (1987) Calculation of survival from double hot-water immersion treatment for papayas infested with oriental fruit flies (Diptera: Tephritidae). *Journal of Economic Entomology* 80, 887–890.

Heard, T.A., Heather, N.W. and Corcoran, R.J. (1991) Dose–mortality relationships for eggs and larvae of *Bactrocera tryoni* (Diptera: Tephritidae) immersed in hot water. *Journal of Economic Entomology* 84, 1768–1770.

Hern, M.D., Penman, D.R. and Chapman, R.B. (1988) Use of fluvalinate for the protection of kiwifruit from contamination by twospotted spider mite (Acari: Tetranychidae). *Journal of Economic Entomology* 81, 863–866.

Hirae, H., Mau, R. and Hara, A. (1987) Experiments with Dursban-impregnated bunch covers. In: *Proceedings, 18th Annual Hawaii Banana Industry Association Conference*, Cooperative Extension Service, University of Hawaii at Manoa, Honolulu, Hawaii, pp. 12–15.

Jang, E.B. (1991) Thermal death kinetics and heat tolerance in early and late third instars of the oriental fruit fly (Diptera: Tephritidae). *Journal of Economic Entomology* 84, 1298–1303.

Japanese Fumigation Engineering Society (1981) The text for the chief of plant quarantine fumigation operation. Japanese Fumigation Engineering Society, Tokyo (in Japanese).

Landolt, P.J., Chambers, D.L. and Chew, V. (1984) Alternative to the use of probit 9 mortality as a criterion for quarantine treatments of fruit fly (Diptera: Tephritidae)-infested fruit. *Journal of Economic Entomology* 77, 285–287.

Maughan, J. (1986) Post harvest treatment of protea cut flowers to eradicate arthropods. In: *Protea National*, June 1986, pp. 13–17.

Moffitt, H.R. (1989) A systems approach to meeting quarantine requirements for insect pests of deciduous fruit. *Proceedings of the Washington State Horticultural Association* 85, 223–225.

Monro, H.A.U. (1969) *Manual of Fumigation for Insect Control*. FAO Agricultural Studies No. 79. Food and Agriculture Organization of the United Nations, Rome.

Moy, J.H. (ed.) (1985) Radiation disinfestation of food and agricultural products. *Proceedings of an International Conference*. 14–18 November 1983, Honolulu, Hawaii Institute of Tropical Agriculture and Human Resources, University of Hawaii at Manoa, Honolulu, Hawaii.

Nakahara, L.M. (1982) Notes and Exhibitions. *Proceedings of the Hawaiian Entomolological Society* 24, 8.

Nascimento, A.S., Malavasi, A., Morgante, J.S. and Duarte, A.L.A. (1992). Hot-water immersion treatment for mangoes infested with *Anastrepha fraterculus*, *A. obliqua* and *Ceratitis capitata* (Diptera: Tephritidae) in Brazil. *Journal of Economic Entomology* 85, 456–460.

Nelson, S.O. (1967) Ionizing radiation. In: Kilgore, W.W. and Doutt, R.L. (eds), *Pest Control: Biological, Physical and Selected Chemical Methods*. Academic Press, New York, pp. 129–139.

Osborne, L.S. (1986) Dip treatment of tropical ornamental foliage cuttings in fluvalinate to prevent spread of insect and mite infestations. *Journal of Economic Entomology* 79, 465–470.

Ouye, M.T. and Gilmore, J.E. (1985) The philosophy of quarantine treatment as related to low-dose radiation. In: Moy, J.H. (ed.), *Radiation Disinfestation of Food and Agricultural Products. Proceedings of an International Conference*. 14–18 November 1983, Hawaii Institute of Tropical Agriculture and Human Resources, University of Hawaii at Manoa, Honolulu, Hawaii, pp. 67–69.

Pesticide and Toxic Chemical News (1992) Pesticide registrants alerted on CFCs, methyl bromide. *Pesticide and Toxic Chemical News* 20, 17–18.

Piriyathamrong, S., Chouvalitvongporn, P. and Sudathit, B. (1985) Disinfestation and vase-life extension of orchids by irradiation. In: Moy, J.H. (ed.), *Radiation Disinfestation of Food and Agricultural Products. Proceedings of an International Conference*. 14–18 November 1983, Hawaii Institute of Tropical Agriculture and Human Resources, University of Hawaii at Manoa, Honolulu, Hawaii, pp. 222–225.

Sanguthai, S. (1991) Marketing orchids from Southeast Asia to Europe and America. In: Leonhardt, K.W., Evans, D.O. and Halloran, J.M. (eds), *The Hawaii Tropical Cut Flower Industry Conference*. 29–31 March 1990, Hilo, Hawaii. Research Extension Series 124, University of Hawaii, College of Tropical Agriculture and Human Resources, Honolulu, Hawaii, pp. 35–39.

Seaton, K.A. and Joyce, D.C. (1988) Post-harvest insect disinfestation treatments for cut flowers and foliage. Western Australia Department of Agriculture, Farmnote No. 89/88. Agdex 280/56, South Perth, Western Australia.

Seaton, K.A. and Joyce, D.C. (1989) Postharvest disinfestation of cut flowers for export. In: *Horticultural Research and Extension Update – 1989*. Western Australia Department of Agriculture, South Perth, Western Australia, pp. 1–7.

Seo, S.T., Hu, B.K.S. , Komura, M., Lee, C.Y.L. and Harris, E.J. (1974) *Dacus dorsalis*: vapor heat treatment in papayas. *Journal of Economic Entomology* 67, 240–242.

Seo, S.T., Akamine, E.K., Goo, T.T.S., Harris, E.J. and Lee, C.Y.L. (1979) Oriental and Mediterranean fruit flies: fumigation of papaya, avocado, tomato, bell pepper, eggplant and banana with phosphine. *Journal of Economic Entomology* 72, 354–359.

Sharp, J.L. (1986) Hot-water treatment for control of *Anastrepha suspensa* (Diptera: Tephritidae) in mangoes. *Journal of Economic Entomology* 79, 706–708.

Sharp, J.L. and Spalding, D.H. (1984) Hot water as a quarantine treatment for Florida mangos infested with Caribbean fruit fly. *Proceedings of the Florida State Horticultural Society* 97, 355–357.

Sharp, J.L., Ouye, M.T., Thalman, R., Hart, W., Ingle, S. and Chew, V. (1988) Submersion of 'Francis' mango in hot water as a quarantine treatment for the West Indian fruit fly and the Caribbean fruit fly (Diptera: Tephritidae). *Journal of Economic Entomology* 81, 1431–1436.

Sinclair, W.B. and Lindgren, D.L. (1955) Vapor heat sterilization of California citrus and avocado fruits against fruit-fly insects. *Journal of Economic Entomology* 48, 133–138.

Soderstrom, E.L., Brandl, D.G., Hartsell, P.L and Mackey, B. (1991) Fumigants as treatments for harvested citrus fruits infested with *Asynonychus godmani* (Coleoptera: Curculionidae). *Journal of Economic Entomology* 84, 936–941.

Spencer, K.A. (1981) A revisionary study of the leafmining flies (Agromyzidae) of California. University of California, Division of Agricultural Sciences, Special Publication 3273.

Spruijt, F.J. and Blanton, F.S. (1933) Vapor-heat treatment for the control of bulb pests and its effect upon the growth of narcissus bulbs. *Journal of Economic Entomology* 26, 613–620.

Tayama, H.K. (1991) Changes in global floriculture crops: Production centers, channels of distribution, and per capita consumption of floriculture products. In: Leonhardt, K.W., Evans, D.O. and Halloran, J.M. (eds), *The Hawaii Tropical Cut Flower Industry Conference.* 29–31 March 1990, Hilo, Hawaii. Research Extension Series 124, University of Hawaii, College of Tropical Agriculture and Human Resources, Honolulu, Hawaii, pp. 2–7.

Tenbrink, V.L., Hansen, J.D. and Hara, A.H. (1990) Postharvest control of banana aphid using dips, 1989. *Insecticide Acaricide Tests* 15, 338–339.

Tenbrink, V.L., Hansen, J.D. and Hara, A.H. (1991a) Postharvest control of mealybugs using dips, 1990. *Insecticide Acaricide Tests* 16, 258.

Tenbrink, V.L., Hansen, J.D. and Hara, A.H. (1991b) Phytotoxicity of Safer's Insecticidal Soap and Mavrik Aquaflow as a postharvest dip, 1990. *Insecticide Acaricide Tests* 16, 261–262.

Tjosvold, S.A. and Ali, A.D. (1989) Fumigation of western flower thrips using banana bags (Polyethylene-D) during simulated shipment of cut flowers. In: *Flower and Nursery Report*, Department of Environmental Horticulture, University of California, Davis, California, pp. 3–4.

Wang, C.L. and Lin, R.T. (1984a) Study on the quarantine treatments of insect pests on chrysanthemum cut flowers – the application of protection bags and the improved dipping methods. *Journal of Agricultural Research China* 33, 325–330.

Wang, C.L. and Lin, R.T. (1984b) Study on the quarantine treatments of insect pests on chrysanthemum cutflowers – fumigation and smoking methods. *Journal of Agricultural Research China* 34, 88–93.

Ware, G.W. (1989) *The Pesticide Book,* 3rd edn. Thomson Publication, Fresno, California.

Wit, A.K.H. and van de Vrie, M. (1985a) Gamma radiation for post harvest control of insects and mites in cutflowers. *Mededelingen van de Faculteit Landbouwwetenschappen Rijksuniversiteit Gent* 50, 697–704.

Wit, A.K.H. and van de Vrie, M. (1985b) Fumigation of insects and mites in cutflowers for post harvest control. *Mededelingen van de Faculteit Landbouwetenschappen Rijksuniversiteit Gent* 50, 705–712.

Wood, J. and Wood, E. (1991) Insect disinfestation of protea cut flowers and foliage. *Protea News* 10, 15–17, Pretoria, Republic of South Africa.

Yathom, S., Padova, R., Tal, S. and Ross, I. (1990) Effects of gamma radiation on the immature-stages of *Liromyza trifolii. Phytoparasitica* 18, 117–124.

Index

Page numbers in *italics* refer to figures and tables

Printed and bound by CPI Group (UK) Ltd, Croydon, CR0 4YY
16/09/2026
14963431-0004